Discovering ANATOMY
A GUIDED EXAMINATION OF THE CADAVER

David A. Morton
University of Utah

John L. Crawley

MORTON
PUBLISHING

925 W. Kenyon Avenue, Unit 12
Englewood, CO 80110

www.morton-pub.com

Book Team

President and CEO	David M. Ferguson
Senior Acquisitions Editor	Marta R. Pentecost
Developmental Editor	Sarah D. Thomas
Project Editors	Trina Lambert, Rayna S. Bailey
Production Manager	Will Kelley
Production Assistants	Joanne Saliger, Amy Heeter
Indexer	Doug Easton
Cover and Illustrations	Imagineering Media Services, Inc.

Copyright © 2018 by Morton Publishing Company

All rights reserved. No part of this publication may be reproduced, stored in a retrieval system, or transmitted, in any form or by any means, electronic, mechanical, photocopying, recording or otherwise, without the prior written permission of the publisher.

Printed in the United States of America

10 9 8 7 6 5 4 3 2 1

ISBN-10: 1-61731-616-4

ISBN-13: 978-1-61731-616-6

Library of Congress Control Number: 2017915760

Preface

Why Did We Write This Book?

As a professor of anatomy, it is my goal to help direct students to become effective health-care providers. Health-care providers repair the human body, much like auto mechanics repair cars. Mechanics can connect an engine with a computer to see where there are problems. However, to become an effective mechanic, it is necessary to learn the parts of the engine and to take the engine apart to better understand the internal relationships and interactions. Similarly, advances in technology enable health-care providers to see the inside of the human body where disease processes are occurring (e.g., CT scans). However, effective health-care providers still must learn the parts of the human body in order to know how to best treat each particular problem. To better understand the internal relationships and interactions, it is particularly important to look at the body in regions or systems.

Discovering Anatomy: A Guided Examination of the Cadaver is written for instructors who teach courses that are fortunate enough to have cadavers at their disposal; however, it may also be used in courses that utilize other means, such as models, to achieve an understanding of anatomical structures.

In the years since I began teaching, I have seen the number and variety of anatomical educational resources skyrocket. However, despite the many print and digital options available, the cadaver remains the most important teaching tool for learning human anatomy. The use of cadavers as a means for learning the human body has been utilized for hundreds of years. The purpose of *Discovering Anatomy: A Guided Examination of the Cadaver* is to provide a reference textbook that enables students to learn the parts, internal relationships, and interactions of the human body using a cadaver. The cadavers used in anatomy labs across the country are from body donors—individuals who make the conscious decision during their lifetimes to altruistically donate their bodies after death so A&P students can study the various systems of the body. This practice exemplifies the Latin saying, "mortui vivos docent," or, in English, "the dead teaching the living."

We recognize that students are often overwhelmed by the vast amount of structures that they must learn in an anatomy course. This lab manual is meant to guide students through the many layers and complex structures that they will encounter in their cadaver-based study of human anatomy. We have designed activities that concisely list and explain each system of the body and all relevant structures, while also providing students with opportunities to put what they've learned into practice.

In addition, to make the content more relatable and accessible, we have included two fun features throughout this manual that provide students with interesting supplemental information related to the organ systems being studied:

Clinical Application

Clinical Applications give practical value to the systems and structures covered in each chapter by showing students examples of health issues or conditions they might encounter as a physician or medical professional, such as ACL tears and kidney stones.

Weird and Wacky boxes provide fun facts and trivia to further engage students in learning about each system of the human body by drawing connections between our body functions and everyday life.

How Do You Use This Book?

Chapters have been organized into five separate units:

- **Unit 1: Inside Out** introduces students to the fundamentals of anatomy and the organ systems at work "beneath the surface" of our bodies.
- **Unit 2: Moving Forward** includes organ systems associated with physical activity.
- **Unit 3: Coming to Your Senses** discusses the systems that transmit signals throughout our body to assist us in reacting to stimuli.
- **Unit 4: Body Highways** focuses on the organs that keep our bodies running.
- **Unit 5: The Next Generation** describes how our bodies reproduce.

Each chapter consists of the following parts:

- **Learning Objectives.** This list provides students with an outline of what they will be learning during the laboratory session.
- **Getting Acquainted.** This portion of the chapter helps students prepare for their time with the cadaver and can be completed prior to going to the lab session. This section includes the following features:
  - Textual overviews of the organ systems and associated structures.
  - Fillable tables of key terms.
  - Activities designed to help students identify and label relevant structures.
- **Observing.** This portion of the chapter provides labeled cadaveric photographs to be used for reference when identifying the structures in the cadaver lab. This material is designed to be done while working directly with the cadaver.
- **Wrapping Up.** This portion of the chapter provides students with an opportunity to reflect on what they have learned. It includes short answer and multiple-choice questions, as well as additional images to label. These review activities can be completed by the student in the lab after observing the cadavers or outside of the lab at a later time.

We hope that *Discovering Anatomy: A Guided Examination of the Cadaver* helps each of you in this amazing journey of exploring the human body.

Sincerely,
David A. Morton
John L. Crawley

Acknowledgments

We thank the following individuals for their contributions to this endeavor:

- Kerry Peterson, Luke Sanders, Geoff Dorius, Chad Decker, and Ashley Morris for their help with dissections. A special thanks to Garrett Clement for assisting in editing the final project.
- The altruistic donation of individuals who continue to teach us even after death.
- The following reviewers for their feedback as we prepared this title for publication:
  - Branko Jablanovic, College of Lake County
  - Eric Katz, Diablo Valley College and Las Positas College
  - John McDaniel, Skyline College
  - Cynthia C. Wingert, M.S., Cedarville University

Photo Credits

- Ash, Dr. John F. and Dr. Sheryl A. Scott, University of Utah School of Medicine, SLC, Utah: Figure 2.2C
- Gustoimages/Science Source: Unit 4 Opener
- Pendarvis, Murray: Unit 1 Opener
- Voisin/Phanie/Science Source: Unit 3 Opener
- Watts, Dr. Gary M.: Figures 1.16, 4.18–4.19, 5.16–5.19, 7.14, 10.16–10.18, 14.7–14.8, 17.18

About the Authors

David A. Morton, Ph.D.

Dr. Morton is a Professor and Vice-Chair of Medical and Dental Education in the Department of Neurobiology and Anatomy at the University of Utah School of Medicine. He serves as curriculum director and teaches gross anatomy, histology, and neuroanatomy to medical, dental, physician assistant, physical therapy, and occupational therapy students. His research interests include the creation and incorporation of active learning activities and use of cadavers in medical education. Dr. Morton has authored numerous textbooks, including *Van De Graaff's Photographic Atlas for the Anatomy & Physiology Laboratory* from Morton Publishing.

John L. Crawley

John spent his early years growing up in Southern California, where he took every opportunity to explore nature and the outdoors. He currently resides in Provo, Utah, where he enjoys the proximity to the mountains, desert, and local rivers and lakes.

He received his degree in Zoology from Brigham Young University in 1988. While working as a researcher for the National Forest Service and Utah Division of Wildlife Resources in the early 1990s, John was invited to work on his first project for Morton Publishing, *A Photographic Atlas for the Anatomy and Physiology Laboratory*. After completion of that title, John has continued to work with Morton Publishing, and, to date, he has completed eight titles with them.

John has spent much of his life observing nature and taking pictures. His photography has provided the opportunity for him to travel widely, allowing him to observe and learn about other cultures and lands. His photos have appeared in national ads, magazines, and numerous publications. He has worked for groups such as Delta Airlines, *National Geographic*, Bureau of Land Management, U.S. Forest Service, and many others. His projects with Morton Publishing have been a great fit for his passion for photography and the biological sciences.

Contents

1 Inside Out 1

CHAPTER 1 Introduction to Anatomy 3
    GETTING ACQUAINTED 4
        Activity 1: Anatomical Position and Planes of Section   4
        Activity 2: Body Regions   7
        Activity 3: Organ Systems   11
    OBSERVING 12
        Activity 1: Body Regions and Anatomical Terms of Position   12
    WRAPPING UP 17

CHAPTER 2 Histology 21
    GETTING ACQUAINTED 21
        Activity 1: Epithelial Tissue   21
        Activity 2: Connective Tissue Proper   26
        Activity 3: Other Types of Connective Tissue   30
        Activity 4: Muscle Tissue   35
        Activity 5: Nervous Tissue   37
    OBSERVING 38
        Activity 1: Epithelial Tissue   38
        Activity 2: Connective Tissue   40
        Activity 3: Muscle and Nervous Tissue   41
    WRAPPING UP 43

CHAPTER 3 Integumentary System 47
    GETTING ACQUAINTED 47
        Activity 1: Integumentary System Structures   50
        Activity 2: Histology of the Skin   52
    OBSERVING 53
        Activity 1: Skin Anatomy   53
    WRAPPING UP 55

UNIT 2 Moving Forward 57

CHAPTER 4 Axial Skeleton 59
GETTING ACQUAINTED 59
- Activity 1: Skeletal Features   59
- Activity 2: Bones of the Skull   60
- Activity 3: Remainder of the Axial Skeleton   65

OBSERVING 72
- Activity 1: Bones and Bony Landmarks of the Axial Skeleton   72

WRAPPING UP 77

CHAPTER 5 Appendicular Skeleton 81
GETTING ACQUAINTED 81
- Activity 1: Pectoral Girdle   81
- Activity 2: Upper Limb   84
- Activity 3: Pelvic Girdle   86
- Activity 4: Lower Limb   88

OBSERVING 90
- Activity 1: Bones and Bony Landmarks of the Appendicular Skeleton   90

WRAPPING UP 95

CHAPTER 6 Arthrology 101
GETTING ACQUAINTED 101
- Activity 1: Classification of Joints   101

OBSERVING 110
- Activity 1: Joint Anatomy   110
- Activity 2: Synovial Joints   114

WRAPPING UP 115

CHAPTER 7 Head and Trunk Muscles 119
GETTING ACQUAINTED 119
- Activity 1: Head and Trunk Muscle Attachments and Actions   121
- Activity 2: Identification of Head and Trunk Muscles   123

OBSERVING 125
- Activity 1: Head and Trunk Muscle Anatomy   125

WRAPPING UP 131

CHAPTER 8 Upper Limb Muscles 135
GETTING ACQUAINTED 135
- Activity 1: Muscles of the Scapulothoracic Joint   135
- Activity 2: Muscles of the Glenohumeral Joint   138
- Activity 3: Muscles of the Elbow Joint   143
- Activity 4: Muscles of the Wrist (Anterior Forearm)   146
- Activity 5: Muscles of the Wrist (Posterior Forearm)   151

OBSERVING 154
- Activity 1: Upper Limb Muscle Anatomy   154

WRAPPING UP 159

CHAPTER 9 **Lower Limb Muscles** **161**
   GETTING ACQUAINTED 161
      Activity 1: Muscles of the Hip Joint   161
      Activity 2: Muscles of the Knee Joint   166
      Activity 3: Muscles of the Ankle (Anterior and Lateral Leg)   169
      Activity 4: Muscles of the Ankle (Posterior Leg)   171
   OBSERVING 175
      Activity 1: Lower Limb Muscle Anatomy   175
   WRAPPING UP 181

3 Coming to Your Senses 185

CHAPTER 10 **Central Nervous System** **187**
   GETTING ACQUAINTED 187
      Activity 1: The Brain   187
      Activity 2: Meninges   191
      Activity 3: Spinal Cord   192
   OBSERVING 194
      Activity 1: Central Nervous System Anatomy   194
   WRAPPING UP 199

CHAPTER 11 **Peripheral and Autonomic Nervous Systems** **203**
   GETTING ACQUAINTED 204
      Activity 1: Tissue Structure of Nerves   204
      Activity 2: Cranial Nerves   205
      Activity 3: Spinal Nerves   207
      Activity 4: Spinal Reflexes   210
      Activity 5: Autonomic Nervous System   211
   OBSERVING 213
      Activity 1: Identification of Structures on the Cadaver   213
      Activity 2: Clinical Tests for Peripheral Nerves   218
         Part A: Pupillary Light Reflex   218
         Part B: Eye Movements   219
         Part C: Tongue   219
         Part D: Ears   220
         Part E: Throat   220
         Part F: Face Sensation   220
         Part G: Shoulder and Neck Movements   220
         Part H: Facial Muscles   221
         Part I: Patellar Ligament Reflex   221
         Part J: Stretch Reflex   221
      WRAPPING UP 223

CHAPTER 12 **Special Senses** **227**
    GETTING ACQUAINTED 227
        Activity 1: Key Terms of the Special Senses   227
        Activity 2: Eye Anatomy   229
        Activity 3: Nasal Anatomy   230
        Activity 4: Ear Anatomy   232
        Activity 5: Tongue Anatomy   233
    OBSERVING 235
        Activity 1: Anatomy of the Special Senses   235
        Activity 2: Tests for the Special Senses   236
            Part A: Blind Spot Test   236
            Part B: Visual Acuity Test   237
            Part C: Weber Test   237
            Part D: Rinne Test   237
            Part E: Romberg Test   238
            Part F: Tongue   238
    WRAPPING UP 239

4 Body Highways 241

CHAPTER 13 **Endocrine System** **243**
    GETTING ACQUAINTED 244
        Activity 1: Key Terms of the Endocrine System   244
    OBSERVING 250
        Activity 1: Microscopic Structures
                     of the Endocrine System   250
        Activity 2: Gross Anatomy of the Endocrine System   252
    WRAPPING UP 255

CHAPTER 14 **Cardiovascular System** **257**
    GETTING ACQUAINTED 257
        Activity 1: Heart Layers   257
        Activity 2: Chambers and Valves   258
        Activity 3: Cardiac Cycle and Coronary Circulation   261
        Activity 4: Conduction System   262
        Activity 5: Structures of the Cardiovascular System   263
        Activity 6: Pulmonary Circulation   265
        Activity 7: Blood Vessels   266
    OBSERVING 275
        Activity 1: Cardiovascular System Anatomy   275
        Activity 2: Blood Flow Tracing   278
    WRAPPING UP 281

CHAPTER 15 **Lymphatic System** **285**
GETTING ACQUAINTED 285
    Activity 1: Lymphatic Structures and Organs   288
OBSERVING 292
    Activity 1: Microscopic Structures
        of the Lymphatic System   292
    Activity 2: Gross Anatomy of the Lymphatic System   293
WRAPPING UP 295

CHAPTER 16 **Respiratory System** **297**
GETTING ACQUAINTED 297
    Activity 1: Conduction Airways   297
    Activity 2: Respiratory Airways   300
    Activity 3: Serous Membranes   304
OBSERVING 305
    Activity 1: Microscopic Structures
        of the Respiratory System   305
    Activity 2: Gross Anatomy of the Respiratory System   307
WRAPPING UP 311

CHAPTER 17 **Digestive System** **315**
GETTING ACQUAINTED 315
    Activity 1: Gastrointestinal Tract   315
    Activity 2: Accessory Digestive Organs   320
    Activity 3: Peritoneum   322
    Activity 4: Tissue Layers of the GI Tract   324
    Activity 5: Retroperitoneal and Intraperitoneal Organs   327
OBSERVING 328
    Activity 1: Digestive Tissue   328
    Activity 2: Gross Anatomy of the Digestive System   330
WRAPPING UP 335

CHAPTER 18 **Urinary System** **339**
GETTING ACQUAINTED 339
    Activity 1: Kidney Anatomy   339
    Activity 2: Structures of the Nephron   343
    Activity 3: Ureters, Urinary Bladder, and Urethra   348
OBSERVING 352
    Activity 1: Microscopic Structures of the Urinary System   352
    Activity 2: Gross Anatomy of the Urinary System   354
WRAPPING UP 355

5 The Next Generation ... 357

CHAPTER 19 Male Reproductive System ... **359**

GETTING ACQUAINTED ... 359
Activity 1: Testes and Associated Structures   359
Activity 2: Additional Organs of the
    Male Reproductive System   361
OBSERVING ... 366
Activity 1: Microscopic Structures of the
    Male Reproductive System   366
Activity 2: Gross Anatomy of the
    Male Reproductive System   367
WRAPPING UP ... 369

CHAPTER 20 Female Reproductive System ... **371**

GETTING ACQUAINTED ... 371
Activity 1: Female Reproductive Organs   371
Activity 2: External Female Genitalia   374
OBSERVING ... 378
Activity 1: Microscopic Structures of the
    Female Reproductive System   378
Activity 2: Gross Anatomy of the
    Female Reproductive System   379
WRAPPING UP ... 381

Index ... **383**

1
Inside Out

CHAPTER 1 Introduction to Anatomy 3

CHAPTER 2 Histology 21

CHAPTER 3 Integumentary System 47

CHAPTER 1

Introduction to Anatomy

At the completion of this laboratory session, you should be able to do the following:
1. Demonstrate proficiency in describing locations of anatomical structures using proper anatomical terms.
2. Describe anatomical position.
3. Identify and describe the planes of sections used in anatomy and radiological imaging.
4. Locate and describe the major body regions.
5. Identify the organ systems, their functions, and the major organs in each system.

This laboratory session introduces the basic language of anatomy, which can be intimidating at times. However, with the guidance and practice provided in this lab, you will begin to demonstrate proficiency in correctly describing anatomical structures and their relationships. Medical professionals all over the world use the same, mainly Latin-based, terms to describe health-related conditions.

Anatomy is the language of medicine, and precision is essential when communicating about the human body. Therefore, to prevent misunderstanding, professionals practicing anatomy and medicine use specialized terms to describe the human body. As such, proficiency in the language of anatomy will not only help you in this course, but also throughout your career.

This chapter will familiarize you with the anatomical terminology associated with the following concepts:

1. **Anatomical terms of position.** Anatomy and medicine use a vocabulary of intricate terms to describe the anatomical location of structures. By using this catalog of words, anatomists aim to reduce ambiguity and increase precision. For example, would you expect a mole described as "on the leg" to be located closer to the ankle or knee? Would you expect to find it on the front, side, or back of the leg? It is difficult to determine using only the description provided. However, by using precise anatomical terminology in describing the location of the mole, ambiguity is significantly reduced (e.g., the mole is located 3 cm above the ankle on the anterolateral surface).

2. **Anatomical position.** The anatomical position is the universally accepted frame of reference for describing and relating different body parts to one another. This position refers to a person standing upright and facing forward, with upper limbs at the side and the palms of the hands facing forward, and with lower limbs hip-width apart and the toes facing anteriorly (facing the front).

3. **Anatomical planes of section.** An anatomical plane is an imaginary section used to divide the human body and to describe the location of structures. The three different planes of section are the sagittal plane, transverse plane, and coronal plane. An understanding of these planes is essential in the interpretation of medical imaging.

4. **Body regions.** To better understand the location of anatomical structures, anatomists use language that divides the parts of the body into regions of the head, trunk (neck, thorax, abdomen, pelvis, and perineum), upper limb, and lower limb.

5. **Organ systems.** A systemic study of anatomy looks at each organ system individually. For example, a systemic study of the digestive system would include a discussion of organs such as the stomach and liver. The human body contains 11 organ systems: integumentary system, skeletal system, muscular system, cardiovascular system, lymphatic system, nervous system, endocrine system, respiratory system, digestive system, urinary system, and reproductive system.

GETTING ACQUAINTED

Complete the following activities to become familiar with anatomical terminology.

MATERIALS
Obtain the following items before beginning the laboratory activities:
- Textbook or access to Internet resources

ACTIVITY 1 Anatomical Position and Planes of Section

1 Using your textbook or online resources, write a definition for each of the following terms in Table 1.1.

TABLE **1.1** Anatomical Terms of Position

Term	Definition
Anterior (ventral)	
Posterior (dorsal)	
Superior	
Inferior	
Medial	
Lateral	
Superficial (external)	
Deep (internal)	
Proximal	
Distal	
Cranial (cephalic)	
Caudal	
Anatomical position	

4 ■ *Discovering Anatomy: A Guided Examination of the Cadaver*

2 Fill in the blank boxes in Figure 1.1 using the terms from Table 1.1.

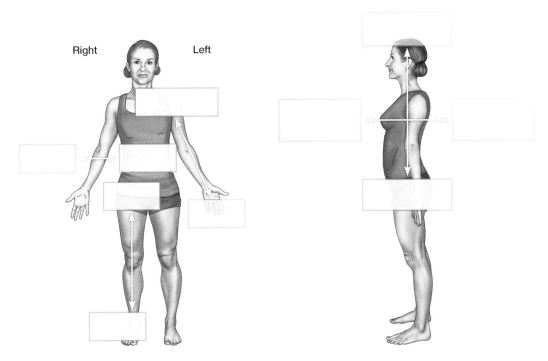

FIGURE **1.1** Anatomical terms of position.

3 Using Figure 1.2 for reference, fill in each of the following statements with the correct directional terms.

a. The skin of the scalp is _____ to the bone of the skull.

b. The brain is _____ to the bone of the skull.

c. The bone of the skull is _____ to the brain.

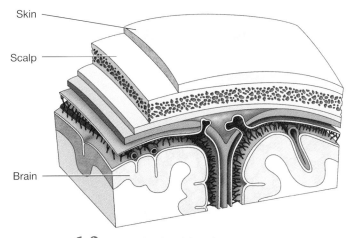

FIGURE **1.2** Superficial and deep layers.

4 Using your textbook or online resources, write a definition for each of the terms in Table 1.2.

TABLE **1.2** Anatomical Planes of Section

Term	Definition
Midsagittal plane	
Coronal (frontal) plane	
Transverse (horizontal/axial) plane	
Longitudinal plane	

5 Label the planes in Figure 1.3 using terms from Table 1.2.

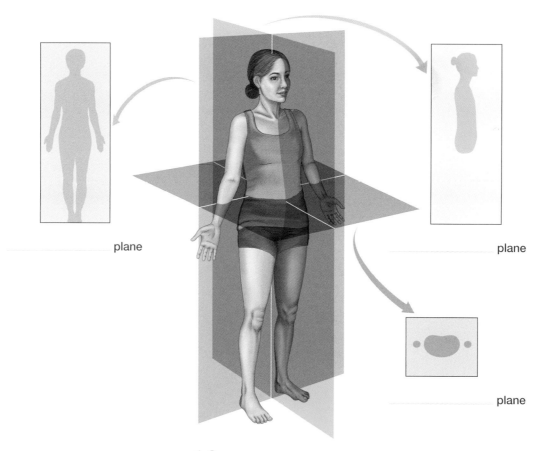

_____ plane

_____ plane

_____ plane

FIGURE **1.3** Anatomical planes of section.

ACTIVITY 2 Body Regions

1 Using your textbook or online resources, write a definition for each region of the head listed in Table 1.3.

TABLE **1.3** Regional Terms for the Head

Term	Definition
Orbital	
Buccal	
Otic	
Frontal	
Nasal	
Mental	
Oral	
Occipital	

2 Label the regions of the head listed in Table 1.3 on Figure 1.4.

FIGURE **1.4** Lateral view of the head.

3 Using your textbook or online resources, write a definition for each region of the trunk listed in Table 1.4.

TABLE **1.4** Regional Terms for the Trunk

Location	Term	Definition
Neck	Cervical	
Thorax	Sternal	
	Mammary	
	Axillary	
	Scapular	
	Costal	
Abdomen	Umbilical	
	Inguinal	
	Lumbar	
	Pubic	
Pelvis/ perineum	Perineal	
	Anal	
	Sacral	

4 Label each region of the trunk listed in Table 1.4 on Figure 1.5.

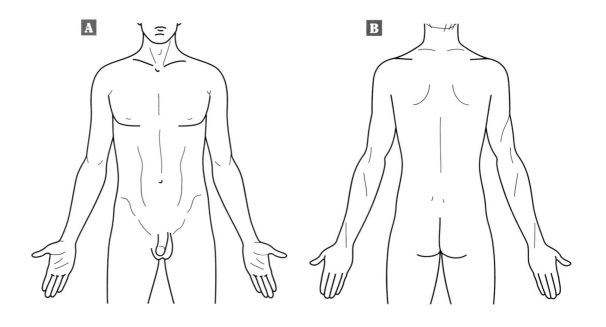

FIGURE **1.5** Trunk: (**A**) anterior; (**B**) posterior.

5 Using your textbook or online resources, write a definition for each region of the upper limb listed in Table 1.5.

TABLE **1.5** Regional Terms for the Upper Limb

Term	Definition
Pectoral area	
Scapular	
Axilla	
Arm (brachial)	
Antecubital	
Forearm (antebrachial)	
Wrist (carpal)	
Palmar	
Digital (phalanges) hand	

6 Label each region of the upper limb listed in Table 1.5 on Figure 1.6.

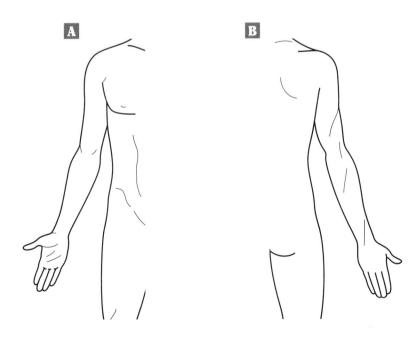

FIGURE **1.6** Upper limb: (**A**) anterior; (**B**) posterior.

7 Using your textbook or online resources, write a definition for each region of the lower limb listed in Table 1.6.

TABLE **1.6** Terms for the Lower Limb

Term	Definition
Pelvic area	
Gluteal	
Thigh	
Popliteal	
Leg (crus)	
Tarsus (ankle)	
Plantar	
Digital (phalanges) foot	

8 Label each region of the lower limb listed in Table 1.6 on Figure 1.7.

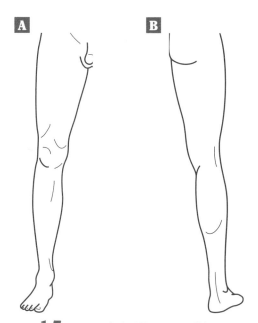

FIGURE **1.7** Lower limb: (**A**) anterior; (**B**) posterior.

ACTIVITY 3 Organ Systems

1 All of our body organs are organized into 11 major organ systems, although some organs are part of more than one organ system. In addition, each organ and organ system possesses a subset of specific functions. Using your textbook or online resources, write down the primary functions for each organ system in Table 1.7.

TABLE **1.7** Major Organs and Functions of Each Body System

Organ System	Primary Function(s)
Integumentary system	
Skeletal system	
Muscular system	
Nervous system	
Cardiovascular system	
Lymphatic system	
Respiratory system	
Endocrine system	
Digestive system	
Urinary system	
Reproductive system	

OBSERVING

Complete the following hands-on laboratory activities to apply your knowledge of anatomical terminology.

Note: Cadavers will be used in the Observing sections throughout this lab manual. When observing the cadaver, note the position of the limbs of the donor as many human donors have their arms pronated. Even though the arms are not in anatomical position, we still use anatomical position as our reference and language of description.

MATERIALS
Obtain the following items before beginning the laboratory activities:
- ❏ Cadavers
- ❏ Gloves
- ❏ Probe
- ❏ Protective gear (lab coat, scrubs, or apron)

ACTIVITY 1 Body Regions and Anatomical Terms of Position

1 Identify the structures listed in Figure 1.8 on the cadaver. Point to each structure as you say it aloud.

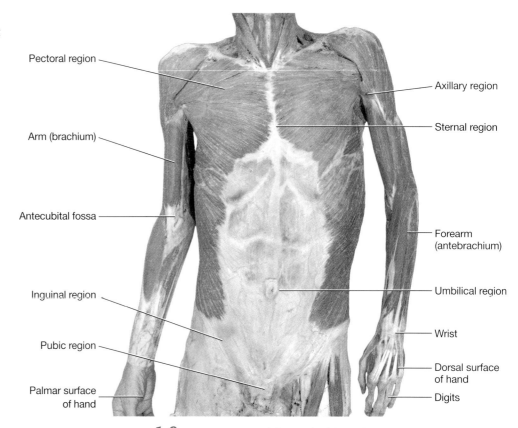

FIGURE **1.8** Anterior view of the trunk of the cadaver.

2 Fill in the blanks using your cadaver as reference.

   a. The pectoral region is _____ to the axillary region.

   b. The forearm is _____ to the arm.

   c. The arm is _____ to the forearm.

   d. The wrist is _____ to the hand.

   e. The wrist is _____ to the forearm.

   f. The sternal region is on the _____ surface of the trunk.

   g. The sternal region is _____ to the umbilical region.

   h. The inguinal region is _____ and _____ to the pubic region.

Discovering Anatomy: A Guided Examination of the Cadaver

3 Identify the structures listed in Figure 1.9 on the cadaver. Point to each structure as you say it aloud.

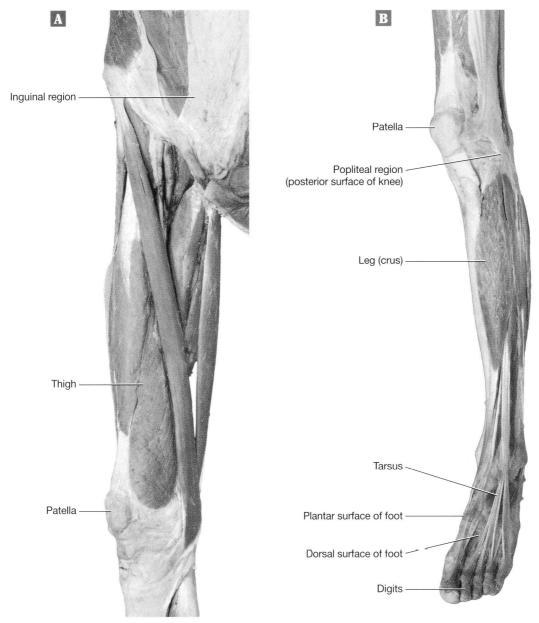

FIGURE **1.9** Anterior view of the leg of the cadaver: (**A**) thigh; (**B**) lower leg.

4 Fill in the blanks using your cadaver as reference.

a. The leg is _____ to the thigh.

b. The tarsus is _____ to the leg.

c. The popliteal region is located on the _____ surface of the lower limb.

d. The patella is _____ to the popliteal region.

5 Identify the structures listed in Figure 1.10 on the cadaver. Point to each structure as you say it aloud.

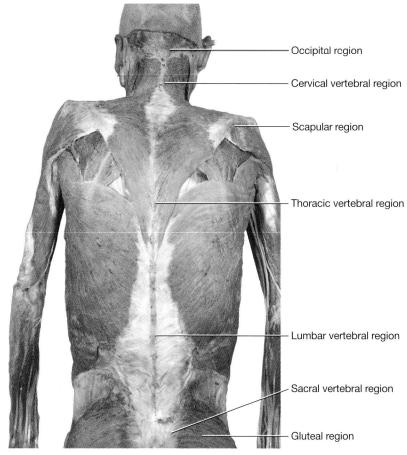

FIGURE **1.10** Posterior view of the trunk of the cadaver.

6 Fill in the blanks using your cadaver as reference.

a. The gluteal region is _____ to the inguinal region.

b. The thoracic region is _____ to the cervical region and _____ to the lumbar region.

c. The sacral region is _____ and _____ to the gluteal region.

d. The scapular region is _____ to the thoracic region.

e. The cervical region is _____ to the thoracic region.

7 Identify the structures labeled in Figure 1.11 on the cadaver. Point to each structure on the cadaver as you say it aloud.

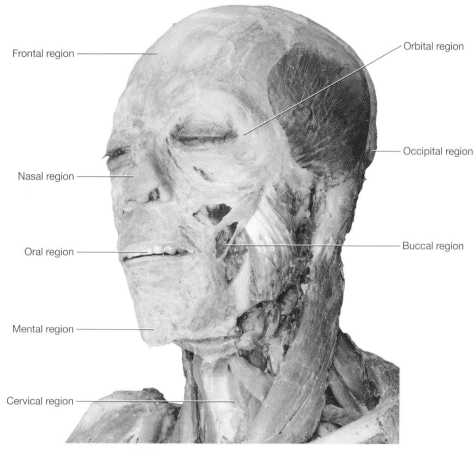

FIGURE **1.11** Anterolateral view of the face of the cadaver.

8 Fill in the blanks with appropriate directional terms.

a. The mental region is _____ to the oral region.

b. The oral region is _____ to the buccal region.

c. The occipital region is _____ to the frontal region and to the cervical region.

d. The nasal region is _____ and _____ to the orbital region.

e. The frontal region is _____ to the orbital region.

9 Identify the structures labeled in Figure 1.12 on the cadaver. Point to each structure on the cadaver as you say it aloud.

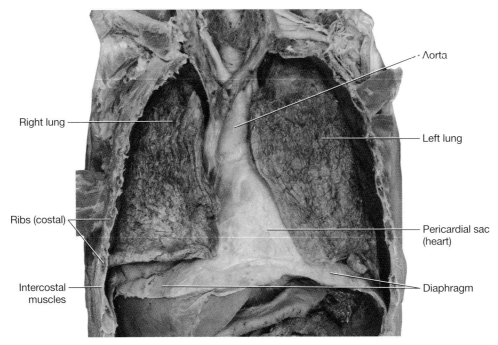

FIGURE **1.12** Interior view of the cadaver.

10 Fill in the blanks with appropriate directional terms.

   a. The lungs are _____ to the heart and _____ to the ribs.

   b. The heart is _____ to the lungs.

   c. The ribs are _____ to the lungs.

   d. The aorta is located on the _____ region of the heart.

   e. The diaphragm is _____ to the lungs and heart.

11 Identify the organ systems associated with each of the following organs.

   a. Lungs:

   b. Heart:

   c. Intercostal muscles:

            and

   d. Diaphragm:

            and

16 ■ *Discovering Anatomy: A Guided Examination of the Cadaver*

WRAPPING UP

Complete the following additional activities to help retain your knowledge of anatomical terminology.

Name _____

Date _____ Section _____

Note: It is not expected that you know the names or functions of the terms listed. These questions are designed for you to practice using directional terms in describing anatomy.

1 Fill in the blanks with appropriate directional terms for Figure 1.13.

a. The gallbladder is

_____

to the liver.

b. The liver is

_____

to the pancreas.

c. The right lobe of the liver is

_____

to the left lobe.

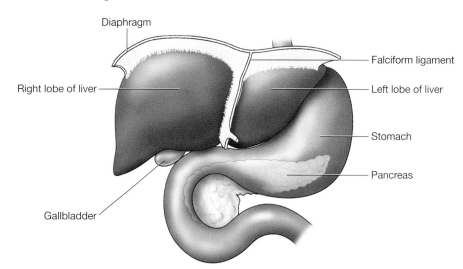

FIGURE **1.13** Liver, gallbladder, and pancreas.

2 Fill in the blanks with appropriate directional terms for Figure 1.14.

a. The lumen is

_____

to the lamina propria.

b. The muscularis mucosae are

_____

to the lamina propria.

c. The serosa is

_____

to the muscularis externa.

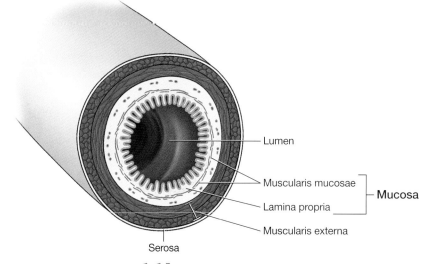

FIGURE **1.14** Tissue layers of the small intestine.

3. Fill in the blanks with appropriate directional terms for Figure 1.15.

   a. The nose is _____ to the mouth.

   b. The ears are _____ and _____ to the eyes.

   c. Hair is _____ to the skull.

   d. The mouth is _____ to the nose.

   e. The skull is _____ to the skin.

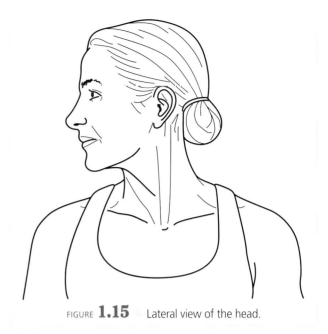

FIGURE **1.15** Lateral view of the head.

4. For each of the radiographic images in Figure 1.16, write the type of plane of section in the space provided.

   a. Plane of section: _____

   b. Plane of section: _____

   c. Plane of section: _____

   d. Plane of section: _____

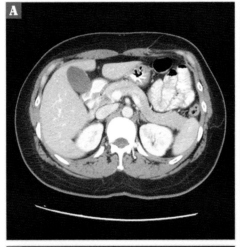
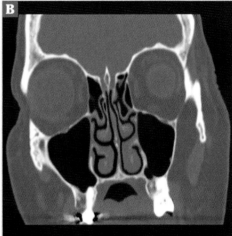
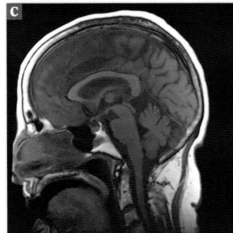
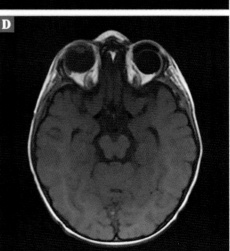

FIGURE **1.16** Various radiographic images of the human body.

Discovering Anatomy: A Guided Examination of the Cadaver

WRAPPING UP
(Continued)

Name _____

Date _____ Section _____

5 The images in Figure 1.17 depict major body organs. In the space provided, write the name of the organ and the organ system to which it belongs.

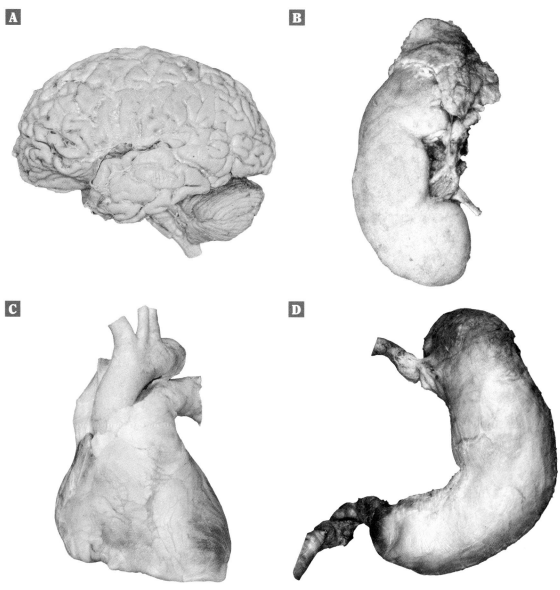

FIGURE **1.17** Various organs of the human body.

a. Organ: _____

   Organ system: _____

b. Organ: _____

   Organ system: _____

c. Organ: _____

   Organ system: _____

d. Organ: _____

   Organ system: _____

CHAPTER 2

Histology

At the completion of this laboratory session, you should be able to do the following:

1. List the basic tissues and describe their position in the hierarchal organization of living matter.
2. Identify and describe the different types of epithelia based upon layers, cell shape, and specialization.
3. List the common types of connective tissues and describe their cellular and extracellular composition.
4. Identify the three types of muscle tissue and describe their distribution throughout the body.
5. Describe nervous tissue and glial cells.

Each of us is composed of seventy trillion cells. All of those cells are organized into groups that form a common purpose. We call those groups "tissues." In other words, a tissue is a group of cells that are similar in structure and organized together for a common purpose. Tissues then combine to form organs, organs combine to form organ systems, and all of the organ systems combine together to make an organism. For example, various tissues combine to form the stomach, the stomach combines with other organs to form the digestive system, and the digestive system combines with other organ systems to complete the human body.

This laboratory session focuses on histology, which is the scientific study of tissues. The primary tissues of the body are the following:

1. Epithelial tissue.
2. Connective tissue.
3. Muscle tissue.
4. Nervous tissue.

Each of these tissues has subclasses that we will explore in this chapter. Tissues possess the following two components:

1. **Cells.** Cells are the smallest structural and functional unit of living matter. For example, epithelial cells make epithelial tissue and muscle cells make muscle tissue.
2. **Extracellular matrix (ECM).** The ECM is a collection of protein fibers (collagen and elastin) and ground substance that reside outside of cells. Ground substance is a gelatinous substance containing water, ions, nutrients, polysaccharides, and glycoproteins. The ECM is produced and secreted by cells and provides structural and biochemical support. Although the ECM is found outside of cells (extracellular) it is contained by the cells of a given tissue. Due to its location between cells (interstitial) it is also called the interstitial fluid.

 GETTING ACQUAINTED

Complete the following activities to become familiar with histology.

ACTIVITY 1 Epithelial Tissue

MATERIALS

Obtain the following items before beginning the laboratory activities:

❏ Textbook or access to Internet resources
❏ Colored pencils

Epithelial tissue predominantly contains epithelial cells with little to no ECM. Without any ECM, there is no room for capillaries between epithelial cells. Therefore, the epithelial tissue is **avascular** because it contains no blood vessels. Epithelium relies on oxygen and nutrients diffusing up from deeper connective tissues. To supply this need, loose connective tissue is always found deep to epithelial basement membranes so its capillaries can provide the necessary nutrient support through diffusion.

Epithelial tissues are anchored to a specialized ECM called the **basement membrane** (referred to as **basal lamina** in electron microscopy). The region of an epithelial cell closest to the basement membrane is termed **basal surface,** and the region closest to the free surface is known as the **apical surface**.

Epithelial tissue is found all throughout the body in the following forms:

1. **Glands.** Epithelium forms glandular tissue for the liver, salivary glands, and pancreas.
2. **Lining.** Epithelium lines the lumen (hollow lining) of all tubular organs (e.g., stomach, uterus, and arteries) and the inside of many body cavities (e.g., abdominal cavity).
3. **Covering.** Epithelium externally covers body surfaces (e.g., skin) and organs (e.g., heart and lungs).

Clinical Application

Carcinomas

Cancer is a disease where a single cell within a tissue, such as an epithelial tissue, has a genetic alteration and starts to replicate uncontrollably. Carcinoma is a type of cancer derived from epithelium. If a carcinoma is detected within the boundaries of the basement membrane it is considered **benign**, or less serious, because the cancer is contained and has not metastasized (spread) to another location. Recall that epithelial tissue is avascular so the cancer cannot spread in the blood stream when corralled by the basement membrane. However, a serious tendency of epithelial cancers is that they fail to respect the basement membrane boundary and penetrate it to invade the tissue beneath. When this occurs, the cancerous cells gain access to blood and lymphatic vessels and are therefore able to spread to other locations. Carcinomas that penetrate the basement membrane are referred to as **malignant**, which are more serious.

Epithelial tissues come in a variety of classifications based on the following characteristics:

1. **Cell layers.** Epithelial cells are found in the following three types of layers:
   a. **Simple.** A single layer of epithelial cells where each cell attaches to the basement membrane.
   b. **Stratified.** Multiple layers of epithelial cells where the upper layers do not attach to the basement membrane.
   c. **Pseudostratified.** Similar to stratified tissue in appearance because the nuclei are at several levels. However, in reality each cell is anchored to the basement membrane.
2. **Cell shape.** Epithelial cells primarily come in the following three shapes:
   a. **Squamous.** Flat cells that resemble a pancake; the nucleus shape is usually flattened and oblong.
   b. **Cuboidal.** Cells that are as wide as they are tall, resembling a cube. Their nuclei have the same volume of cytoplasm to their sides or above and below them.
   c. **Columnar.** Cells that are tall and thin with a nucleus sharing the same shape.
3. **Cellular specializations.** The most common epithelial specializations are as follows:
   a. **Keratin.** Intermediate filament primarily found in epithelial tissue (e.g., epidermis and hair).
   b. **Cilia.** Long, apical projections from the cell's surface that assist in movement of particles.
   c. **Microvilli.** Tiny, fingerlike projections of the cell membrane along the apical surface of the epithelial cell.

To classify epithelium, the following formula is used:

1	Cell layers
2	Cell shape
3	Cellular specialization (if any)
+ 4	The word "epithelium"
Total	Epithelial classification

For example:

1	The tissue has one layer of cells = simple
2	The cells are flat = squamous
3	There are not specializations
+ 4	The word "epithelium"
Total	Simple squamous epithelium

The structure of epithelial tissue is connected to its function. For example, stratified squamous epithelium lines the outside of the body and protects against abrasion and pathogens entering the body. In contrast, simple columnar epithelium lines the internal surface of the small intestines and mediates transport from the lumen, through the epithelial wall, and into the underlying capillaries. Epithelial cells are often arranged as continuous sheets of cells with tight junctions binding the individual epithelial cells together to keep the sheets physically intact.

1 Using your textbook or online resources for reference, draw the listed epithelial tissues in the space provided. Label the following on each sketch: basement membrane, underlying loose connective tissue, apical surface, basal surface, epithelial cell types, and nucleus.

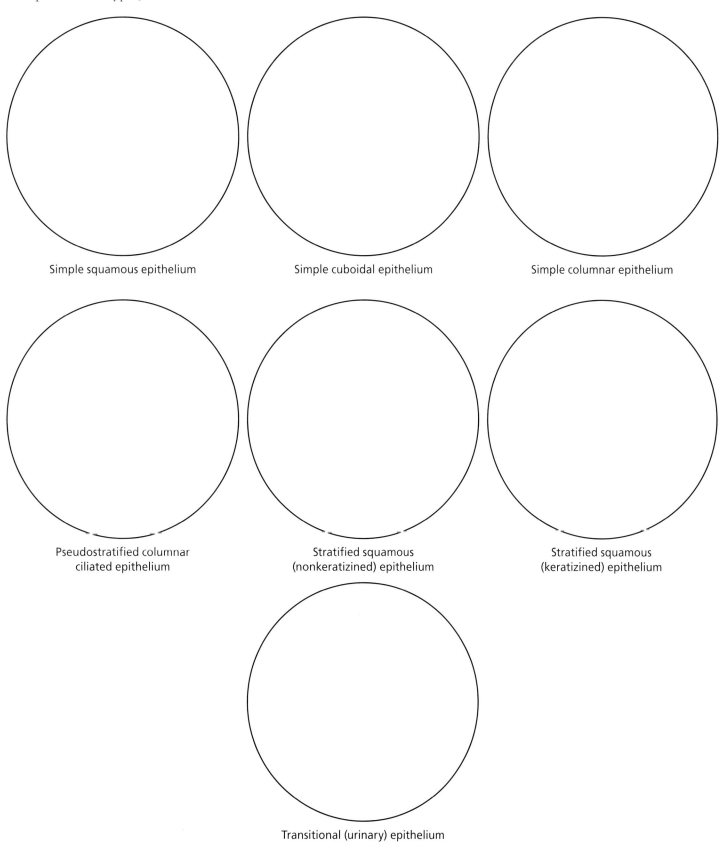

Simple squamous epithelium

Simple cuboidal epithelium

Simple columnar epithelium

Pseudostratified columnar ciliated epithelium

Stratified squamous (nonkeratinized) epithelium

Stratified squamous (keratinized) epithelium

Transitional (urinary) epithelium

2 Table 2.1 contains the primary classifications of epithelium. Descriptions and micrographs are provided in no particular order in Table 2.2 and Figure 2.1, respectively. Match the epithelial classifications with their correct description and micrograph and record your answers in Table 2.1.

TABLE **2.1** Classification of Epithelial Tissue

Type	Description	Micrograph
Simple squamous epithelium		
Simple cuboidal epithelium		
Simple columnar epithelium		
Stratified squamous keratinized epithelium		
Stratified squamous nonkeratinized epithelium		
Pseudostratified columnar ciliated epithelium		
Transitional epithelium		

TABLE **2.2** Epithelial Tissue Descriptions

1	Several layers of epithelial cells; basal cells are cuboidal; surface cells are squamous, lack keratin and, therefore, have nuclei
2	Single layer of tall cells with tall, basally located nuclei
3	Single layer of flattened cells with central, flat nuclei
4	Resembles stratified cuboidal epithelium with surface cells; dome-shaped or squamous-shaped depending on degree of organ stretch
5	Single layer of cube-like cells with central, oval nuclei
6	Single layer of cells of differing heights, with some not reaching the free surface; nuclei located at different levels; contains cilia
7	Several layers of epithelial cells; basal cells are cuboidal; surface cells are squamous, full of keratin, and, therefore, without nuclei

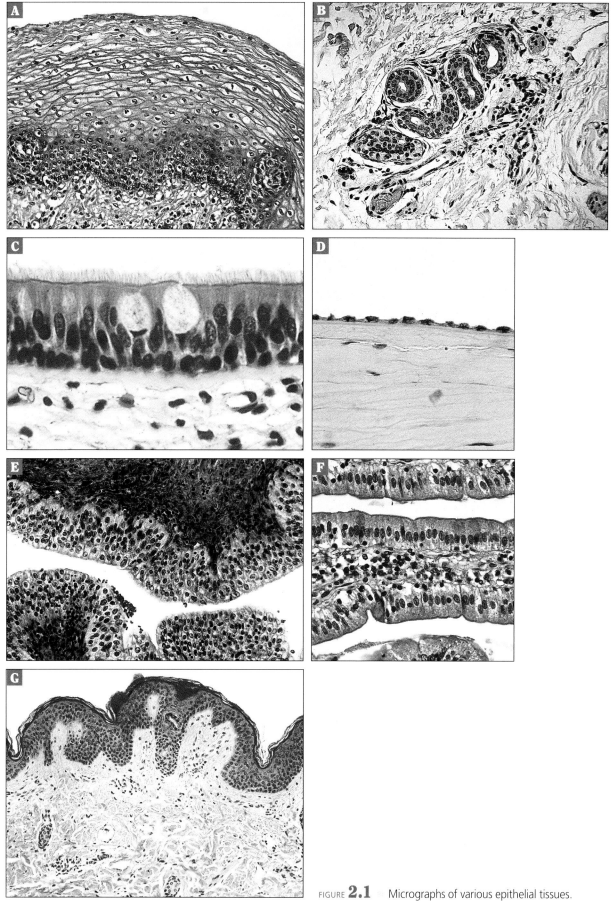

FIGURE **2.1** Micrographs of various epithelial tissues.

ACTIVITY 2 Connective Tissue Proper

A second type of tissue found in the body is connective tissue. It is found in locations all over the body. Connective tissues are classified in the following ways:

1. Connective tissue proper (loose and dense).
2. Fluid CT (lymph and blood).
3. Support CT (bone and cartilage).

Note: Fluid CT will be covered in greater detail in Chapter 14.

The name "connective tissue" is appropriate because they connect other tissues together. For example, loose connective tissue connects the epidermis to the underlying dermis. Connective tissues also aid the body in immune defense, storage, tissue repair, and nutrition. All connective tissues possess a small number of cells relative to their extracellular matrix, are derived from the same embryonic tissue (mesenchyme), and share the following three components in common:

1. Specialized cells.
2. Ground substance.
3. Fibers.

The proportion of each of these components differs in the three types. For example, blood looks very different than bone because bone has higher amounts of collagen fiber, while blood has more ground substance (plasma) and specialized cells. As such, some connective tissues are flexible, tough, and leathery (tendons) in contrast to some that are solid and inflexible (bone) and others that are fluid (blood).

Connective tissue proper consists of loose connective tissues and dense connective tissues based upon the concentration of fibers within the extracellular matrix. Loose (areolar) connective tissues are the most widespread and are located deep to all epithelial basement membranes, surround and cushion most organs, and play a role in the inflammatory process. Dense connective tissue consists of a higher concentration of collagen fibers within a regular or irregular arrangement. Dense regular connective tissue knits muscles and bones together. Dense irregular connective tissue is located all throughout the body in the walls of tubular organs (e.g., GI tract), the skin, nerves, vessels, and musculoskeletal system.

1 This activity will focus on connective tissue proper, specifically. Using your textbook or online resources, write a definition for each of the terms listed in Table 2.3.

TABLE 2.3 Terms Describing Connective Tissue Proper

Term	Definition
Connective tissue proper cells	
Fibrocytes (fibroblasts)	
Adipocytes	
Macrophages	
Mast cells	
Plasma cells	
Extracellular matrix	
Ground substance	
Collagen fiber	
Elastin fiber	
Reticular fiber	

2 Table 2.4 contains the primary classifications of connective tissue proper. Descriptions and micrographs are provided in no particular order in Table 2.5 and Figure 2.2, respectively. Match the classifications with their correct description and micrograph and record your answers in Table 2.4. Identify one or two places in the body where this particular tissue is located.

TABLE **2.4** Classification of Connective Tissue Proper

Classification	Description	Micrograph	Location
Loose connective tissue			
Dense regular collagenous connective tissue			
Dense irregular collagenous connective tissue			
Elastic tissue			
Adipose tissue			

TABLE **2.5** Descriptions of Connective Tissue Proper

1	Fibroblasts and dense collection of collagen fibers in parallel arrangement
2	Closely packed adipocytes with nuclei pushed to the side by large fat droplet
3	Ground substance is more abundant than fibers; located deep to all basement epithelial membranes
4	Dense connective tissue with high proportion of elastic fibers
5	Fibroblasts and dense collection of collagen fibers in irregular arrangement

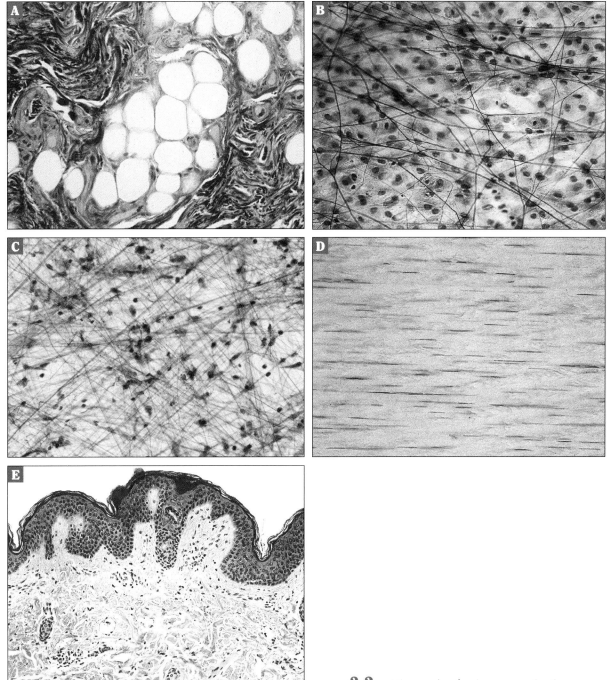

FIGURE 2.2 Micrographs of various connective tissues.

ACTIVITY 3 Other Types of Connective Tissue

This activity will focus on the three other types of connective tissue:

1. **Cartilage.** Cartilage is tough yet flexible and resists tension, twisting, and compressive forces. Cartilage consists of **chondrocytes** (cartilage cells) located in cavities, called **lacunae**, which are embedded in the ECM. Cartilage lacks nerve endings and is avascular. Therefore, it receives its nutrients by diffusion from blood vessels located in the **perichondrium**, its surrounding connective tissue. The three types of cartilage are:
   a. **Hyaline cartilage.** Contains mostly chondrocytes scattered in ground substance with few visible protein fibers. This lack of protein fibers gives hyaline cartilage a smooth, glassy appearance. Hyaline cartilage covers the ends of bones, connects ribs to the sternum, and forms the framework for the respiratory system.
   b. **Fibrocartilage.** Contains a high concentration of collagen fibers, which makes it tough and strong. Fibrocartilage is located in intervertebral discs, the pubic symphysis, and the menisci.
   c. **Elastic cartilage.** Contains a high concentration of the fiber elastin, which enables it to stretch and recoil. Elastic cartilage is found in the ear and epiglottis.
2. **Bone.** Bone is composed of hard osseous tissue that supports and protects softer tissue (e.g., the skull protecting the brain). Bones provide attachments for tendons and ligaments, as well as cavities for fat storage and synthesis of blood cells. Bone tissue consists of the following cells:
   a. **Osteoblasts.** Secrete bone matrix and build new bone.
   b. **Osteoclasts.** Secrete enzymes that catalyze the breakdown of bone matrix.
   c. **Osteocytes.** Mature osteoblasts that reside in lacunae and monitor and maintain the extracellular matrix.

Clinical Application

Strains and Sprains
Athletes are put to the physical test on a regular basis. Sometimes the strain of training or competition overwhelms the body and an injury occurs. One example of this type of athletic injury occurs with CT proper. Ligaments are a type of CT proper that connects bone to bone. When too much stress or strain is placed upon a ligament (e.g., rolling an ankle or lateral stress on the knee), overstretching or tearing of the ligament may occur. This type of injury is known as a "sprain." In other words, a "sprained ankle" is a tearing of the CT proper connecting ankle bones together. For example, an anterior cruciate ligament (ACL) sprain is a tearing of the ligament connecting the tibia and femur bones together.

Clinical Application

Cartilage Damage
Because cartilage is an avascular tissue, a lack of blood supply makes it difficult for the body to send nutrients to help heal injury. This feature of cartilage is compounded by the fact that aging cartilage cells lose their ability to divide. As such, sports injuries that include damage to cartilage (e.g., disc rupture or meniscus injuries in the knees) are very slow to heal or fail to heal altogether. Over time, damaged cartilage may die or become ossified.

Anterior view of right knee as it is flexed. — Medial meniscus

The ECM of bone tissue is arranged in concentric layers called **lamellae**, with the osteocytes sandwiched between them. This structure makes bone the hardest tissue in the body and the most resistant to mechanical stresses. The ECM consists of inorganic calcium salts and abundant collagen fibers.

With few exceptions, long bones have the following general structures:
   a. **Diaphysis.** The shaft of a long bone. From superficial to deep, the diaphysis consists of compact bone, spongy bone, and a medullary cavity filled with bone marrow.
   b. **Epiphysis.** The proximal and distal rounded ends of a long bone.
   c. **Periosteum.** A thin layer of dense irregular connective tissue surrounding and anchored to the external surface of bone via **Sharpey's fibers**, providing an attachment point for tendons and ligaments.
   d. **Compact (cortical) bone.** The external surface of a bone that is hard and dense due to repeating, densely packed subunits called **osteons**. Osteons contain the following features:
      i. **Lamellae.** Concentric rings of bone matrix providing compact bone with a great deal of strength (akin to the annular rings of a tree trunk).

ii. **Central (Haversian) canal.** A canal running down the center of the osteon filled with blood vessels and nerves; lined with endosteum.
iii. **Lacunae.** Microscopic cavities between lamellae that contain osteocytes. Tiny canals, called **canaliculi,** connect adjacent lacunae and their osteocyte occupants to each other.
iv. **Perforating (Volkmann's) canals.** Located perpendicular to the osteon. These canals carry blood vessels into the bone from the periosteum.
v. **Spongy (cancellous) bone.** Located deep to compact bone. This bone resembles the look of a sponge, hence its name. It consists of a latticework-type structure, called **trabeculae.** Bone marrow lies within the spaces of the trabeculae. Red bone marrow produces blood cells and yellow bone marrow is adipose tissue.

3. **Blood.** Blood is different from the other connective tissues in that it does not physically connect anything together and it is a fluid. Blood, however, follows the same pattern as the other connective tissues in that there are cells (red and white blood cells) surrounded by an extracellular matrix, or plasma.

1 Using your textbook or online resources, write a definition for each of the terms listed in Table 2.6.

TABLE **2.6** Osteological Terms

Term	Definition
Osteoblasts	
Osteoclasts	
Osteocytes	
Diaphysis	
Epiphysis	
Periosteum	
Osteon	
Lamellae	
Central (Haversian) canal	
Endosteum	
Lacunae	
Canaliculi	
Perforating (Volkmann's) canal	
Bone marrow	
Medullary cavity	

2 Identify, label, and color Figure 2.3 using the terms from Table 2.6.

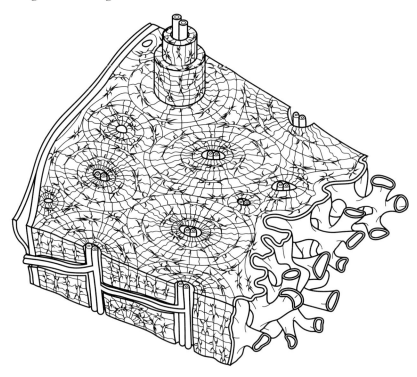

FIGURE **2.3** Microscopic anatomy of compact bone.

3 Identify, label, and color Figure 2.4 using the terms from Table 2.6.

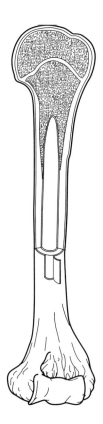

FIGURE **2.4** Anatomy of a long bone.

32 ■ *Discovering Anatomy: A Guided Examination of the Cadaver*

4 Using your textbook or Internet resources, write a definition for each of the terms listed in Table 2.7.

TABLE **2.7** Terms Describing Other Types of Connective Tissue

Term	Definition
Chondrocytes	
Lacunae	
Perichondrium	
Calcium hydroxyapatite	
Red blood cell	
White blood cell	
Plasma	

5 Table 2.8 contains different types of cartilage, bone, and blood. Descriptions and micrographs are provided in no particular order in Table 2.9 and Figure 2.5, respectively. Match the classifications with their correct description and micrograph and record your answers in Table 2.8.

TABLE **2.8** Classification of Other Connective Tissues

Classification	Description	Micrograph
Hyaline cartilage		
Fibrocartilage		
Elastic cartilage		
Bone		
Blood		

TABLE **2.9** Descriptions of Other Connective Tissues

1	Cartilage possessing a high concentration of elastin
2	Fluid connective tissue consisting of RBCs, WBCs, and plasma
3	Cartilage possessing a high concentration of cells with few fibers; glassy appearance
4	Cartilage possessing a high concentration of collagen
5	The hardest tissue in the body

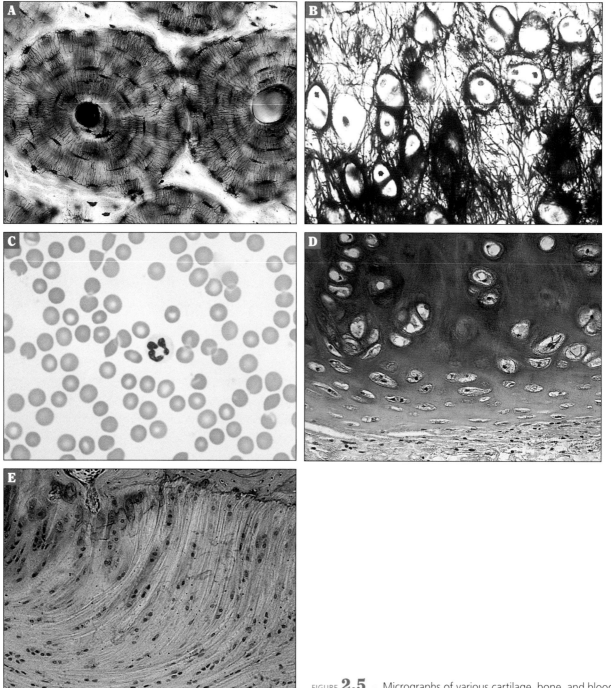

FIGURE 2.5 Micrographs of various cartilage, bone, and blood tissues.

ACTIVITY 4 Muscle Tissue

Muscle tissue is formed from long, thin, muscle cells called **myocytes**. Myocytes are considered striated or smooth depending on the presence or absence of repeating arrangements of contractile proteins called **myosin** and **actin**.

The three types of muscle tissue are:

1. **Skeletal muscle tissue.** Skeletal muscle tissues are voluntary muscle fibers that are long, striated, and multinucleated. This type of tissue attaches voluntary muscles to the skeleton.
2. **Cardiac muscle tissue.** Cardiac muscle tissues are involuntary muscle fibers that are striated and short with specialized cell junctions called intercalated discs, which link myocytes together. This type of tissue is only located in the myocardium of the heart.
3. **Smooth muscle tissue.** Smooth muscle tissues are involuntary muscle fibers that possess a centrally located nucleus and are smooth in appearance because of the irregularly arranged contractile proteins. This type of tissue is primarily located within the walls of blood vessels, hollow organs, skin, and eyes.

1 Using your textbook or Internet resources, write a definition for each of the terms listed in Table 2.10.

TABLE **2.10** Terms Describing Muscle Tissue

Term	Definition
Myofilaments	
Myocyte	
Myofiber	
Endomysium	
Intercalated discs	
Striated muscle	
Nonstriated muscle	

2 Table 2.11 contains different types of muscle tissue. Descriptions and micrographs are provided in no particular order in Table 2.12 and Figure 2.6, respectively. Match the classifications with their correct description and micrograph, and record your answers in Table 2.11.

TABLE **2.11** Classification of Muscle Tissues

Type	Cell Shape	Position of Nucleus	Presence of Striations	Presence of Intercalated Discs
Skeletal muscle				
Cardiac muscle				
Smooth muscle				

TABLE **2.12** Descriptions of Muscle Tissues

1	Muscle tissue comprised of branching, uninucleated, striated fibers with intercalated discs; under involuntary control
2	Muscle tissue comprised of long, cylindrical, multinucleated, striated fibers; under voluntary control
3	Muscle tissue comprised of spindle-shaped cells with central nuclei and nonstriated fibers, nonstriated; under involuntary control

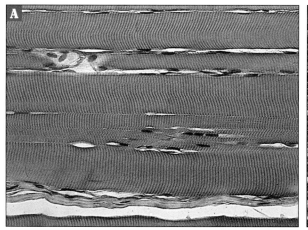

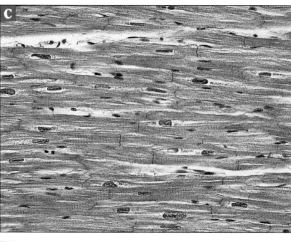

FIGURE **2.6** Micrographs of various muscle tissues.

ACTIVITY 5 Nervous Tissue

Nervous tissue is composed of neurons (nerve cells) and neuroglia (supporting cells).

1. **Neurons.** Neurons are the functional cells of nervous tissue in that they are responsible for sending and receiving messages. Unlike the other three tissue types (epithelial, muscle, and connective tissue cells), neurons are not attached to their neighboring neurons. Neurons are separated by a space called a synapse and communicate across this synapse via neurotransmitters. Most neurons contain the following components:
   a. **Cell body (soma).** Spherical part of the neuron that contains the nucleus and connects to dendrites and axon.
   b. **Dendrites.** Extensions of the cell body that receive messages from adjacent cells at synapses throughout the body.
   c. **Axon.** The long extension of a neuron where impulses are conducted from the cell body to other neurons, muscle cells, or glands.
   d. **Synapses.** Neurons are separated via a synaptic cleft or synapse, and neurons communicate across this synaptic cleft via neurotransmitters.
2. **Neuroglia.** Neuroglia are the smaller, more numerous cells that surround and support the neurons by insulating, providing scaffolding, or contributing to the blood brain barrier. The neuroglia include the following:
   a. **Oligodendrocytes.** Produce myelin for neurons in the central nervous system.
   b. **Schwann cells.** Produce myelin for neurons in the peripheral nervous system.
   c. **Astrocytes.** Star-shaped glial cells that regulate the transmission of electrical impulses in neurons and support the blood brain barrier.
   d. **Ependymal cells.** Line the ventricular system of the brain.
   e. **Microglia.** Immune cells that serve as macrophages of the central nervous system.

1 Using your textbook or online resources, write a definition for each of the terms listed in Table 2.13.

TABLE **2.13** Terms Describing Nervous Tissue

Term	Definition
Neuron	
Neuroglia	
Cell body	
Dendrite	
Axon	
Synapse	
Neurotransmitter	

👁 OBSERVING

Complete the following hands-on laboratory activities to apply your knowledge of histology.

> **Note:** Digital images of the tissues may be used if light microscopes and prepared slides are not available.

During this activity you will identify the four types of tissues and their subclasses on the slides provided, using the microscope or digital images provided by your instructor. Here are a few suggestions to consider when examining microscopic images:

- **Examine the slide with the naked eye.** What different colors, shapes, and structures can you see before you place the slide on the microscope?
- **View the slide on low power first.** Once you scan the image on low power (4×), advance to higher magnifications and scan the slide to see more details.
- **Use example micrographs to guide you.** Use the micrographs from the activities in this chapter's Getting Acquainted section to help you identify each tissue type. If at first your slide does not resemble the images in this text, try scrolling around the slide and keep looking. Note that you should not completely rely on the images in the pre-class work because the slides you are provided with may be from different tissues or prepared with different stains. Many slides are sections from whole organs and may contain multiple tissues. As such, you may have to hunt for the specific tissues being studied.
- **Study other images.** It sometimes takes a little while, but you will get your "histology eyes" eventually. Initially, epithelial tissues may just look like "purple dots." After viewing numerous examples, however, you will start to form your histology eyes and distinguish basement membranes, layers, and shapes of cells.

MATERIALS
Obtain the following items before beginning the laboratory activities:

- ❏ Colored pencils
- ❏ Light microscope
- ❏ Prepared microscope slides of the following:
    - ☐ Simple squamous epithelium
    - ☐ Simple cuboidal epithelium
    - ☐ Simple columnar epithelium
    - ☐ Adipose tissue
    - ☐ Hyaline cartilage
    - ☐ Fibrocartilage
    - ☐ Loose connective tissue
    - ☐ Dense irregular collagenous connective tissue
    - ☐ Dense regular collagenous connective tissue
    - ☐ Dense elastic tissue
    - ☐ Bone (osseous) tissue
    - ☐ Blood tissue
    - ☐ Muscle tissue
    - ☐ Skeletal muscle tissue
    - ☐ Cardiac muscle tissue
    - ☐ Smooth muscle tissue
    - ☐ Nervous tissue

ACTIVITY 1 Epithelial Tissue

A few helpful hints to consider while looking at epithelial tissues include the following:

- **Epithelium is highly cellular.** You will find that epithelial tissues are comprised mostly of cells. Many of the purple dots in epithelium are nuclei. The cell membranes are often difficult to see so you need to impose cell membranes around the nuclei.
- **Epithelium is avascular.** As a result, you will not see any vascular elements like capillaries within the epithelial tissue.
- **Epithelium is peripherally located.** Therefore, you will typically find epithelium on the outer edge of the slide. Because most slides have several tissues in each section, you may need to scroll around the slide to find epithelium.

1 Look at the epithelial tissue slides under the microscope and draw and label what you see. Use your coloring pencils to differentiate between cells, tissues, and structures. Write a description of each epithelial tissue's appearance and list some hints to yourself on how to distinguish structures.

a. Simple squamous epithelium

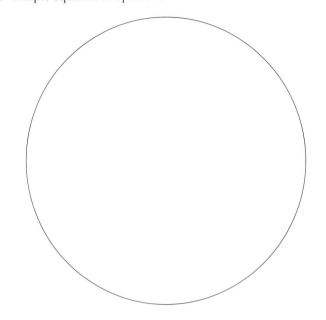

Description

Location(s)

b. Simple cuboidal epithelium

Description

Location(s)

c. Simple columnar epithelium

Description

Location(s)

d. Stratified squamous keratinized epithelium

Description

Location(s)

ACTIVITY 2 Connective Tissue

The following principles will assist you in identifying and differentiating between the various types of connective tissues:

1. **Cells.** Connective tissues typically have spread-out cells (look for the nuclei) surrounded by a large amount of space called extracellular matrix (ECM); ECM contains ground substance and protein fibers.
   a. One exception is adipose tissue, which consists of densely packed adipocytes filled with a large lipid droplet.
2. **CT Fibers.** Connective tissues can be distinguished by the variety of fibers they contain.
   a. **Collagen fibers.** Thick fibers that often stain pink and are located in fibrocartilage, dense regular collagenous CT, dense irregular collagenous CT, and loose CT.
   b. **Elastic fibers.** Fibers that are thinner than collagen, usually stain black or purple, are wavy in appearance, and are located in elastic tissue, elastic cartilage, and loose CT.
3. **Cartilage.** Cartilage fibers are the easiest to discern from CT proper by looking at the shape of the cells. CT proper cells (fibroblasts) are generally small and flat. In contrast, cartilage cells (chondrocytes) are larger and round and usually sit in lacunae.
4. **Blood and Bone.** The easiest of the tissues to identify are blood and bone. They should look just like the images in Activity 3 of this chapter's Getting Acquainted section.

1 Look at the connective tissue slides under the microscope and draw and label what you see. Use your coloring pencils to differentiate between cells, tissues, and structures. Write a description of each CT's appearance and list some hints to yourself on how to distinguish structures.

a. Loose connective tissue

b. Dense irregular collagenous connective tissue

Description

Location(s)

Description

Location(s)

c. Dense regular collagenous connective tissue

d. Adipose tissue

Description

Description

Location(s)

Location(s)

ACTIVITY 3 Muscle and Nervous Tissue

The following principles will assist you in identifying and differentiating between the various types of muscle and nervous tissues:

1. **Skeletal muscles.** Skeletal muscle tissue has long, cylindrical, multinucleated cells with a striated appearance.
2. **Cardiac muscle.** Cardiac muscle tissue has cells that branch, are striated, and have a centrally located nucleus; look for intercalated discs between adjacent cardiac muscle cells.
3. **Smooth muscles.** Smooth muscle tissue has spindle-shaped cells and no striations. Their nuclei are centrally located and look like corkscrews at higher power.

1 Look at the muscle and nervous tissue slides under the microscope and draw and label what you see. Use your coloring pencils to differentiate between cells, tissues, and structures. Write a description of each muscle and nervous tissue's appearance and list some hints to yourself on how to distinguish structures.

a. Skeletal muscle tissue

Description

Location(s)

b. Cardiac muscle tissue

Description

Location(s)

c. Smooth muscle tissue

Description

Location(s)

d. Nervous tissue

Description

Location(s)

WRAPPING UP

Complete the following additional activities to help retain your knowledge of histology.

Name _____

Date _____ Section _____

1. An organ is a group of tissues organized for a common purpose. Therefore, an organ is composed of a variety of tissues knitted together. Determine the main tissue types that constitute each of the organs listed in Table 2.14, and record the information in the space provided. Be specific about the type of epithelial tissue, connective tissue, or muscle tissue.

TABLE **2.14** Tissues in Various Organs

Organ	Epithelial Tissue	Connective Tissue	Muscle Tissue
Left ventricle			
Aorta			
Skin			
Stomach			
Uterus			
Primary bronchus			
Urinary bladder			

UNIT 1: Inside Out

CHAPTER 2: Histology

2. Identify and classify the tissues in Figure 2.7 using the answer lines provided.

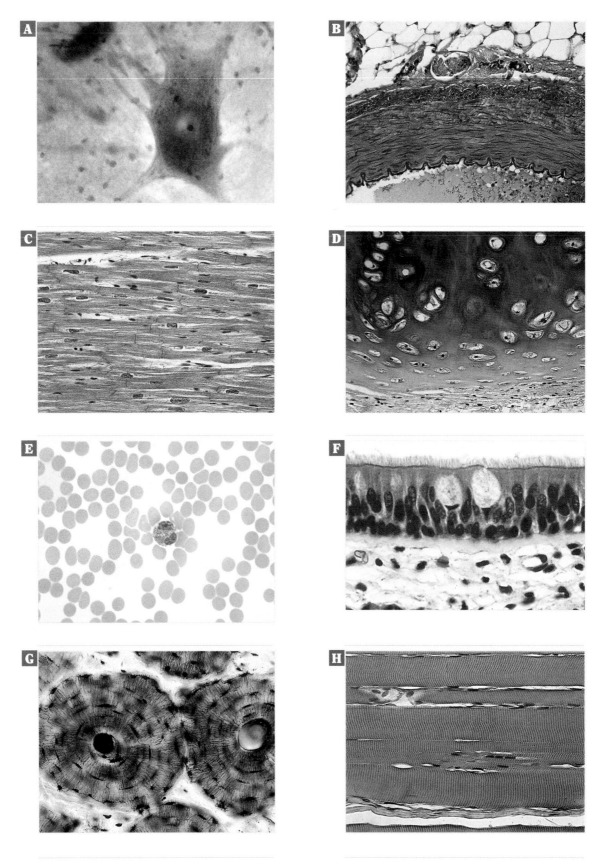

FIGURE 2.7 Various tissues.

WRAPPING UP
(Continued)

Name _____

Date _____ Section _____

3 Match the tissue type in the left-hand column with its correct location in the body in the right-hand column.

_____ Dense irregular collagenous connective tissue	a. Gluteus maximus muscle
_____ Simple cuboidal epithelium	b. Lining the lumen of the trachea
_____ Cardiac muscle	c. Biceps tendon
_____ Pseudostratified ciliated columnar epithelium	d. Tunica media of an artery
_____ Nervous tissue	e. Deep to epithelial basement membrane
_____ Fibrocartilage	f. Lining the lumen of the jejunum
_____ Skeletal muscle	g. Spinal cord
_____ Simple squamous epithelium	h. Proximal convoluted tubule in nephron
_____ Smooth muscle	i. Intervertebral disc
_____ Dense regular collagenous connective tissue	j. Dermis of the skin
_____ Hyaline cartilage	k. Right atrium
_____ Bone	l. Epiphyseal (growth) plate
_____ Simple columnar epithelium	m. Femur
_____ Loose connective tissue	n. Alveoli

4 Identify the structures in Figure 2.8 and list them with their functions in Table 2.15.

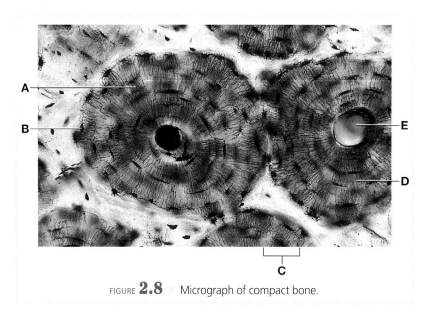

FIGURE **2.8** Micrograph of compact bone.

TABLE **2.15** Structures and Functions of Structures in Compact Bone

	Structure	Function
A		
B		
C		
D		
E		

CHAPTER 3

Integumentary System

At the completion of this laboratory session, you should be able to do the following:

1. Name the specific tissue types composing the layers of the integument (epidermis, dermis, and hypodermis), describe their histologic composition, and list their functions.
2. Identify appendages of the epidermis (hair, sebaceous glands, sweat glands, and nails) and describe their functions.
3. Describe the gross and microscopic structure of thin and thick skin.

The integument, more commonly referred to as skin, is the largest organ in the body. It covers 2 square meters, weighs 4 kilograms, and accounts for 7 percent of total body weight. The integument (Latin for "covering") covers our body and forms a protective barrier between our body's internal environment and the outside world. Without our integument we would dehydrate, become hypothermic, and quickly fall prey to countless germs. Additional functions of the integumentary system include protection (against desiccation, germs, and UV radiation), thermoregulation, synthesis of vitamin D, and reception of sensory stimuli.

🖐 GETTING ACQUAINTED

Complete the following activities to become familiar with the integumentary system.

The integument consists of the following three layers:

1. **Epidermis.** The epidermis is the outer layer of skin that is in continual contact with the outside world. It is formed by stratified squamous keratinized epithelium. The two principal cells comprising the epidermis include the following:

    a. **Keratinocytes.** Keratinocytes comprise approximately 90 percent of the epidermal cells. These cells produce the protein keratin, which forms a barrier against germs (bacteria, fungi, and viruses) and prevents water loss. Keratinocytes are tightly attached together via anchoring proteins, called desmosomes, and arise from the basal layer of the epidermis where they undergo continuous mitotic division. Daughter cells are continually pushed toward the skin surface by the production of new cells beneath. Migrating keratinocytes produce keratin that will eventually dominate their cytoplasm. By the time the keratinocytes reach the superficial layer of the epidermis, they are dead and little more than scalelike cells filled with keratin, having been too far from the nutrients supplied by the capillaries in the dermis. Millions of these dead cells rub off every day, resulting in a new epidermis every month.

    b. **Melanocytes.** Melanin is made and bundled in tiny packages called melanosomes. Melanocytes extend cytoplasmic projections to adjacent keratinocytes and transfer the melanosomes to the cell's apical surface to protect the body from the harmful effects of UV radiation. Sunlight stimulates melanocytes to greater activity and, therefore, prolonged sun

MATERIALS
Obtain the following items before beginning the laboratory activities:

❏ Textbook or access to Internet resources
❏ Colored pencils

Clinical Application

Psoriasis
Psoriasis is a chronic inflammatory skin disease. In individuals with normal skin, the proliferation and sloughing off of the epidermis is in a state of balance called homeostasis. However, in individuals with psoriasis, the proliferation of keratinocytes is greatly accelerated and their maturation is abnormal. This imbalance of proliferation and sloughing off results in the accumulation of incompletely differentiated keratinocytes in the stratum corneum. As a result, the skin in areas where this occurs (often scalp, elbow, and knees) possesses itchy, scaly, red plaques.

exposure results in melanin buildup to protect against the harmful exposure to the sun's UV radiation. As a result, the skin becomes darker, producing what we know as a tan. Ironically, melanocytes are lighter in color than keratinocytes when viewed under a microscope. Pigmented moles and freckles are simply accumulations of melanin. People have different skin color because each person produces different amounts and kinds of melanin. For example, an African-American's melanocytes produce more and darker melanin, and their melanin is retained longer than melanin from a Caucasian individual.

The epidermis consists of the following layers:
  a. **Stratum corneum.** The most superficial layer of dead keratinocytes that are sloughed off from abrasion.
  b. **Stratum lucidum.** The clear layer observed in thick skin only.
  c. **Stratum granulosum.** The granular layer.
  d. **Stratum spinosum.** The spiny layer.
  e. **Stratum basale/germinativum.** The bottom layer anchored to the basement membrane that mitotically divides to supply the more superficial layers of the epidermis; melanocytes usually reside here as well.

2. **Dermis.** The dermis is the layer of integument deep to the epidermis composed of loose connective tissue and dense irregular collagenous connective tissue. It supports and nourishes the epidermis. The dermis consists of two layers:
    a. **Papillary layer.** Superficial layer composed of loose connective tissue; located deep to the basement membrane of the epidermis.
        i. **Tactile (Meissner's) corpuscle.** A cutaneous mechanoreceptor within the papillary layer responsible for sensing light touch concentrated in hairless skin (e.g., fingertips).
    b. **Reticular layer.** Layer consisting of dense irregular collagenous connective tissue.
        i. **Lamellar (Pacinian) corpuscle.** A cutaneous mechanoreceptor within the reticular layer responsible for sensing vibration and pressure.
3. **Hypodermis.** The hypodermis (also referred to as the subcutaneous layer or superficial fascia) is the deepest layer of the integument and consists of loose connective tissue and adipose tissue. It stores fat and attaches the skin to underlying tissues.

Skin contains a variety of appendages derived from epidermis such as sweat glands, hair follicles and their associated sebaceous glands, and nails.

Clinical Application

Malignant Melanoma

Malignant melanoma accounts for only 5 percent of skin cancers. However, it is the most aggressive type of skin cancer and, if it metastasizes, can spread unpredictably to any organ. Melanomas first appear in the epidermis, often in an area of normal skin, where they initially progress in a radial growth phase (think east to west). They may progress to a more dangerous vertical growth phase (think north to south) where the melanoma breaks through the epithelial basement membrane. Once it progresses into the dermis, blood vessels and lymphatics can transport the cancer cells to any organ in the body. To distinguish a "happy" mole from a suspicious or "unhappy" mole, remember the A, B, C, D, and E's of melanomas:

A – Asymmetry. Happy moles are often symmetrical. Unhappy moles are often irregular and asymmetrical in shape.

B – Border irregularity. Happy moles often have smooth, even borders. Unhappy moles usually have irregular borders that are difficult to define.

C – Color. Happy moles usually have a single shade of brown or tan. Unhappy moles often have more than one color in an uneven distribution (blue, black, brown, tan, etc.).

D – Diameter. Happy moles are usually smaller than 6 mm. Unhappy moles are usually greater than 6 mm.

E – Evolution. The evolution of the mole is the most important factor to consider when it comes to diagnosing a melanoma. If a mole undergoes changes in size or color, it should be brought to the attention of a physician immediately.

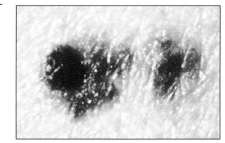

Cancerous mole.

 **Weird and Wacky**

Tattoos

The word "tattoo" is defined as a mark upon the skin with pigment. The act of tattooing has been practiced for centuries. In fact, the oldest recorded human tattoo was found on a 5,000-year-old mummy. When considering tattoos, it is important to remember the first two layers of the skin: the epidermis and dermis. If a tattoo is placed on the epidermis (think of the cheap ones purchased at convenience stores), then the tattoo lasts only days to weeks because of the continual shedding of the layers of the epidermis. However, a dermal tattoo is permanent because a needle from a tattoo machine (moving up and down 50 to 150 times per second) traverses the epidermis and injects ink one drop at a time into the papillary layer of the dermis. The dermis does not turn over like the epidermis and, therefore, the ink is there permanently. Over the years, the tattoo fades and becomes blurred because the ink particles move deeper into the reticular layer of the dermis. In summary, epidermal tattoos are temporary, whereas dermal tattoos are forever.

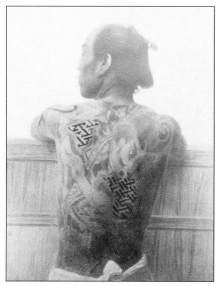

Japanese man with tattoos (c. 1890–1909).

 Clinical Application

Hypodermic Injections

Injections into the hypodermis are a common and preferred route for administering certain medicines, such as insulin. This is due to the high concentration of blood vessels, which enables the injected drug to be easily absorbed into the blood stream and carried throughout the body. The hypodermic needle is also a clean way of injecting substances into the body with minimum risk of contamination. The smooth surface of the needle prevents germs from accumulating on its surface, and the sharp point creates a very small opening, preventing germs from entering.

 Clinical Application

First-, Second-, and Third-Degree Burns

Burns are characterized by the damage caused to the skin, rather than the burning sensation associated with this injury. There are three primary types of burns, each based on the severity of damage to the skin:

- **First-degree burns.** The burn extends only into the top layer of the epidermis. The skin is usually red, non-blistered, and heals in a few days without scarring.
- **Second-degree burns.** The burn extends beneath the epidermis and into the dermis where plasma leaks out and forms blisters. The second-degree burns are divided into superficial partial thickness burns (extending through the papillary layer of the dermis) and deep thickness burns (extending into the reticular layer of the dermis).
- **Third-degree burns.** The burn extends through the epidermis and dermis and into the hypodermis. These burns are usually white and leathery in appearance. They have a high risk for infection and usually result in hospital stays.

ACTIVITY 1 Integumentary System Structures

1 Using your textbook or Internet resources, write a definition for each of the structures listed in Table 3.1.

TABLE **3.1** Integumentary System Terms

Layer	Term	Definition
Epidermis	Epidermis	
	Stratified squamous keratinized epithelium	
	Keratinocyte	
	Melanocyte	
	Stratum corneum	
	Stratum lucidum	
	Stratum granulosum	
	Stratum spinosum	
	Stratum basale	
	Sweat gland	
	Hair follicle	
	Sebaceous gland	
Dermis	Dermis	
	Dermal papillae	
	Loose connective tissue	
	Reticular layer of dermis	
	Dense irregular collagenous connective tissue	
	Tactile (Meissner's) corpuscle	
	Lamellar (Pacinian) corpuscle	
	Thermoreceptor	
	Arrector pili muscle	
Hypodermis	Hypodermis	
	Adipose tissue	

2 Identify, label, and color the following structures on Figure 3.1.

- ☐ Epidermis
- ☐ Stratum corneum
- ☐ Stratum basale
- ☐ Sweat gland and duct
- ☐ Hair follicle
- ☐ Arrector pili muscle
- ☐ Sebaceous gland
- ☐ Dermis
- ☐ Meissner's corpuscle
- ☐ Pacinian corpuscle
- ☐ Hypodermis
- ☐ Adipose tissue
- ☐ Cutaneous blood vessels

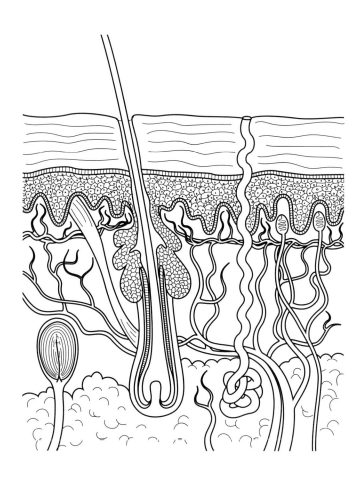

FIGURE **3.1** Anatomy of the skin.

ACTIVITY 2 Histology of the Skin

1 The following list contains anatomical structures associated with the skin and a list of associated micrographs, in no particular order. Match the anatomical structure with its associated micrograph in Figure 3.2.

Epidermis   Hypodermis   Sebaceous gland

Dermis (papillary layer)   Hair follicle   Pacinian corpuscle

Dermis (reticular layer)   Sweat gland

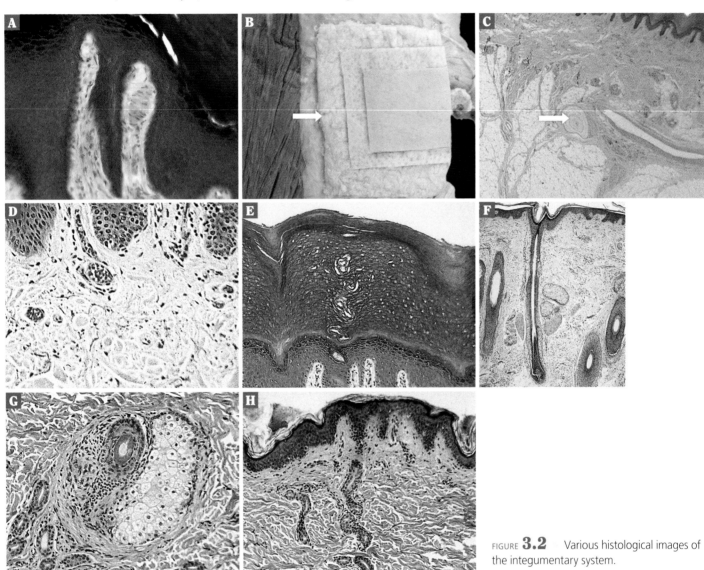

FIGURE **3.2** Various histological images of the integumentary system.

Weird and Wacky

Dust Bunnies
There are statistics floating around that dust is comprised of 80 percent dead human cells. Yikes! This supposed fact might make you want to move out and live in your front yard. However, before you start packing, rest assured that this is not true. Dust particles accumulating throughout the house primarily come from outside sources such as soil tracked in from our shoes and outdoor air particles that float in through open windows and doors. Approximately one-third of dust volume comes from inorganic sources such as carpet fibers and furniture lint. And yes, a very small percentage does come from dead flakes of sloughed-off human skin as well.

Dust "bunnies."

👁 OBSERVING

Complete the following hands-on laboratory activities to apply your knowledge of the integumentary system.

Note: Digital images of the tissues may be used if light microscopes and prepared slides are not available.

During this activity you will study micrographs of the skin using a microscope or through digital images. Refer to the hints and tips provided on page 38 in Chapter 2 for guidelines to assist you when examining microscopic images.

MATERIALS
Obtain the following items before beginning the laboratory activities:

❑ Colored pencils
❑ Light microscope
❑ Prepared microscope slides of thin skin and thick skin

ACTIVITY 1 Skin Anatomy

1. Examine prepared slides of thick and thin skin.

2. Use colored pencils to draw what you see under the microscope, and label the indicated structures. Write at least two examples of where that region of skin is found in the body.

   a. Thick skin
      Draw and label the following structures:
      ☐ Epidermis
         - Stratum corneum
         - Stratum lucidum
         - Stratum granulosum
         - Stratum spinosum
         - Stratum basale
      ☐ Epidermis appendages
         - Sweat gland
      ☐ Dermis
         - Papillary layer
         - Reticular layer

   Location(s):

   1.

   2.

UNIT 1: Inside Out   CHAPTER 3: Integumentary System ■ 53

b. Thin skin

Draw and label the following structures:

- ☐ Epidermis
    - Stratum corneum
    - Stratum lucidum
    - Stratum granulosum
    - Stratum spinosum
    - Stratum basale
- ☐ Epidermis appendages
    - Sweat gland
    - Hair follicle
    - Sebaceous gland
- ☐ Dermis
- ☐ Papillary layer
    - Reticular layer

Location(s):

1.

2.

WRAPPING UP
Complete the following additional activities to help retain your knowledge of the integumentary system.

Name _____

Date _____ Section _____

1. Label the following parts of the skin on Figure 3.3.
   - ☐ Arrector pili muscle
   - ☐ Cutaneous blood vessels
   - ☐ Dermal papillae
   - ☐ Dermis
   - ☐ Epidermis
   - ☐ Hair follicle
   - ☐ Hair shaft
   - ☐ Hypodermis
   - ☐ Sebaceous gland
   - ☐ Sweat gland

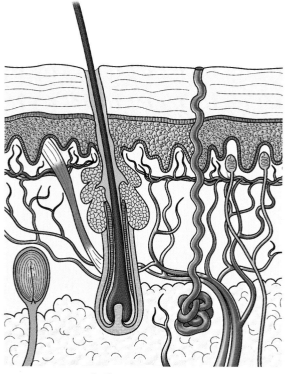

FIGURE **3.3** Anatomy of the skin.

2. Match the integumentary structure in the left-hand column with its correct description in the right-hand column.

   _____ Stratum corneum
   _____ Stratum basale
   _____ Melanocyte
   _____ Keratinocyte
   _____ Dermis: papillary layer
   _____ Dermis: peticular layer
   _____ Hypodermis
   _____ Sebaceous gland
   _____ Sweat gland
   _____ Pacinian corpuscle

   a. Functions in thermal regulation
   b. Epidermal cell that produces the structural protein keratin
   c. Pressure receptor in the dermis
   d. Projections of the dermis that indent the epidermis
   e. Deep layer of dense irregular collagenous connective tissue
   f. Secretes sebum
   g. Produces a pigment protecting skin from the sun
   h. Deepest layer of the epidermis containing a single row of mitotically active cells
   i. Layers of epidermal cells that protect underlying tissue from infection, dehydration, and abrasion
   j. Deepest layer of the skin containing adipose tissue

3 Using your knowledge of the integumentary system, provide explanations for the following:

a. The palm of the hand contains more Pacinian corpuscles than the arm. Why?

b. What factor(s) determine why some paper cuts bleed while others do not?

c. Why does a deep cut increase the risk of an infection?

d. Very few drugs can be delivered transdermally (through the skin). Why?

e. The skin produces sweat when we exercise. Why?

4 A 29-year-old man sustains multiple burns to his lower limbs as a result of a fire. In the emergency room, his burns are assessed for the amount of skin damaged and grading. The burn injuries to his left thigh involve the epidermis and dermis and do not blanch with manual pressure. What grading depth is most appropriate for the burns on his left thigh?

a. First-degree burn.
b. Second-degree burn.
c. Third-degree burn.
d. Both a and c.

5 An 18-year-old woman presents with frostbite, which is similar to a thermal injury but is characterized by ice crystals within cells and subsequent interruption of blood. Identify the type of epithelium that is most likely no longer nourished.

a. Simple columnar epithelium.
b. Simple cuboidal epithelium.
c. Simple squamous epithelium.
d. Stratified squamous epithelium.
e. Pseudostratified epithelium.

6 In psoriasis, dividing keratinocytes are located in several of the epidermal strata, causing a highly accelerated production of epithelial cells. Identify the epithelial layer that houses the mitotic keratinocytes that account for cellular division in normal skin.

a. Stratum basale.
b. Stratum corneum.
c. Stratum granulosum.
d. Stratum lucidum.
e. Stratum spinosum.

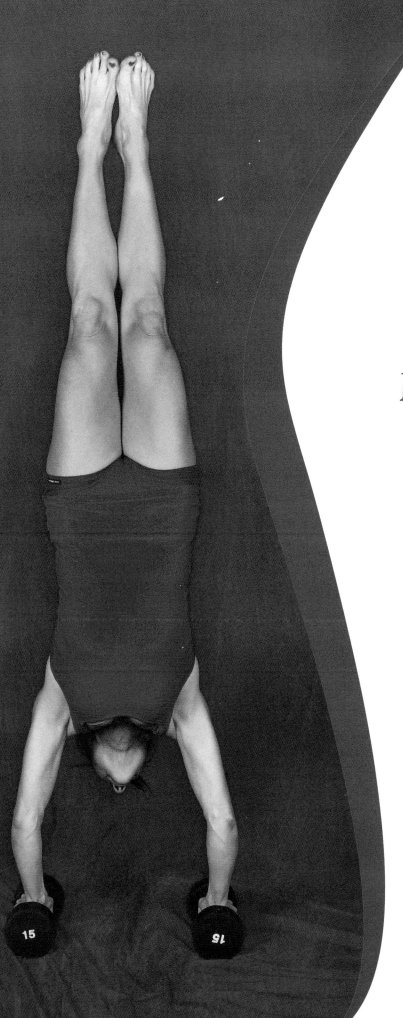

2

Moving Forward

CHAPTER 4	Axial Skeleton	**59**
CHAPTER 5	Appendicular Skeleton	**81**
CHAPTER 6	Arthrology	**101**
CHAPTER 7	Head and Trunk Muscles	**119**
CHAPTER 8	Upper Limb Muscles	**135**
CHAPTER 9	Lower Limb Muscles	**161**

CHAPTER 4

Axial Skeleton

At the completion of this laboratory session, you should be able to do the following:

1 Identify bones and bony landmarks of the axial skeleton.
2 Compare and contrast sutures in adults and fontanelles in newborns.

The human skeleton consists of 206 bones divided into the following two divisions:

1. **Axial skeleton.** The axial skeleton is composed of the bones of the head and the trunk. The head is comprised of cranial and facial bones. The trunk includes the hyoid, vertebral column, sternum, and ribs.
2. **Appendicular skeleton.** The appendicular skeleton is composed of the bones of the pectoral girdle, upper limbs, pelvic girdle, and lower limbs. This division will be covered in Chapter 5.

In this chapter we will examine the bones and primary bony landmarks of the axial skeleton, which are important to know when studying the brain, spinal cord, and thoracic organs.

GETTING ACQUAINTED

Complete the following activities to become familiar with the axial skeleton.

ACTIVITY 1 Skeletal Features

MATERIALS
Obtain the following items before beginning the laboratory activities:
- Textbook or access to Internet resources
- Colored pencils

1 Using your textbook or Internet resources, write a definition for each of the structures listed in Table 4.1.

TABLE **4.1** Osteology Glossary Terms

Term	Definition
Suture	
Foramen	
Fissure	
Meatus	
Spine	
Process	
Tubercle	
Tuberosity	
Fossa	
Groove	
Condyle	
Epicondyle	

ACTIVITY 2 Bones of the Skull

The skull is composed of two general types of bones: cranial bones and facial bones.

Cranial Bones

The cranial bones encase the brain. In addition, cranial bones form the cranial base, which contains depressions called the anterior, middle, and posterior cranial fossae that accommodate the brain. The cranial bones are as follows:

1. **Frontal bone.** Located on the front of the skull, the frontal bone possesses the following landmarks:
   a. **Frontal sinuses.** Hollow spaces within the frontal bone.
   b. **Coronal suture.** Articulates with the parietal bones.
   c. **Glabella.** Smooth surface between the supraorbital ridge.
   d. **Supraorbital foramen.** Opening on the superior aspect of the orbit that transmits the supraorbital nerve, artery, and vein.
   e. **Supraciliary arches.** Bony prominence on the supraorbital margin.
2. **Parietal bones.** Parietal bones are paired bones that form the lateral walls of the skull and possesses the following landmarks:
   a. **Sagittal suture.** Articulates the parietal bones.
   b. **Lambdoid suture.** Articulates with the occipital bone.
   c. **Squamous suture.** Articulates with the temporal bone.
3. **Temporal bones.** Temporal bones are paired bones that form the lateral walls of the skull and possesses the following landmarks:
   a. **Internal acoustic meatus.** Transmits nerves for hearing, balance, and facial expression.
   b. **External acoustic meatus.** Passageway for sound waves to reach the tympanic membrane (eardrum).
   c. **Petrous region.** The densest portion of the skull that houses the cochlea (hearing) and vestibular apparatus (balance).
   d. **Mastoid process.** Rounding protuberance on the posteroinferior temporal bone.
   e. **Styloid process.** Pointy projection of temporal bone for muscle and ligament attachment.
   f. **Zygomatic process.** Projection that articulates with the zygomatic bone creating the zygomatic arch.
   g. **Stylomastoid foramen.** Opening between the styloid and mastoid processes that transmits the facial nerve.
4. **Occipital bone.** The occipital bone forms the posterior region of the skull and possesses the following landmarks:
   a. **Foramen magnum.** Large opening at the base of the skull, which transmits the spinal cord, vertebral arteries, and spinal accessory nerve.
   b. **External occipital protuberance.** A prominent bump on the posterior aspect of the skull.
   c. **Occipital condyles.** Articulates with C1 vertebra (atlas) to make the atlanto-occipital joint.
   d. **Hypoglossal canal.** Located adjacent the occipital condyles and transmits the hypoglossal nerve.
   e. **Jugular foramen.** Opening between the occipital and temporal bones and transmits the internal jugular vein, glossopharyngeal nerve, vagus nerve, and spinal accessory nerve.
5. **Sphenoid bone.** The butterfly-shaped sphenoid bone is located internally and articulates with most of the other cranial bones. It possesses the following landmarks:
   a. **Sella turcica.** Depression that houses the pituitary gland.
   b. **Sphenoid sinus.** Hollow chamber inferior to the sella turcica.
   c. **Lesser wings.** Bony platform that makes a small contribution to the anterior cranial fossa.
   d. **Greater wings.** Bony platform that makes a large contribution to the middle cranial fossa.
   e. **Pterygoid processes.** Perpendicular processes that descend at the junction of the greater wing and sphenoid body. Each process possesses a medial and lateral pterygoid plate for attachment of the medial and lateral pterygoid muscles.
   f. **Superior orbital fissure.** Transmits oculomotor, trochlear, ophthalmic, and abducens nerves, as well as the superior ophthalmic vein.
   g. **Optic canal.** Transmits the optic nerve and ophthalmic artery.
   h. **Foramen ovale.** Transmits the mandibular nerve.

    i. **Foramen spinosum.** Transmits the middle meningeal artery.
    j. **Foramen rotundum.** Transmits the maxillary nerve.
    k. **Foramen lacerum.** A foramen in a dried skull but is filled with cartilage in a living subject.
  6. **Ethmoid bone.** The ethmoid bone is located inferior to the frontal bone, posterior to the nasal bones, and anterior to the sphenoid bone. As such, this particular bone is difficult to see from standard views. The ethmoid bone possesses the following landmarks:
    a. **Cribriform foramina.** Holes that transmit the olfactory nerves from the nasal cavity into the olfactory bulb.
    b. **Crista galli.** Projection that serves as an attachment for the falx cerebri.
    c. **Ethmoid air cells (sinuses).** Hollow air cells within the ethmoid bone.
    d. **Superior nasal conchae.** Projections in the nasal cavity.
    e. **Middle nasal conchae.** Projections in the nasal cavity (*Note:* The inferior nasal conchae are separate bones and do not articulate with the ethmoid bone).
    f. **Perpendicular plate.** Forms the nasal septum along with the vomer bone and nasal cartilage.

Skull bones articulate together via an immovable fibrous joint called a **suture**. The four primary sutures are as follows:
  1. **Coronal suture.** Connects the frontal and parietal bones.
  2. **Sagittal suture.** Connects the parietal bones.
  3. **Squamous suture.** Connects the temporal and parietal bones.
  4. **Lambdoid suture.** Connects the occipital and parietal bones.

Weird and Wacky

Coneheads

The skull of an infant contains five bones (two frontal, two parietal, and one occipital). These bones are knitted together by dense collagenous connective tissue with gaps of tissue between them. The gaps in the skull, which are called **fontanelles**, enable the skull to be flexible. In other words, the skull bones in a newborn can mold to make accomodations during the squeezing through the narrow birth canal during delivery. The more squeezing that occurs, the greater the chance for a newborn to sport a temporarily cone-shaped head for a few days of family photos. Fontanelles also serve as a window to the internal health of a skull. For example, a fontanelle that bulges outward could indicate raised intracranial pressure (i.e., hydrocephalus) or a sunken fontanelle could mean the baby is dehydrated.

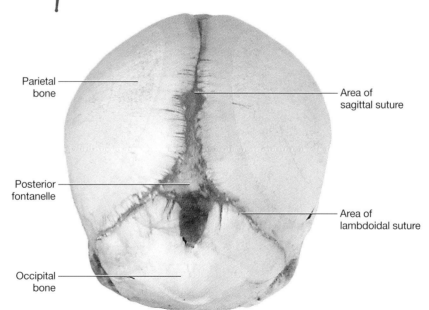

Posterior view of the fetal skull.

Facial Bones

The facial bones are as follows:
  1. **Mandible.** The mandible forms the lower jaw; the perpendicular branch of the mandible is called the mandibular ramus, which possesses the following two superior projections:
    a. **Condyloid process.** A posterior projection from the mandibular ramus. The condyloid process articulates with the mandibular fossa of the temporal bone, forming the **temporomandibular joint** (TMJ).
    b. **Coronoid process.** An anterior process from the mandibular ramus serving as an attachment for the temporalis muscle.

2. **Maxillae.** Forming the upper jaw and hard palate, the maxillae possess the following landmarks:
   a. **Palatine processes.** Projections that form the hard palate.
   b. **Maxillary sinuses.** Hollow cavities that flank the nasal cavity; mucous drains into the nasal cavity via openings inferior to the middle nasal conchae.
   c. **Infraorbital fissure.** Large opening along the inferior floor of the orbit that transmits branches of the maxillary nerve, artery, and vein.
   d. **Infraorbital foramen.** Opening below the orbit that transmits infraorbital nerves, arteries, and veins to the face.
3. **Lacrimal bones.** Small lacrimal bones are located in the medial part of the orbit and form part of the **nasolacrimal duct,** which drains tears from the eye into the nasal cavity.
4. **Nasal bones.** Nasal bones form the bony bridge of the nose (where glasses sit).
5. **Vomer.** The vomer forms the nasal septum, along with the perpendicular plate of the ethmoid bone and nasal cartilage.
6. **Inferior nasal conchae.** The inferior nasal conchae form part of the lateral wall of the nasal cavity.
7. **Palatine bones.** Palatine bones form the posterior portion of the hard palate and posterolateral walls of the nasal cavity.
8. **Zygomatic bones.** Zygomatic bones form the cheekbone of the skull.

1 Identify, label, and color the following structures on Figure 4.1.

- ☐ Frontal bone
- ☐ Parietal bones
- ☐ Temporal bones
- ☐ Zygomatic bones
- ☐ Maxillae
- ☐ Mandible
- ☐ Vomer bone
- ☐ Nasal conchae
- ☐ Sphenoid bone
- ☐ Lacrimal bone
- ☐ Nasal bone
- ☐ Superior orbital fissure
- ☐ Optic canal
- ☐ Inferior orbital fissure
- ☐ Supraorbital foramen
- ☐ Infraorbital foramen

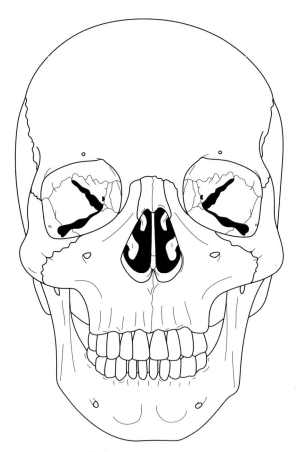

FIGURE **4.1** Anterior view of the skull.

2 Identify, label, and color the following structures on Figure 4.2.
- ☐ Parietal bones
- ☐ Temporal bones
- ☐ Occipital bone
- ☐ Maxillae
- ☐ Mandible
- ☐ Vomer bone
- ☐ Palatine bone
- ☐ Sphenoid bone
- ☐ Sagittal suture
- ☐ Squamous suture
- ☐ Lambdoid suture
- ☐ Sutural bones

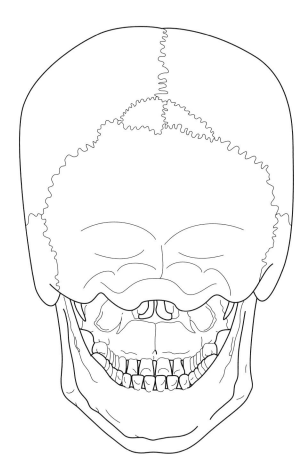

FIGURE **4.2** Posterior view of the skull.

3 Identify, label, and color the following structures on Figure 4.3.
- ☐ Temporal bones
  - Styloid process
  - Stylomastoid foramen
  - Mastoid process
  - Carotid canal
  - Mandibular fossa
- ☐ Occipital bone
  - Foramen magnum
  - Occipital condyles
  - Hypoglossal canal
- ☐ Zygomatic bones
- ☐ Maxillae
- ☐ Vomer bone
- ☐ Sphenoid bone
- ☐ Lambdoid suture
- ☐ Jugular foramen

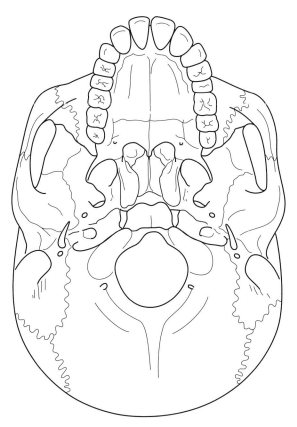

FIGURE **4.3** Inferior view of the skull.

4 Identify, label, and color the following structures on Figure 4.4.
- ☐ Frontal bone
- ☐ Parietal bones
- ☐ Temporal bones
  - Internal acoustic meatus
- ☐ Occipital bone
  - Foramen magnum
- ☐ Ethmoid bone
  - Crista galli
  - Cribriform foramina
- ☐ Sphenoid bone
  - Foramen spinosum
  - Foramen ovale
  - Foramen rotundum
  - Sella turcica
  - Optic canal

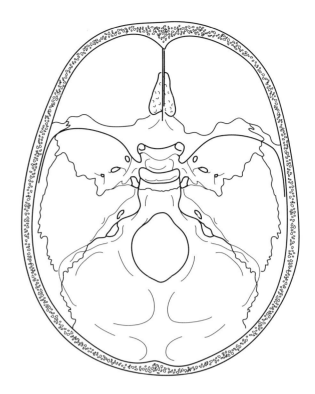

FIGURE **4.4** Superior view of the skull with the calvarium removed.

5 Identify, label, and color the following structures on Figure 4.5.
- ☐ Frontal bone
- ☐ Parietal bone
- ☐ Temporal bone
  - External acoustic meatus
  - Mastoid process
  - Styloid process
- ☐ Occipital bone
- ☐ Zygomatic bone
- ☐ Maxilla
- ☐ Mandible
  - Coronoid process
  - Mandibular notch
  - Condyloid process
  - Mental foramen
  - Ramus of the mandible
- ☐ Sphenoid bone
- ☐ Lacrimal bone
- ☐ Nasal bone
- ☐ Coronal suture
- ☐ Squamous suture
- ☐ Lambdoid suture

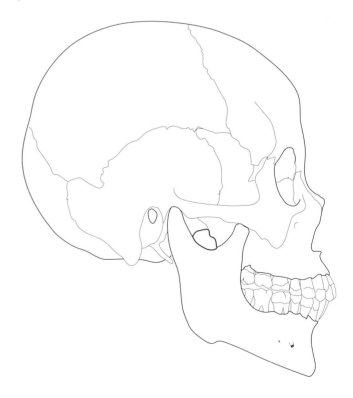

FIGURE **4.5** Lateral view of the skull.

6 Identify, label, and color the following on Figure 4.6.
- ☐ Frontal bone
  - Frontal sinus
- ☐ Parietal bone
- ☐ Temporal bone
- ☐ Occipital bone
- ☐ Sphenoid bone
  - Sella turcica
  - Sphenoid sinus
- ☐ Ethmoid bone
  - Perpendicular plate
- ☐ Maxilla
- ☐ Mandible
- ☐ Vomer bone
- ☐ Nasal bone
- ☐ Coronal suture
- ☐ Squamous suture
- ☐ Lambdoid suture

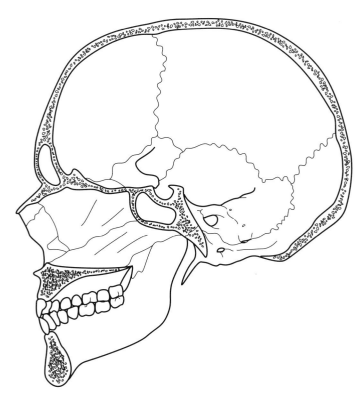

FIGURE **4.6** Sagittal section of the skull, medial view.

ACTIVITY 3 Remainder of the Axial Skeleton

Vertebral Column

Another bony element that forms the axial skeleton is the vertebral column, which consists of seven cervical, twelve thoracic, five lumbar, five fused sacral, and three or four fused coccygeal vertebrae. For abbreviation, often the first letter of each vertebra is used along with its number. For example, the fourth cervical vertebra would be referred to as C4 and the eleventh thoracic vertebra as T11. The vertebral column forms a solid, protective, flexible tube. This surrounds the spinal cord and its meninges, maintains posture, and supports the weight of the head and body. All of these functions are a trade-off between vertebral column stability and flexibility. The cervical, thoracic, and lumbar vertebrae are mobile, while the sacral and coccygeal are fused. A typical vertebra consists of the following features:

1. **Vertebral body.** Located anteriorly, the vertebral body serves as the main weight-bearing element of the vertebra.
2. **Pedicles.** The pedicle is the region of the vertebral arch that connects the transverse process to the vertebral body.
3. **Laminae.** The lamina is the region of the vertebral arch that connects the spinous process to the transverse process.
4. **Vertebral foramen.** Formed by the vertebral arch and vertebral body, the vertebral foramen contains the spinal cord. The stacked vertebral foramina of the entire vertebral column form the vertebral canal.
5. **Intervertebral (neural) foramina.** The intervertebral foramina are the bilateral openings between every pair of adjacent vertebrae where the spinal nerves communicate between the spinal cord and body tissues.
6. **Vertebral arch.** The **vertebral arch** consists of paired pedicles and laminae.
7. **Spinous process.** The spinous process is the posteriorly projecting tip of the vertebral arch that is easily palpated through the skin.
8. **Transverse processes.** The transverse processes are laterally projecting processes that extend from the junction of the pedicle and lamina.
9. **Superior articular process and inferior articular process.** Articular surfaces that project superiorly and inferiorly from the vertebral arch are called superior and inferior articular processes, respectively. They articulate with adjacent vertebrae, forming facet (zygopophyseal) synovial joints.

10. **Intervertebral disc.** The intervertebral disc is a cartilage cushion located between adjacent vertebral bodies to function as a shock absorber. It consists of a tough fibrocartilaginous outer ring called the **annulus fibrosus** and a gelatinous core called the **nucleus pulposus**.

The basic properties of each region of the vertebral column are as follows:

1. **Cervical vertebrae (7).** Located in the neck, the cervical vertebrae possess **transverse foramina,** which transport the vertebral arteries and veins to and from the posterior region of the brain. The spinous processes of the cervical vertebrae are often bifid or forked. The cervical vertebrae include the following:
    a. **Atlas (C1).** Also known as the **atlas** because the C1 vertebra articulates with the occipital bone and "holds" the weight of the skull, much like the mythological Atlas held the weight of the world.
    b. **Axis (C2).** Also known as the **axis** because its articulation with the atlas, via the **dens (odontoid process),** enables rotation of the skull. In other words, the skull rotates because of the dens, hence the name "axis."
    c. **Vertebral prominens (C7).** The most prominent spinous process at the base of the neck.
2. **Thoracic vertebrae (12).** The thoracic vertebrae are located in the thoracic or chest region. The spinous processes are thin and point inferiorly. Costal facets are present for articulation with the 12 pairs of ribs; the lateral view of this structure resembles the head of a giraffe.
3. **Lumbar vertebrae (5).** The lumbar vertebrae are located in the lower back. Vertebral bodies are large and blocklike. Spinous processes are thick and point posteriorly. The lateral view of this structure resembles the head of a moose.
    a. **Superior and inferior articulating facets.** These structures articulate close to the sagittal plane, which enables flexion and extension but not side-to-side movements.
4. **Sacrum (5 fused vertebrae).** The sacrum is located between os coxae. Spinal nerves traverse the **sacral foramina** that flank the vertebral bodies. The lateral surfaces of the sacrum articulate with the ilium of the os coxae to form **sacroiliac joints.**
5. **Coccyx (3 to 4 small, fused vertebrae).** Located below the sacrum, the coccyx is also called the "tailbone."

Carrying the Weight of the World
In Greek mythology, Atlas and his brother joined the Titans in the war against the Olympians. However, the Titans were defeated and many were confined to Tartarus. Zeus condemned Atlas to stand on the edge of the earth and hold the sky upon his shoulders for eternity! Through some misconceptions, Atlas was shown in art to hold the "earth" upon his shoulders instead of the sky. This misconception stuck so much that early anatomists named the C1 vertebra after Atlas because it holds the skull (the world) upon its superior articulating facets (shoulders).

Crouching Figure of Atlas,
*Baldassare Tommaso Peruzzi
(1481–1536).*

Clinical Application

Why are Slipped Discs Painful?
There are many names associated with a slipped disc: herniated disc, pinched nerve, and bulging disc are just a few examples. A herniated disc occurs when the nucleus pulposus region of an intervertebral disc partly protrudes through the annulus fibrosis and compresses an adjacent spinal nerve. When the slipped disc pushes on the adjacent spinal nerve, a radiating pain along the distribution of the nerve is often produced.

1 Identify, label, and color the following skeletal structures in Figure 4.7.
- ☐ Cervical vertebrae
- ☐ Thoracic vertebrae
- ☐ Lumbar vertebrae
- ☐ Sacral vertebrae
- ☐ Coccygeal vertebrae

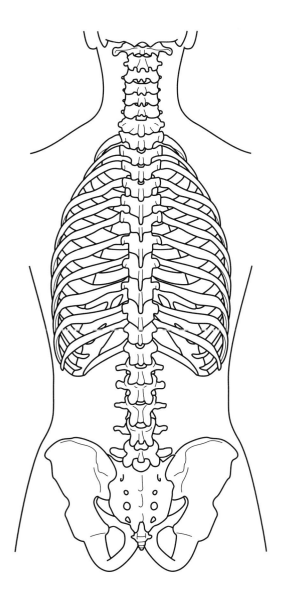

FIGURE **4.7** Bones of the upper skeleton, posterior view.

2 Identify, label, and color the following structures of the vertebral column in Figure 4.8.
- ☐ Cervical vertebrae
- ☐ Thoracic vertebrae
- ☐ Lumbar vertebrae
- ☐ Sacral vertebrae
- ☐ Coccygeal vertebrae
- ☐ C4, T4, and S4 vertebrae

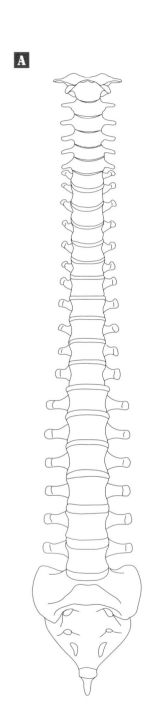
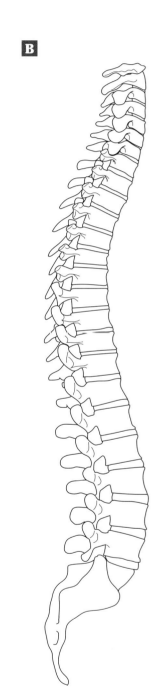

FIGURE **4.8** Vertebral column: (**A**) anterior; (**B**) lateral.

3 Identify, label, and color the following structures of vertebrae in Figures 4.9 and 4.10.
- Cervical vertebrae:
  - Bifid spinous process
  - Lamina
  - Vertebral foramen
  - Vertebral body
  - Transverse foramen
  - Superior articular facet
- Thoracic vertebrae:
  - Spinous process
  - Lamina
  - Transverse process
  - Pedicle
  - Vertebral foramen
  - Vertebral body
  - Superior articular facet
- Lumbar vertebrae:
  - Spinous process
  - Lamina
  - Transverse process
  - Pedicle
  - Vertebral foramen
  - Vertebral body
  - Superior articular facet

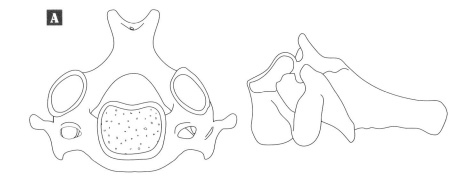

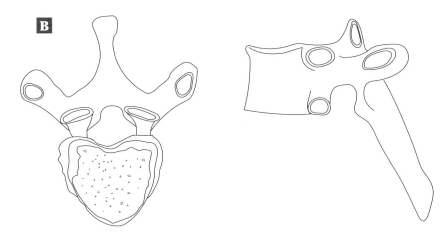

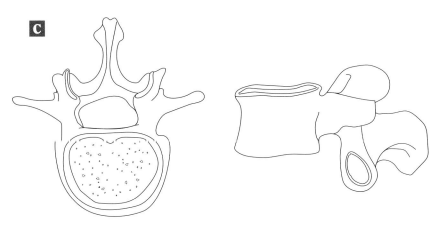

FIGURE **4.9** Superior and lateral views of vertebrae: (**A**) cervical; (**B**) thoracic; (**C**) lumbar.

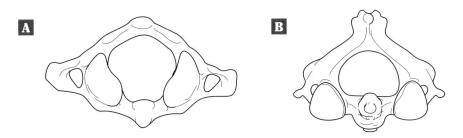

FIGURE **4.10** Superior view of vertebrae: (**A**) atlas [C1]; (**B**) axis [C2].

UNIT **2**: Moving Forward   CHAPTER **4**: Axial Skeleton

Rib Cage

The rib cage forms the remainder of the axial skeleton. The following structures comprise the rib cage:

1. **Sternum.** The sternum, also known as the breastbone, is divided into the manubrium, sternal angle, sternal body, and xiphoid process.
2. **Ribs.** There are 12 pairs of ribs, most of which articulate to the thoracic vertebrae posteriorly and the sternum anteriorly. They are divided into the following classifications according to how they attach to the sternum:
   a. **True ribs (ribs 1–7).** Considered true ribs because they attach directly to the sternum by their own cartilage.
   b. **False ribs (ribs 8–10).** Considered false ribs because they attach to the cartilage of the true ribs, rather than directly to the sternum.
   c. **Floating ribs (ribs 11–12).** Considered floating ribs because they do not attach to the sternum.

Weird and Wacky

Waist Not, Want Not

Some urban legends have claimed that various movie stars have removed their floating ribs so their waists could be thinner. This is not true. However, the desire to have a thinner waist did cause the invention of the corset to become so popular. Victorian women used to wear a corset, a tight-laced item of clothing, around the abdominal region for the purpose of constricting the size of the waist. Some were so tight that breathing became difficult (who needs to breathe?). One woman allegedly whittled her waist size down to measure 15 inches through the use of corsets.

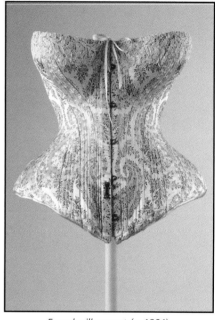

French silk corset (c. 1891).

Clinical Application

Why Do You Push on the Chest During CPR?

Cardiopulmonary resuscitation (CPR) is a lifesaving technique used in emergency situations. When someone's heart has stopped, blood is no longer circulating through the body and, therefore, the oxygen in the blood is not being used. However, their blood should still contain oxygen, assuming the person was breathing normally seconds before. As such, it is crucial to deliver the oxygenated blood to the cells until medical professionals arrive. The rib cage is composed of both bone and cartilage and is therefore flexible enough that chest compressions will manually compress the heart. This forces blood out of the heart chambers into systemic arteries where blood flows into the capillaries and oxygen is delivered to body cells. In a nutshell, chest compressions create forced heartbeats.

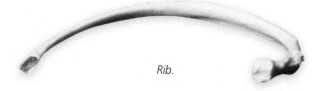

Rib.

4 Identify, label, and color the following structures on Figure 4.11.
- ☐ Suprasternal notch
- ☐ Manubrium
- ☐ Sternal angle
- ☐ Sternal body
- ☐ Xiphoid process
- ☐ True ribs
- ☐ False ribs
- ☐ Floating ribs
- ☐ Costal cartilage
- ☐ Costal angle
- ☐ T1 vertebra
- ☐ T12 vertebra

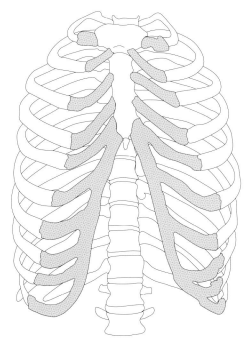

FIGURE **4.11** Rib cage.

Hyoid Bone

The hyoid bone is located in the neck and is held in place on the C3 vertebra by muscles and ligaments. It helps stabilize the larynx and serves as an attachment site for the tongue and pharyngeal muscles. Due to its location, the act of choking will often leave the victim with a broken hyoid bone.

Clinical Application

Hyoid Bone Fractures

The "U"-shaped hyoid bone is situated at the midline of the neck and in a prime location to be injured during strangulation. If the hyoid bone reveals a fracture during post-mortem (examination after death), the death will be ruled a homicide from strangulation until proven otherwise. As fingers grasp the throat of the victim, the ends of the "U" squeeze inward toward each other, and this displacement often results in a fracture.

👁 OBSERVING

Complete the following hands-on laboratory activities to apply your knowledge of the axial skeleton.

ACTIVITY 1 Bones and Bony Landmarks of the Axial Skeleton

1 Identify the bones and bony landmarks on the skull labeled in Figure 4.12.

MATERIALS

Obtain the following items before beginning the laboratory activities:

- ❑ Articulated axial skeleton
- ❑ Disarticulated axial skeleton (skull, vertebrae, ribs, sternum, and hyoid)
- ❑ Probe (use metal probe to point to various structures; DO NOT use pens or pencils, which will mark up the bones)

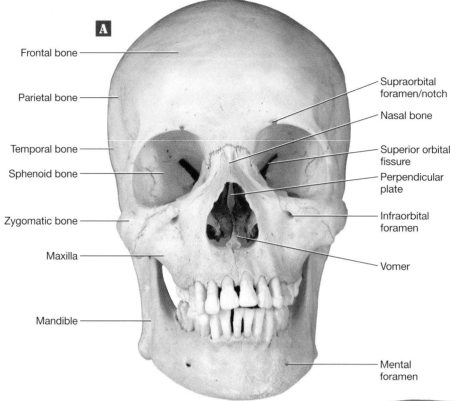

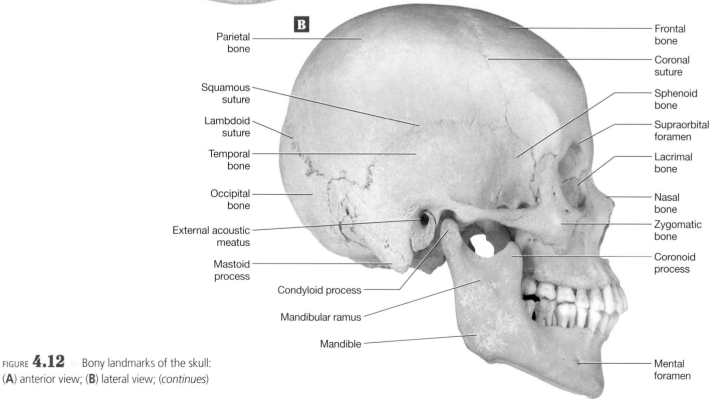

FIGURE **4.12** Bony landmarks of the skull: (**A**) anterior view; (**B**) lateral view; (*continues*)

72 ■ *Discovering Anatomy: A Guided Examination of the Cadaver*

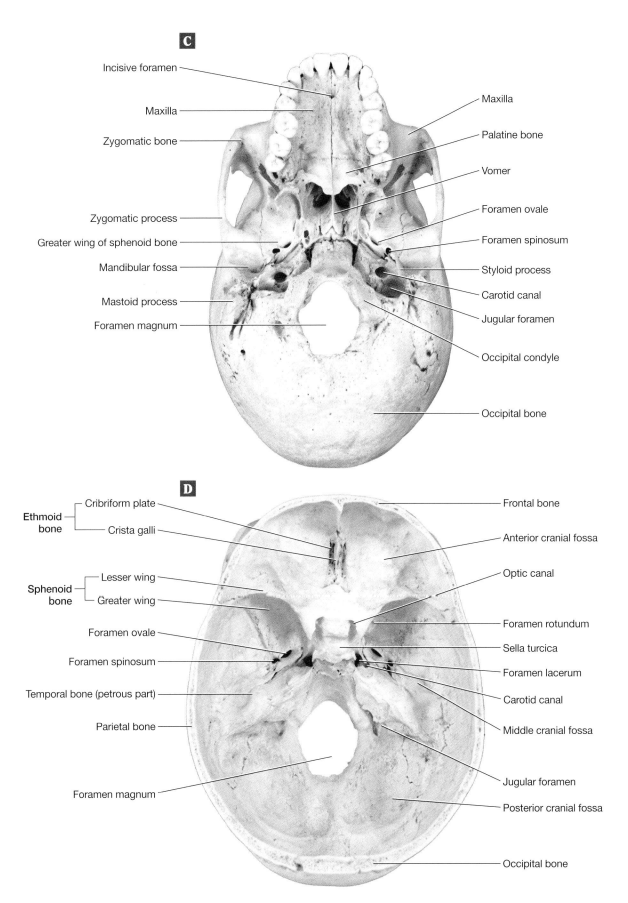

FIGURE **4.12** (cont.)   Bony landmarks of the skull: (**C**) inferior view; (**D**) superior view with calvarium removed; (continues)

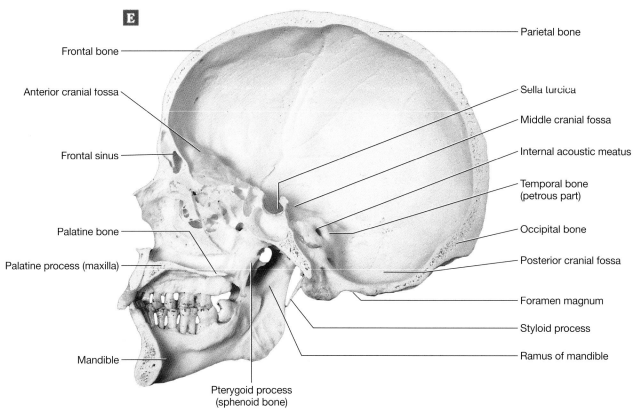

FIGURE 4.12 (cont.) Bony landmarks of the skull: (E) sagittal section with medial view.

2 Identify the bones and bony landmarks on the skeleton in Figure 4.13.

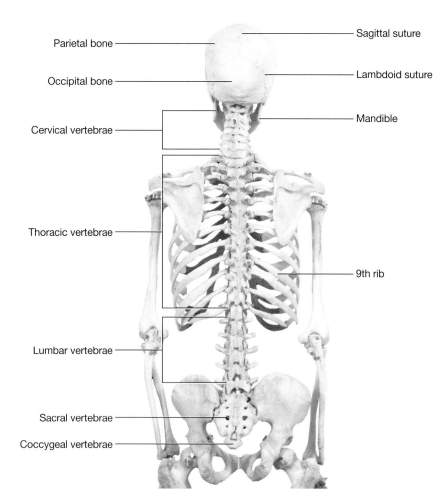

FIGURE 4.13 Posterior view of the bones of the upper skeleton.

3 Identify the bony landmarks on the vertebrae in Figure 4.14.

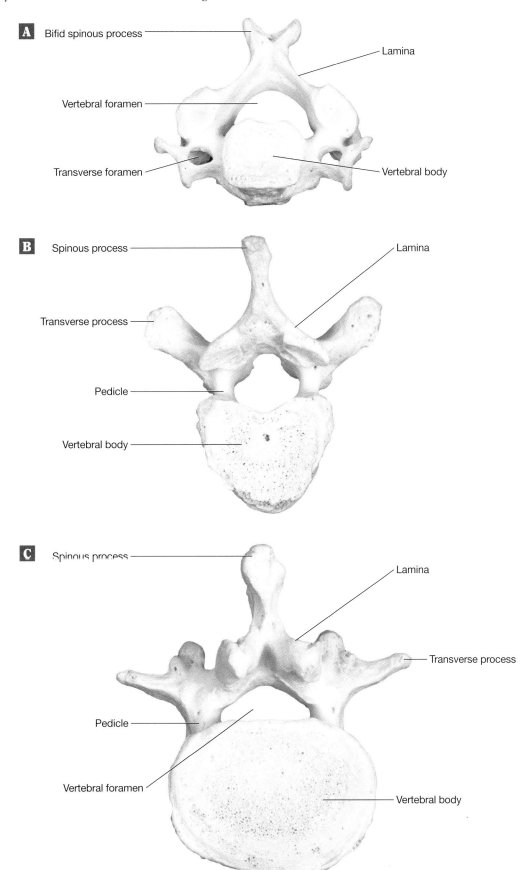

FIGURE **4.14** Superior view of vertebrae: (**A**) cervical; (**B**) thoracic; (**C**) lumbar.

4 Identify the bony landmarks on the vertebrae in Figure 4.15.

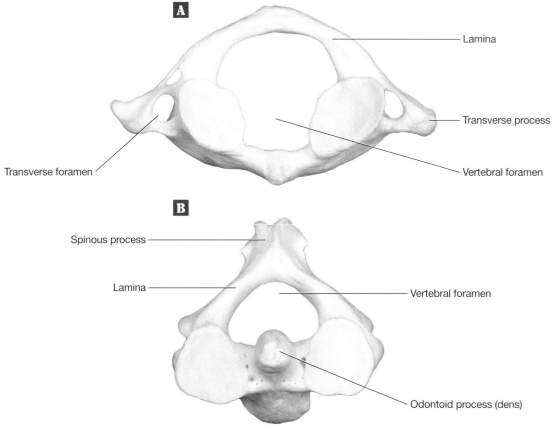

FIGURE **4.15** Superior view: (**A**) atlas [C1]; (**B**) axis [C2].

5 Identify the bones and bony landmarks on the rib cage in Figure 4.16.

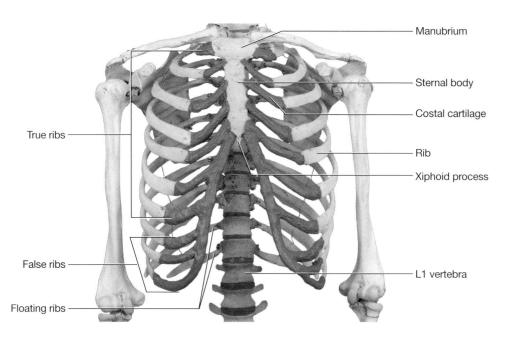

FIGURE **4.16** Anterior view of the rib cage.

WRAPPING UP

Complete the following additional activities to help retain your knowledge of the axial skeleton.

Name _____

Date _____ Section _____

1 Identify the following bones and bony landmarks on the disarticulated skull in Figure 4.17.

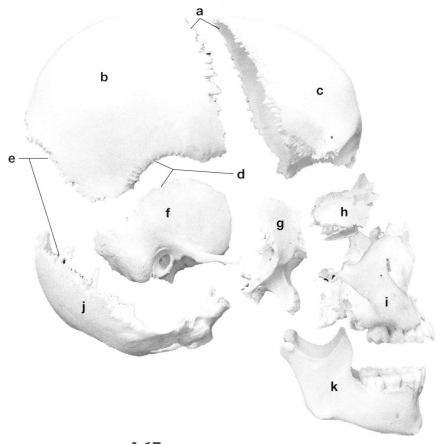

FIGURE **4.17** Lateral view of a disarticulated skull.

a. _____

b. _____

c. _____

d. _____

e. _____

f. _____

g. _____

h. _____

i. _____

j. _____

k. _____

2. Identify the following bones on the anterior X-ray of the face in Figure 4.18.

a. _____
b. _____
c. _____
d. _____
e. _____
f. _____

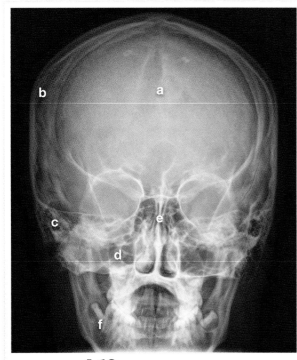

FIGURE 4.18 Anterior X-ray of the skull.

3. Identify the following bones on the lateral X-ray of the face in Figure 4.19.

a. _____
b. _____
c. _____
d. _____
e. _____
f. _____
g. _____
h. _____
i. _____

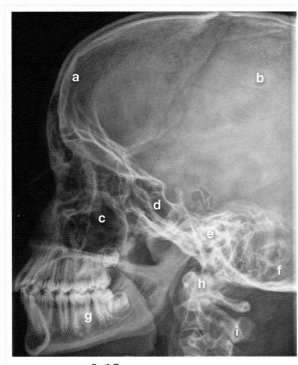

FIGURE 4.19 Lateral X-ray of the face.

WRAPPING UP
(Continued)

Name _____

Date _____ Section _____

Scientists uncover human bones during an archeology dig. Answer the following questions associated with this scenario:

4 How many bones would need to be counted to ensure they found the entire skeleton?
   a. 202.
   b. 204.
   c. 206.
   d. 208.
   e. 210.

5 Identify a distinguishing feature that contrasts a cervical vertebra from a thoracic vertebra.
   a. Intervertebral disc.
   b. Lamina.
   c. Pedicle.
   d. Superior articular facet.
   e. Transverse foramen.

6 The mandible was uncovered during the dig with part of another bone articulating with the condyloid process. Identify this articulating bone fragment.
   a. Frontal bone.
   b. Mandible.
   c. Maxilla.
   d. Temporal bone.
   e. Zygomatic bone.

7 Identify a distinguishing feature ensuring that the occipital bone was located.
   a. Foramen lacerum.
   b. Foramen magnum.
   c. Intervertebral foramen.
   d. Medial pterygoid plate.
   e. Zygomatic process.

8 When counting the vertebrae in order from cranial to caudal, the lumbar vertebrae would be numbered _____ .
   a. 1–7
   b. 8–19
   c. 20–24
   d. 25–29
   e. 30–32

9 The bone articulating with the skull is uncovered during the dig. Identify the name of the bone.
   a. Atlas.
   b. Axis.
   c. Occipital.
   d. Sphenoid.
   e. Vomer.

10 Identify the reason the eleventh rib differs from the other ribs.
   a. Articulates with the vertebral column.
   b. Attaches to the sternum.
   c. Distal ends lack costal cartilage.
   d. Located in the abdominal wall.

11 The bone articulating with the lateral sides of the sacrum is uncovered during the dig. Identify the name of this bone.
   a. Coccyx.
   b. Ilium.
   c. Ischium.
   d. Lumbar vertebra.
   e. Pubis.

12 A bone articulating anteriorly with the temporal bone is identified. Identify the name of this bone.
   a. Frontal bone.
   b. Mandibular bone.
   c. Occipital bone.
   d. Parietal bone.
   e. Zygomatic bone.

13 You overhear an anthropologist describing a bone she uncovered and how it has a very distinct "external acoustic meatus." Identify the name of the bone she uncovered.
   a. Frontal bone.
   b. Maxillary bone.
   c. Parietal bone.
   d. Sphenoid bone.
   e. Temporal bone.

CHAPTER 5

Appendicular Skeleton

At the completion of this laboratory session, you should be able to do the following:

1 Identify bones and bony landmarks of the appendicular skeleton.
2 Describe how adjacent appendicular bones articulate together.

The appendicular skeleton is composed of the bones of the pectoral girdle (scapula and clavicle), the upper limb (humerus, radius, ulna, carpals, metacarpals, and phalanges), the pelvic girdle (os coxae), and the lower limb (femur, patella, tibia, fibula, tarsals, metatarsals, and phalanges).

 GETTING ACQUAINTED

Complete the following activities to become familiar with the appendicular skeleton.

MATERIALS

Obtain the following items before beginning the laboratory activities:

❏ Textbook or access to Internet resources
❏ Colored pencils

ACTIVITY 1 Pectoral Girdle

The pectoral girdle consists of the clavicle and the scapula, forming the skeletal base of the upper limb. The only bony attachment between the upper limb and axial skeleton is through the clavicle of the pectoral girdle. As such, the upper limb relies on large muscular slings to suspend it from the trunk. Many important muscles (e.g., pectoralis major) that cross and move the shoulder joint attach to the bones of the pectoral girdle. The pectoral girdle consists of the following bones:

1. **Clavicle (collar bone).** The clavicle is an "S"-shaped bone located on the anterior, superior aspect of the thoracic cage. The medial end of the clavicle articulates with the sternum and the lateral end articulates with the acromion of the scapula. The clavicle braces the glenohumeral joint and upper limb laterally away from the rib cage and is easily palpable throughout its length.

 Clinical Application

Breaking Your Fall
A function of the clavicle is to prop the glenohumeral joint and upper limb laterally away from the narrow superior portion of the rib cage. This bracing function becomes immediately obvious when a clavicle is fractured, resulting in the entire shoulder region collapsing medially. The clavicle also transmits compression forces from the upper limbs to the axial skeleton. However, they are not very strong and are likely to fracture. For example, a clavicle fracture is likely to occur when a person uses outstretched arms to break a fall. The way the clavicle is curved ensures that it will fracture anteriorly (outward) rather than collapse posteriorly (inward), which would damage the subclavian artery that serves the upper limb.

 Clinical Application

Curiosities of the Clavicle
Cleidocranial dysostosis is a congenital condition where the individual has a partial or complete absence of the clavicle. This defect typically occurs bilaterally. An absent clavicle enables hypermobility of the glenohumeral joint including the ability to touch the shoulders together in front of the chest.

Weird and Wacky

Make a Wish

The furcula is a forked bone formed by the union of the two clavicles in birds and enables the thoracic skeleton to withstand the stresses of wing movement during flight. Superstition has surrounded the furcula for centuries including its ability to foresee the victor of battle, predict the weather, and forecast good fortune. During the early 1600s, a custom originated whereby two individuals pull on either side of a dried furcula bone until it breaks. The individual that receives the larger part of the furcula is granted his or her wish, hence its common name of the "wishbone." Considering that dinosaurs are ancestors of birds, you can imagine that the furcula of the Tyrannosaurus rex would have been a beast to make a wish with after Thanksgiving dinner!

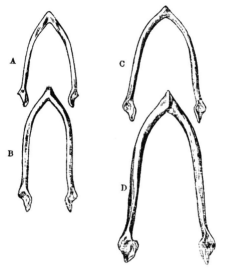

Furculae: (A) Short-faced tumbler; (B) and (C) Fantails; (D) Pouter. From Charles Darwin's Variation of Animals and Plants under Domestication, *Volume 1 (1868).*

2. **Scapula (shoulder blade).** The scapula is a large, flat, triangular bone that spans ribs 2 through 7 and possesses the following landmarks:
   a. **Spine.** A ridge along the posterior aspect of the scapula.
   b. **Acromion.** A superolateral projection formed by the continuation of the spine.
   c. **Supraspinous fossa.** A deep concave surface superior to the spine.
   d. **Infraspinous fossa.** A shallow concave surface inferior to the spine.
   e. **Glenoid cavity.** A lateral depression that articulates with the humeral head to form a ball-and-socket joint (glenohumeral joint).
   f. **Supraglenoid tubercle.** A small, bony projection above the glenoid cavity that serves as an attachment site for the long head of the biceps brachii muscle.
   g. **Infraglenoid tubercle.** A small, bony projection below the glenoid cavity that serves as an attachment site for the long head of the triceps brachii muscle.
   h. **Coracoid process.** The anterior projection inferior to the acromioclavicular (AC) joint.
   i. **Subscapular fossa.** Shallow concave surface on the anterior surface of the scapula.
   j. **Lateral border.** The border of the scapula inferior to the glenoid cavity.
   k. **Medial border.** The medial border of the scapula.

1 Identify, label, and color the following parts of the scapula in Figure 5.1.

- ☐ Spine
- ☐ Infraspinous fossa
- ☐ Subscapular fossa
- ☐ Acromion
- ☐ Glenoid cavity
- ☐ Lateral border
- ☐ Supraspinous fossa
- ☐ Coracoid process
- ☐ Medial border

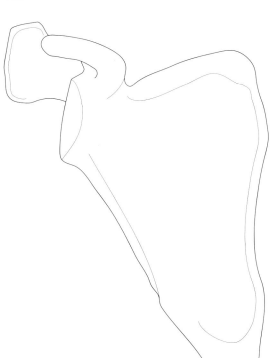
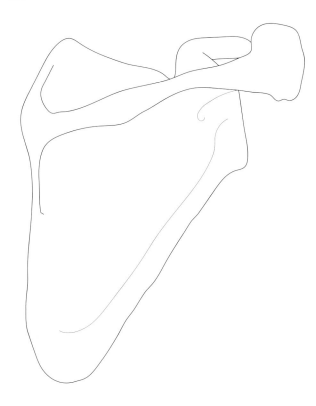

FIGURE **5.1** Scapula: (**A**) anterior; (**B**) posterior.

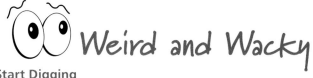 Weird and Wacky

Start Digging
The etymology of the word scapula is Latin for "spades" or "shovels." The scapulae of animals were used as tools for digging and shoveling. For example, Native Americans sometimes bound the scapula of a buffalo to a piece of wood to be used as a hoe to till the soil.

Hidatsa woman cultivating maize and squashes with a bone hoe.

UNIT **2: Moving Forward** — CHAPTER **5:** Appendicular Skeleton ■ **83**

ACTIVITY 2 Upper Limb

The upper limb consists of the arm, forearm, wrist, and hand. The following list contains the bones of the upper limb, including the primary landmarks of each bone:

1. **Humerus.** The humerus is a long bone located in the arm between the shoulder and elbow joints; it possesses some of the following landmarks:
   a. **Head.** A proximal portion of the humerus that articulates with the glenoid cavity of the scapula to form the glenohumeral joint.
   b. **Lesser tubercle.** A small prominence on the anterior, proximal surface of the humerus.
   c. **Greater tubercle.** A large prominence on the lateral, proximal surface of the humerus.
   d. **Intertubercular (bicipital) groove.** A groove between the lesser and greater tubercles in which the tendon of the long head of the biceps brachii muscle courses.
   e. **Deltoid tuberosity.** A projection of the mid-lateral humerus where the deltoid muscle attaches.
   f. **Medial epicondyle.** A bony protuberance on the distal, medial surface.
   g. **Lateral epicondyle.** A bony protuberance on the distal, lateral surface.
   h. **Trochlea.** The distal condyle that articulates with the trochlear notch of the ulna.
   i. **Capitulum.** The distal, ball-shaped condyle that articulates with the head of the radius.
   j. **Coronoid fossa.** An indentation on the distal anterior surface of the humerus.
   k. **Olecranon fossa.** An indentation on the distal posterior surface of the humerus.

2. **Ulna.** The ulna is located medial to the radius in the forearm, distal to the humerus, and proximal to the carpals. The ulna is thick and notched proximally (looks like a hook), thin distally, and possesses some the following features:
   a. **Coronoid process.** A beak-like process that forms the inferior border of the trochlear notch.
   b. **Olecranon process.** A knob-like projection the forms the superior border of the trochlear notch. The lay term "elbow" is associated with the olecranon, the bony landmark that rests on the table when leaning on the elbows.
   c. **Trochlear notch.** The deep, "C"-shaped, anterior facing notch that forms a synovial hinge joint with the trochlea of the humerus. It is formed by the coronoid and olecranon processes. The trochlear notch forms the "U" shape (for ulna), which makes it easy to distinguish this feature from the radius.
   d. **Styloid process.** A distal pointy projection.

 Weird and Wacky

That's Not Funny
The "funny bone" is neither funny (like when you accidentally hit it against the back of your chair), nor is it a bone (whether you hit it or not). When you "hit your funny bone," you're actually hitting the ulnar nerve as it courses deep to the medial epicondyle of the humerus. The ulnar nerve provides motor innervation to forearm and hand muscles and sensory innervation to the skin in the medial portion of the hand—the fifth digit (the "pinkie") and half of the fourth digit (the "ring finger"). Similar to the rest of the nerves in the body, the ulnar nerve is protected by bone and muscles. However, as the ulnar nerve courses posterior to the medial epicondyle, it is only covered by skin and fat and thus exposed for a good "bumping." Therefore, when you hit your funny bone, you are actually hitting the nerve against the bone, causing pain and tingling sensations to shoot to the hand. What a hoot!

Anterior view of the right hand.

3. **Radius.** The radius is located lateral to the ulna in the forearm, distal to the humerus, and proximal to the carpals. This structure is thin proximally, expands distally, and possesses some the following features:
   a. **Radial head.** A bony, wheel-like landmark that articulates with the ulna to form a synovial pivot joint enabling pronation and supination.
   b. **Radial tuberosity.** A proximal swelling that serves as an attachment site for the biceps brachii muscle.
   c. **Styloid process.** A pointy, bony landmark on the distal radius.
4. **Wrist and hand.** The bones of the wrist and hand include the following:
   a. **Carpal bones.** At the base of the hand are the eight short carpal bones that compose the region defined as the wrist. The carpal bones are the scaphoid, lunate, triquetrum, pisiform, trapezium, trapezoid, capitate, and hamate.
   b. **Metacarpal bones.** Distal to the carpal bones are the five metacarpal bones that form the digital skeleton of the palm. The metacarpals 1 through 5 are numbered in order from the thumb to the pinkie.
   c. **Phalanges.** The fourteen slender columns of phalangeal bones (phalanges = plural; phalanx = singular) form the mobile parts of the digits (fingers and thumb). This is an extremely mobile region of the skeleton's upper limb that produces a variety of movements that provide the incredible array of functional activity associated with the hands.

Clinical Application

Carpal Tunnel Syndrome

The eight carpal bones are organized into two rows of four bones in a concave (bowl) shape. A strong ligament called the flexor retinaculum spans the top of this concavity (forming the "lid" of the "bowl"). Thus the carpal bones and flexor retinaculum form the carpal tunnel. Nine flexor tendons and the median nerve traverse the carpal tunnel. Carpal tunnel syndrome is a condition caused when the synovial sheaths surrounding the tendons become swollen or inflamed (tenosynovitis), from repetitive motions, vibrational damage infection, or fluid retention. The pressure resulting from the chronic inflammation of the flexor tendon sheets is trapped beneath the flexor retinaculum (bracelet-like ligament that prevents "bowing" of the flexor tendons) compressing the median nerve against the wrist bones. This results in pressure on the median nerve. The result is tingling and numbness in the thumb, index, and middle fingers primarily. In more severe cases, atrophy of the thumb muscles may occur.

1 Identify, label, and color the following parts of the upper limb in Figure 5.2.
- ☐ Head of humerus
- ☐ Capitulum
- ☐ Trochlear notch
- ☐ Greater tubercle
- ☐ Trochlea
- ☐ Head of the radius
- ☐ Lesser tubercle
- ☐ Olecranon process
- ☐ Styloid process of radius
- ☐ Radial tuberosity
- ☐ Radial notch
- ☐ Deltoid tuberosity
- ☐ Coronoid process

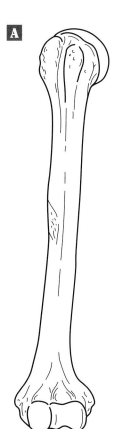

FIGURE **5.2** Anterior view: (**A**) right humerus; (**B**) right radius and ulna.

2 Identify, label, and color the following parts of the hand in Figure 5.3.
- ☐ Scaphoid
- ☐ Pisiform
- ☐ Capitulum
- ☐ Proximal phalanx
- ☐ Lunate
- ☐ Trapezium
- ☐ Hamate
- ☐ Middle phalanx
- ☐ Triquetrum
- ☐ Trapezoid
- ☐ Metacarpals
- ☐ Distal phalanx

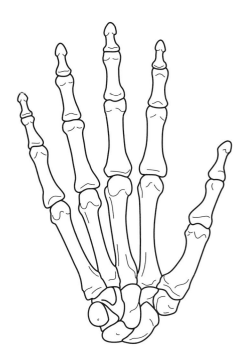

FIGURE **5.3** Anterior view of the right hand.

ACTIVITY 3 Pelvic Girdle

The pelvic girdle consists of three distinct bones on each side (ilium, ischium, and pubis), which fuse during development to form a single bone called the os coxa. The paired os coxae bond anteriorly at the pubic symphysis to form the pelvic girdle. This strong girdle of bone fuses with the sacrum to form the pelvic skeleton. The os coxae form a strong skeletal base for the transfer of forces generated by the lower limb during locomotion. All three bones come together on the lateral side of each os coxa to form a deep socket called the **acetabulum**, which articulates with the head of the femur to form the ball-and-socket hip joint. The **obturator foramen** is a large hole formed where the ischium and pubis meet. Note that in a living person this hole is covered with a membrane and only the obturator nerve, artery, and vein pass through.

The three os coxa bones have the following three features:

1. **Ilium**. The largest of the os coxa bones, the ilium contains the following primary bony landmarks:
    a. **Body**. The main portion of the ilium closest to the acetabulum.
    b. **Ala**. The superior wing of the body.
    c. **Iliac fossa**. Concave surface on the anterior region of the ilium that serves as an attachment for the iliacus muscle.
    d. **Iliac crest**. The superior ridge of the ala where you rest your hands on your hips.
    e. **Anterior superior iliac spine**. The anterior end of the iliac crest that serves as an attachment for the sartorius muscle and inguinal ligament.
    f. **Anterior inferior iliac spine**. A bony prominence inferior to the anterior superior iliac spine that serves as an attachment for the rectus femoris muscle.
    g. **Posterior superior iliac spine**. The posterior end of the iliac crest.
    h. **Posterior inferior iliac spine**. A bony prominence inferior to the posterior superior iliac spine.
    i. **Greater sciatic notch**. A notch in the posterior ilium, which transmits the sciatic nerve between the pelvis and thigh.
2. **Ischium**. The ischium forms the posteroinferior pelvis and contains the following primary features:
    a. **Ischial spine**. A pointed, triangular projection on the posterior margin of the ischium.
    b. **Lesser sciatic notch**. A small notch situated below the ischial spine and above the ischial tuberosity; serves as a passageway for the pudendal nerve and internal pudendal artery and vein.
    c. **Ischial tuberosity**. A large, rough bump on the posterior aspect of the ischium; known as the "sit bones" as they bear the weight of the body when you sit down.
    d. **Ischial ramus**. Anteromedial extension of the ischium.

3. **Pubis.** The pubis forms the anterior portion of the pelvis and contains the following primary features:
   a. **Superior ramus.** Superior extension of the pubic body.
   b. **Inferior ramus.** Inferior extension of the pubic body.
   c. **Pubic symphysis.** Formed by the bodies of both pubic bones at a pad of fibrocartilage.
   d. **Pubic angle.** The angle formed by both inferior pubic rami, which can help to determine the gender of a skeleton.

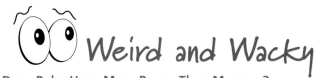 Weird and Wacky

Does Baby Have More Bones Than Mommy?
Well, kind of. The skeleton of an adult has 206 bones, while the skeleton of a newborn contains 300. However, most of the bones of newborns are actually cartilage. Anatomically speaking, an adult has the same number of bones as a baby; however, some of these bones will fuse together and be considered a single bone as the infant develops. For example, adults have an os coxa on the left and an os coxa on the right. However, the pelvis in an infant is actually made of six bones (ilium, ischium, and pubis on each side). Furthermore, the ilium articulates with the sacrum. The sacrum is one bone in an adult. However, the sacrum begins as five separate bones that later fuse together during development. Therefore, a baby has more bones than mommy until they grow up.

1 Identify, label, and color the following parts of the os coxa in Figure 5.4.
- ☐ Iliac crest
- ☐ Body of the ilium
- ☐ Anterior superior iliac spine
- ☐ Anterior inferior iliac spine
- ☐ Posterior superior iliac spine
- ☐ Posterior inferior iliac spine
- ☐ Acetabulum
- ☐ Greater sciatic notch
- ☐ Ischial spine
- ☐ Lesser sciatic notch
- ☐ Ischial tuberosity
- ☐ Ischial ramus
- ☐ Obturator foramen
- ☐ Superior pubic ramus
- ☐ Inferior pubic ramus

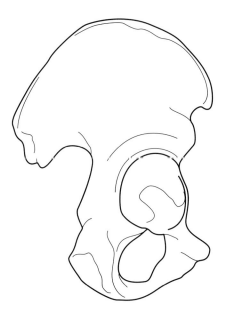

FIGURE **5.4**   Lateral view of the right os coxa.

ACTIVITY 4 Lower Limb

The lower limb consists of the thigh, leg, ankle, and foot. The following list contains the bones of the lower limb, including the primary landmarks of each bone:

1. **Femur.** The femur is the largest bone of the body and is located between the hip and knee joints. It possesses the following landmarks:
   a. **Head.** The rounded proximal end of the femur that articulates with the acetabulum forming the hip joint.
   b. **Neck.** The cantilevered extension of the femur that is capped by the head.
   c. **Greater trochanter.** The lateral bony projection of the femur that serves as the insertion point for the gluteal and external hip rotator muscles.
   d. **Lesser trochanter.** The medial bony projection that serves as the insertion point for the iliopsoas muscles.
   e. **Linea aspera.** A rough, longitudinal ridge along the posterior surface.
   f. **Medial and lateral condyles of the femur.** Projections on the distal femur for articulating with the tibia to make the knee joint and attachment site for the gastrocnemius muscle.
   g. **Medial and lateral epicondyle of the femur.** Small prominences above the medial and lateral condyles.
   h. **Adductor tubercle.** A bony prominence proximal to the medial epicondyle that serves as an attachment for the adductor magnus muscle.
2. **Patella (knee cap).** The patella is a flat bone within the quadriceps tendon anterior to the knee joint. It functions as a compression strut in the quadriceps tendon and increases the mechanical advantage of the quadriceps. The patella is the largest sesamoid bone in the body (a bone embedded within a tendon or muscle).

I "Knee" That Bone
Did you know that babies are born without their patellae? Technically, newborn babies do have their kneecaps (patellae) but they consist of cartilage and do not begin to ossify into the bony patellae until about three years of age. The patella is classified as a sesamoid bone. A sesamoid bone is embedded in a tendon and forms in response to mechanical strain from the attaching muscles.

3. **Tibia.** The tibia is the larger medial bone of the leg that possesses the following landmarks:
   a. **Medial and lateral condyles of the tibia.** Articulates with the medial and lateral condyles of the femur to make the knee joint.
   b. **Tibial tuberosity.** Roughened protrusion from the anterior surface of the proximal tibia that serves as an attachment site for the patellar tendon from the quadriceps femoris muscles.
   c. **Anterior border.** The vertical bony ridge along the anterior surface.
   d. **Medial malleolus.** Projection at the distomedial end of the tibia that forms the medial wall of the mortise-like ankle joint.
4. **Fibula.** A slender, splint-like bone that serves as an important site of muscle attachment but does not play a weight-bearing role. The fibula attaches to the tibia via the interosseous membrane and is divided into two parts:
   a. **Head.** The proximal end of the fibula.
   b. **Lateral malleolus.** The projection at the distolateral end of the fibula that forms the lateral wall of the mortise-like ankle joint.

Clinical Application

Shin Splints
Shin splint pain is a common exercise-related problem and usually refers to pain along the medial region of the tibia (shinbone). Shin splints typically develop when a tendon from a shin muscle, such as the tibialis anterior and its attachment to the tibial periosteum, becomes overworked by repetitive injury caused by activities like running. Symptoms often include razor-sharp pain and mild swelling along the medial region of the tibia.

 Weird and Wacky

The Two-Faced Bone

Most of our fancy-sounding anatomical terms come from either a Latin or Greek etymology. However, sometimes an anatomical structure received multiple names from both Latin and Greek origins. For example, the Latin etymology for the word "fibula" is a clasp, pin, or brooch; however, the Greek word "peroneus" also means clasp, pin, or brooch. For some reason, both the Latin and Greek terms have persisted throughout the centuries, making it difficult for all anatomy students to learn this region of the body. This is why there are two names for the muscles and nerves associated with the fibula as shown in the table provided:

Latin	Greek
Fibularis brevis	Peroneus brevis
Fibularis longus	Peroneus longus
Fibularis tertius	Peroneus tertius
Common fibular nerve	Common peroneal nerve
Superficial fibular nerve	Superficial peroneal nerve
Deep fibular nerve	Deep peroneal nerve

5. **Foot.** The foot consists of the following three bony subdivisions:
   a. **Tarsal bones.** At the base of the foot are the short tarsal bones that compose the region defined as the ankle. The seven tarsal bones are talus, calcaneus (heel bone), navicular, medial cuneiform, intermediate cuneiform, lateral cuneiform, and cuboid.
   b. **Metatarsal bones.** Distal to the tarsal bones are the five metatarsal bones that form the digital skeleton of the foot.
   c. **Phalanges.** The slender columns of 14 phalangeal bones form the mobile parts of the digits, or toes.

1 Identify, label, and color the following parts of the lower limb in Figure 5.5.

- ☐ Head of the femur
- ☐ Neck of the femur
- ☐ Greater trochanter
- ☐ Lesser trochanter
- ☐ Patellar surface
- ☐ Tibial tuberosity
- ☐ Medial malleolus
- ☐ Head of the fibula
- ☐ Lateral malleolus
- ☐ Medial and lateral condyles (femur)
- ☐ Medial and lateral condyles (tibia)
- ☐ Anterior tibial border

 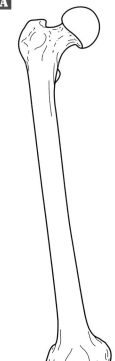 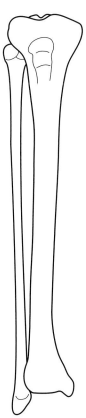

FIGURE **5.5** Anterior view: (**A**) right femur; (**B**) right tibia and fibula.

2 Identify, label, and color the following parts of the foot in Figure 5.6.
- ☐ Calcaneus
- ☐ Talus
- ☐ Navicular
- ☐ Cuboid
- ☐ Medial cuneiform
- ☐ Intermediate cuneiform
- ☐ Lateral cuneiform
- ☐ Metatarsals
- ☐ Proximal phalanx
- ☐ Middle phalanx
- ☐ Distal phalanx

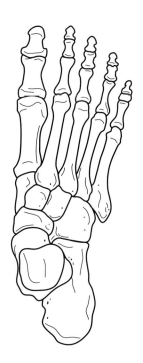

FIGURE **5.6** Dorsal surface of the right foot.

👁 OBSERVING

Complete the following hands-on laboratory activities to apply your knowledge of the appendicular skeleton.

ACTIVITY 1 Bones and Bony Landmarks of the Appendicular Skeleton

MATERIALS
Obtain the following items before beginning the laboratory activities:
- ❑ Articulated appendicular skeleton
- ❑ Disarticulated appendicular skeleton
- ❑ Metal probe

1 Identify the bony landmarks on the scapula labeled in Figure 5.7.

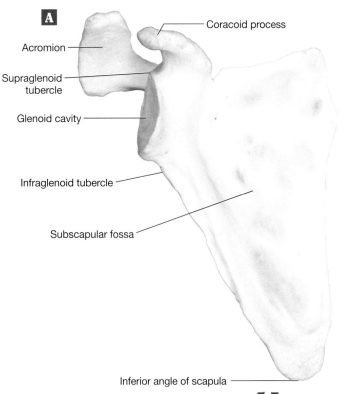
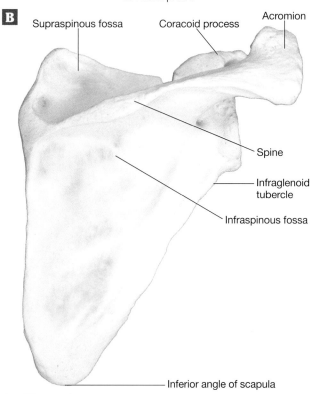

FIGURE **5.7** Right scapula: (**A**) anterior; (**B**) posterior.

90 ■ *Discovering Anatomy: A Guided Examination of the Cadaver*

2 Identify the bony landmarks on the humerus labeled in Figure 5.8.

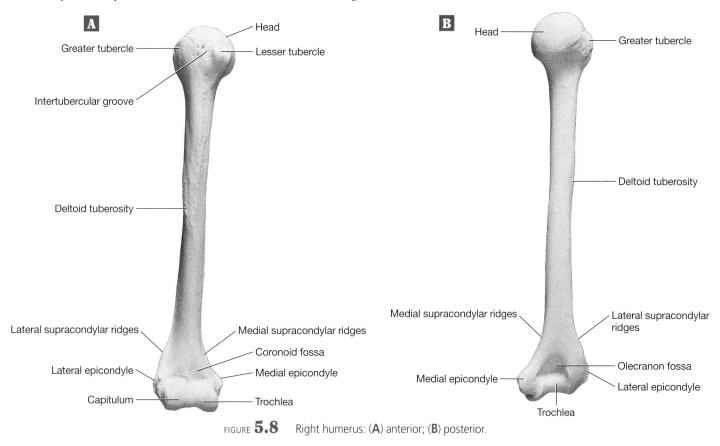

FIGURE **5.8** Right humerus: (**A**) anterior; (**B**) posterior.

3 Identify the bony landmarks on the radius and ulna labeled in Figure 5.9.

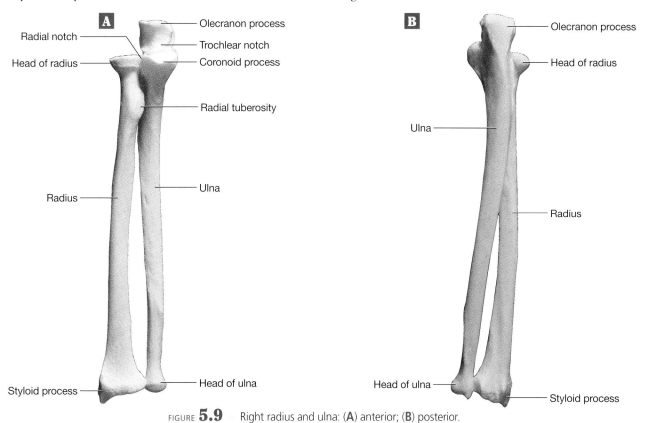

FIGURE **5.9** Right radius and ulna: (**A**) anterior; (**B**) posterior.

4 Identify the bones of the hand labeled in Figure 5.10.

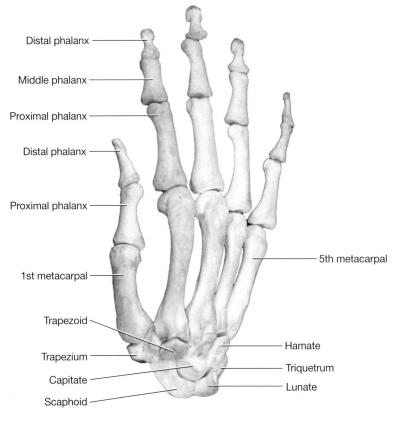

FIGURE **5.10** Posterior view of the bones of the left hand.

5 Identify the bony landmarks on the os coxa labeled in Figure 5.11.

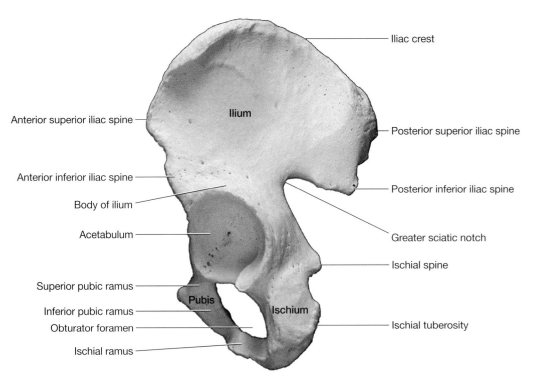

FIGURE **5.11** Lateral view of the right os coxa.

6 Identify the bony landmarks on the os coxa labeled in Figure 5.12.

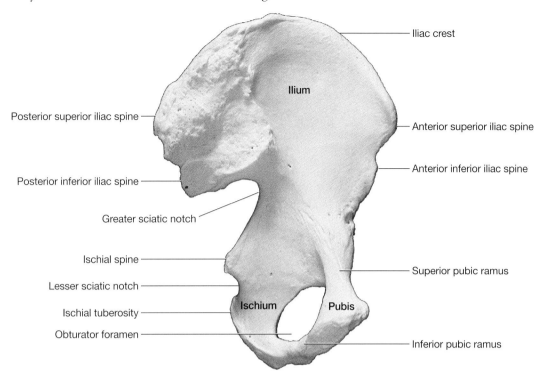

FIGURE **5.12**   Medial view of the right os coxa.

7 Identify the bony landmarks on the femur labeled in Figure 5.13.

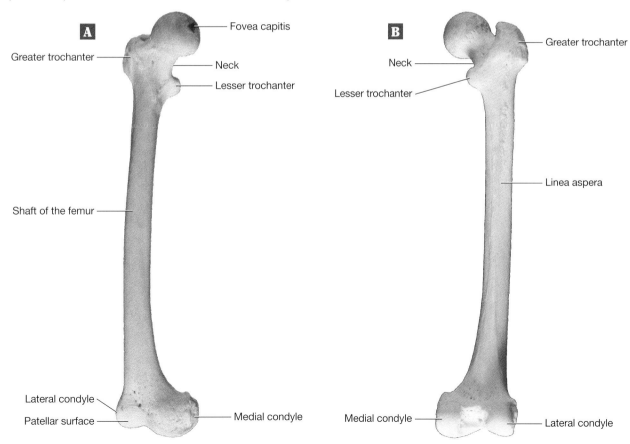

FIGURE **5.13**   Right femur: (**A**) anterior; (**B**) posterior.

8 Identify the bony landmarks on the tibia and fibula labeled in Figure 5.14.

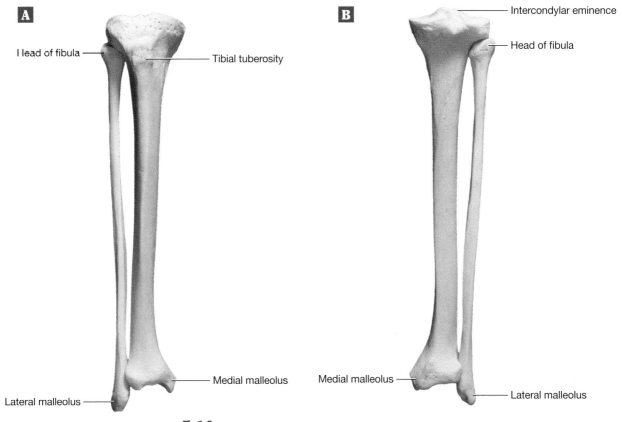

FIGURE **5.14** Right tibia and fibula: (**A**) anterior; (**B**) posterior.

9 Identify the bones of the foot labeled in Figure 5.15.

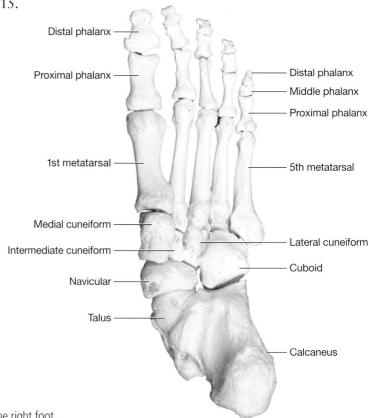

FIGURE **5.15** Plantar surface of the bones of the right foot.

WRAPPING UP
Complete the following additional activities to help retain your knowledge of the appendicular skeleton.

Name _____

Date _____ Section _____

1 Identify the bones and bony landmarks on the X-ray of the pelvis in Figure 5.16.

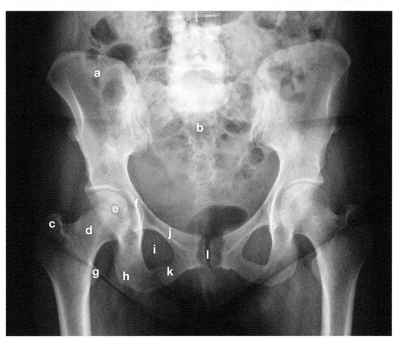

FIGURE **5.16** X-ray of the pelvis.

a. _____

b. _____

c. _____

d. _____

e. _____

f. _____

g. _____

h. _____

i. _____

j. _____

k. _____

l. _____

2. Identify the bones and bony landmarks on the X-ray of the left glenohumeral joint in Figure 5.17.

a. _____
b. _____
c. _____
d. _____
e. _____
f. _____
g. _____
h. _____

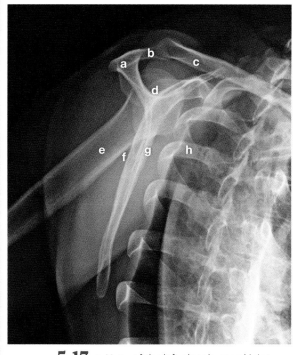

FIGURE **5.17** X-ray of the left glenohumeral joint.

3. Identify the bones and bony landmarks on this lateral X-ray of the right elbow joint in Figure 5.18.

a. _____
b. _____
c. _____
d. _____
e. _____
f. _____
g. _____
h. _____

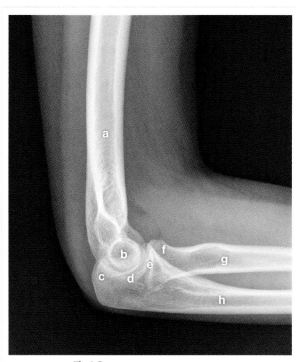

FIGURE **5.18** X-ray of the right elbow joint.

WRAPPING UP
(Continued)

Name _____

Date _____ Section _____

4. Identify the bones and bony landmarks on the X-ray of the right ankle joint in Figure 5.19.

   a. _____

   b. _____

   c. _____

   d. _____

   e. _____

   f. _____

   g. _____

   h. _____

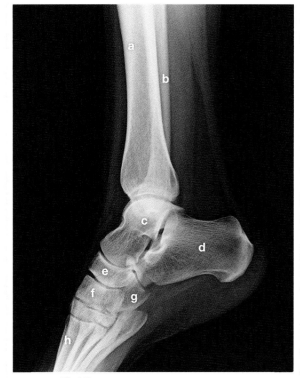

FIGURE **5.19** X-ray of the medial view of the right ankle joint.

5. Identify the bony landmark that participates in the formation of joint between the clavicle and the scapula.

   a. Acromion.

   b. Coracoid process.

   c. Deltoid tuberosity.

   d. Glenoid cavity.

   e. Subscapular fossa.

6. Identify the bony marking that participates in the formation of the joint between the scapula and humerus.

   a. Acromion.

   b. Coracoid process.

   c. Deltoid tuberosity.

   d. Glenoid cavity.

   e. Subscapular fossa.

7 The articulations primarily responsible for the elbow hinge joint are the _____.

   a. capitulum and radial head
   b. capitulum and trochlear notch
   c. coronoid process and coronoid fossa
   d. coronoid process and olecranon fossa
   e. olecranon process and coronoid fossa
   f. olecranon process and olecranon fossa
   g. trochlea and radial head
   h. trochlea and trochlear notch

8 A massage therapist palpating the greater trochanter is touching the _____.

   a. femur
   b. fibula
   c. ilium
   d. ischium
   e. pubis
   f. tibia

9 The medial malleolus is fractured during a bike accident. What bone is injured?

   a. Femur.
   b. Fibula.
   c. Humerus.
   d. Radius.
   e. Tibia.
   f. Ulna.

10 The acetabulum is formed by the ilium, pubis, and _____.

   a. femur
   b. fibula
   c. ischium
   d. tibia
   e. patella

WRAPPING UP
(Continued)

Name _____

Date _____ Section _____

11. The vastus intermedius muscle arises from the anterior surface of the femur and attaches to the tibial tuberosity. Identify the primary action of the vastus intermedius muscle.

    a. Knee extension (straighten the knee).

    b. Knee flexion (bend the knee).

    c. Hip abduction (spread legs apart).

    d. Hip adduction (bring legs together).

    e. Hip extension (straighten the hip).

    f. Hip flexion (bend the hip).

12. The bone articulating with the medial surface of the ilium is shown in an X-ray. Identify the name of this bone.

    a. Coccyx.

    b. Ilium.

    c. Ischium.

    d. Lumbar vertebra.

    e. Pubis.

    f. Sacrum.

13. The femur is to the acetabulum as the humerus is to the _____.

    a. acromion

    b. coracoid process

    c. glenoid cavity

    d. infraspinous fossa

    e. subscapular fossa

    f. supraspinous fossa

14. The deltoid muscle attaches to the clavicle, scapula, and the _____.

    a. capitulum

    b. humerus

    c. radius

    d. trapezium

    e. ulna

CHAPTER 6

Arthrology

At the completion of this laboratory session, you should be able to do the following:
1. Distinguish between joints classified by structure and function.
2. Identify the various fibrous joints (sutures, gomphoses, syndesmoses) on a skeleton.
3. Identify the various cartilaginous joints (symphysis, intervertebral discs) on a skeleton.
4. Describe the different types of synovial joints.
5. Identify the synovial joints on the upper limb (shoulder, elbow, wrist, and hand) and lower limb (hip, knee, ankle, and foot) and describe their actions.

The study of joints and articulations is known as arthrology ("arthro" is the Greek word for joints). A joint is a place within the skeleton where at least two bones articulate with each other. A ligament is a band of dense collagenous connective tissue that binds joints together. Joints are constructed to enable or restrict movement and to provide mechanical support.

GETTING ACQUAINTED
Complete the following activities to become familiar with arthrology.

ACTIVITY 1 Classification of Joints

Joints are described by either the type of tissue connecting the bones together (structure) or by the range of motion permitted by the joint (function):

MATERIALS
Obtain the following items before beginning the laboratory activities:
- ❏ Textbook or access to Internet resources
- ❏ Colored pencils

1. **Structure.** Joints may be classified based upon the structural tissues binding the articulating bones. There are three ways of classifying a joint based upon structure:
   a. **Fibrous joint.** Dense irregular collagenous connective tissue binds the bones together. The three different types of fibrous joints are:
      i. **Sutures.** Bind skull bones together (e.g., coronal suture).
      ii. **Gomphosis.** Binds teeth to the skull (i.e., periodontal ligaments).
      iii. **Syndesmosis.** A broad ligament that binds bones together (e.g., interosseous membrane with radius and ulna).
   b. **Cartilaginous joint.** Cartilage binds bones together. There are two types of cartilaginous joints:
      i. **Symphysis.** A cartilaginous joint comprised of fibrocartilage (e.g., pubic symphysis or intervertebral disc).
      ii. **Synchondrosis.** A cartilaginous joint comprised of hyaline cartilage (e.g., growth plate).
   c. **Synovial joint.** A joint capsule (consisting of dense irregular collagenous connective tissue and synovial membrane) is filled with synovial fluid and surrounds a joint (e.g., shoulder/glenohumeral joint).

Listen to Your Elders
Before bad weather, like a rainstorm, a drop in atmospheric pressure is often recorded. This slight drop in atmospheric pressure may cause body tissue, which includes synovial joints, to expand. This expansion of synovial joints may produce achy pain in people with inflamed joints. As such, next time Grandpa recommends taking an umbrella with you to school, you may want to take his advice.

 Weird and Wacky

Cracking Up!
Most people know someone in their social circle that cracks his or her knuckles. There are a number of theories as to why our synovial joints crack. The most accepted one is that "cracking" occurs as a result of negative pressure pulling nitrogen gas temporarily into the synovial fluid of the joint. The cracking sounds occur from either the formation of this gas bubble or the bursting of this gas bubble. Either way, cracking knuckles does not cause arthritis or cause finger joints to enlarge. Dr. Donald Unger gained fame when he cracked the knuckles of his left hand every day for more than sixty years, but not his right. He reported that no arthritis or joint swelling occurred in either hand.

2. **Function**. Joints may be classified based upon the range of motion permitted within the articulation. Joints are classified according to function in the following manner:
   a. **Synarthrotic**. Non-moveable joint (e.g., suture).
   b. **Amphiarthrotic**. Only a little movement occurs (e.g., intervertebral disc).
   c. **Diarthrotic**. Freely moving (i.e., any synovial joint).

For the sake of simplicity, this chapter is focused on describing joints according to structure.

Motions of Synovial Joints

Each time you move your body you are producing movement at a number of synovial joints. Many possible motions can occur at synovial and cartilaginous joints. These motions include the following:

- **Extension**. Describes a straightening motion that increases the angle of a synovial joint. For example, extending (straightening) the elbow or knee (Fig. 6.1A).
- **Flexion**. Describes a bending motion that decreases the angle of a synovial joint; for example, flexing (bending) the elbow or knee (Fig. 6.1B).
- **Abduction**. Describes a motion where a structure moves away from the midline; for example, moving a limb away from the trunk (Fig. 6.2A).
- **Adduction**. Describes a motion where a structure moves toward the midline; for example, moving a limb toward the trunk (Fig. 6.2B).
- **Medial (internal) rotation**. Describes a motion where a structure rotates toward the midline (Fig. 6.3A).
- **Lateral (external) rotation**. Describes a motion where a structure rotates away from the midline (Fig. 6.3B).
- **Elevation**. Describes a motion where a structure moves in the superior direction; for example, shrugging the shoulders (Fig 6.4A).
- **Depression**. Describes a motion where a structure moves in the inferior direction; for example, shoulders moving from an elevated to a neutral position (Fig. 6.4B).
- **Supination**. Describes a rotational movement where the hand is turned outward (faces anteriorly in anatomical position) (Fig. 6.5A).
- **Pronation**. Describes a rotational movement where the hand is turned inward (faces posteriorly in anatomical position) (Fig. 6.5B).
- **Dorsiflexion**. Describes the decrease in the angle of the ankle on the dorsal (top) surface of the foot; for example, lifting your foot off the ground (Fig 6.6A).
- **Plantar flexion**. Describes the decrease in the angle of the ankle on the plantar (bottom) surface of the foot; for example, pressing on the car brakes (Fig. 6.6B).

Clinical Application

Bumper Bones
Arthritis describes pain or disease in a joint (usually a synovial joint). Degenerative arthritis is the most common type of arthritis and involves the erosion or breakdown of the articular cartilage found in the joint capsule. Think of the bone like a car and the articular cartilage is the bumper. When the front of Car One bumps into the back of Car Two, the metal from each vehicle remained intact. Only the bumpers collided. Similarly, when a bone in a synovial joint bumps against the other bone, only the cartilage touched, not the bone. As such, when degenerative arthritis causes deterioration of the articular cartilage, bone starts to wear on bone. This often causes pain, swelling, redness, and limited function.

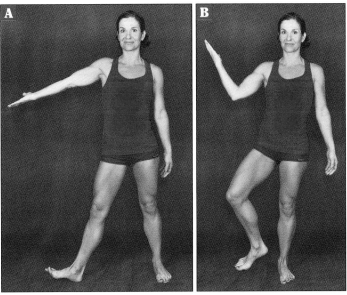

FIGURE **6.1** (**A**) Extension; (**B**) flexion.

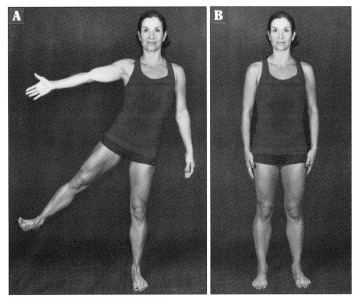

FIGURE **6.2** (**A**) Abduction; (**B**) adduction.

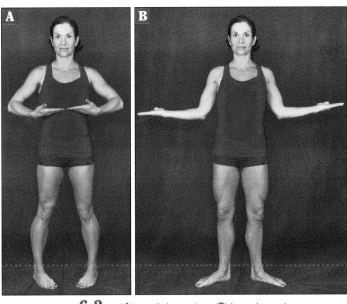

FIGURE **6.3** (**A**) Medial rotation; (**B**) lateral rotation.

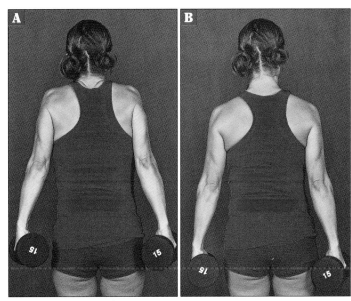

FIGURE **6.4** (**A**) Elevation; (**B**) depression.

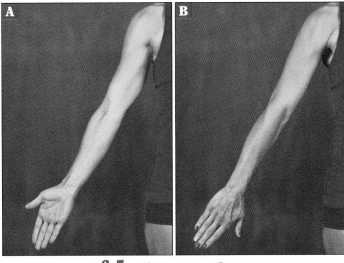

FIGURE **6.5** (**A**) Supination; (**B**) pronation.

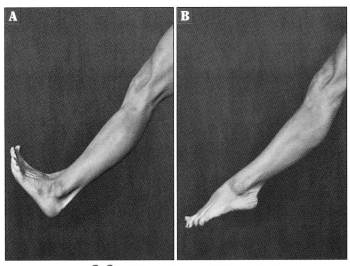

FIGURE **6.6** (**A**) Dorsiflexion; (**B**) plantar flexion.

1 Using your textbook or Internet resources, write a definition for each term listed in Table 6.1 prior to coming to lab.

TABLE **6.1** Descriptions of Various Joint Classifications

Joint Classification	Definition
Joints Classified According to Structure	
Fibrous joint	
Suture	
Gomphosis	
Syndesmosis	
Cartilaginous joint	
Symphysis	
Intervertebral disc	
Synovial joint	
Plane joint	
Condyloid joint	
Saddle joint	
Hinge joint	
Pivot joint	
Ball-and-socket joint	
Joints Classified According to Function	
Synarthrotic	
Amphiarthrotic	
Diarthrotic	

2 Identify, label, and color the following structures in Figure 6.7.

- ☐ Dense irregular collagenous CT
- ☐ Parietal bone
- ☐ Occipital bone
- ☐ Sagittal suture
- ☐ Frontal bone
- ☐ Temporal bone
- ☐ Coronal suture
- ☐ Lambdoid suture

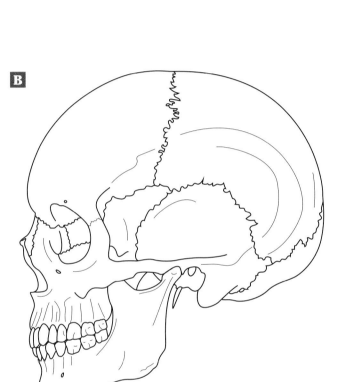
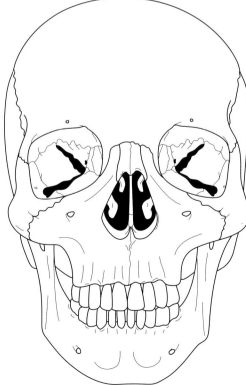

FIGURE **6.7** Fibrous joints in the skull: (**A**) superior; (**B**) lateral; (**C**) anterior.

UNIT 2: Moving Forward CHAPTER 6: Arthrology ■ **105**

3 Identify, label, and color the following structures associated with fibrous joints in Figures 6.8 through 6.10.

- ☐ Dense irregular collagenous CT
- ☐ Ulna
- ☐ Fibula
- ☐ Radius
- ☐ Tibia
- ☐ Syndesmoses joints

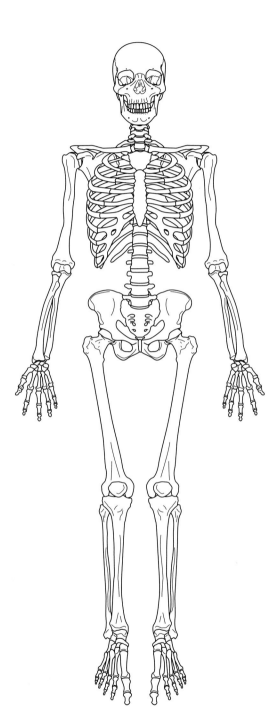

FIGURE **6.8** Fibrous joints in the skeleton.

FIGURE **6.9** Fibrous joints in the radius and ulna.

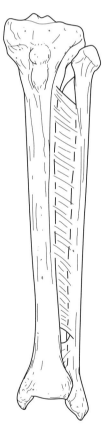

FIGURE **6.10** Fibrous joints in the tibia and fibula.

4 Identify, label, and color the following structures associated with cartilaginous joints in Figures 6.11 and 6.12.
- ☐ Symphysis joint
- ☐ Intervertebral discs
- ☐ L1 vertebra
- ☐ Pubic bones
- ☐ T12 vertebra

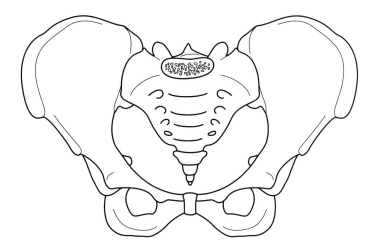

FIGURE **6.11** Cartilaginous joints in the os coxa.

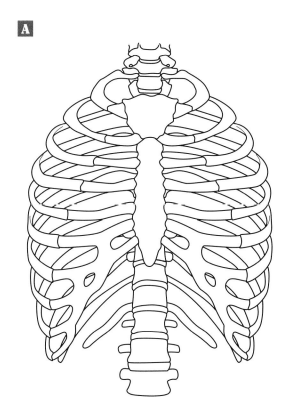
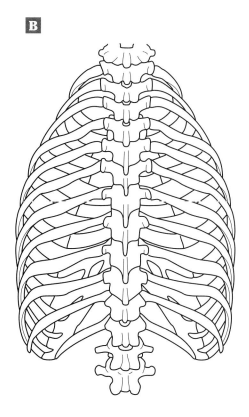

FIGURE **6.12** Cartilaginous joints in the rib cage: (**A**) anterior; (**B**) posterior.

5 Identify, label, and color the following structures associated with synovial joints in Figure 6.13.
- ☐ Bone
- ☐ Joint capsule
- ☐ Articular cartilage
- ☐ Synovial membrane

6 Identify and label the following structures associated with synovial joints in Figure 6.14.
- ☐ Plane joint
- ☐ Condyloid joint
- ☐ Saddle joint
- ☐ Hinge joint
- ☐ Pivot joint
- ☐ Ball-and-socket joint

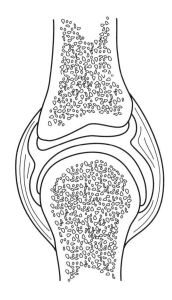

FIGURE **6.13** Synovial joint in the knee.

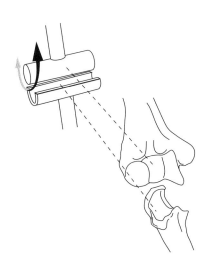

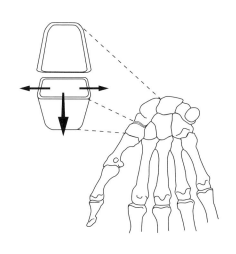

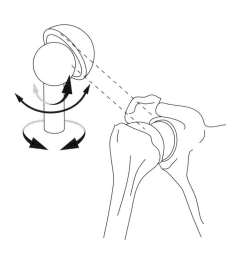

_____

_____

_____

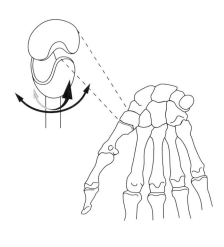

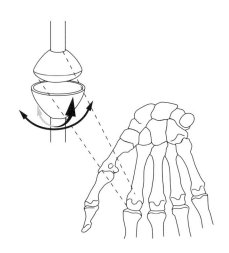

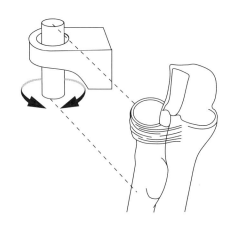

_____

_____

_____

FIGURE **6.14** Types of synovial joints.

7 Identify and label the following actions associated with synovial joints in Figure 6.15.

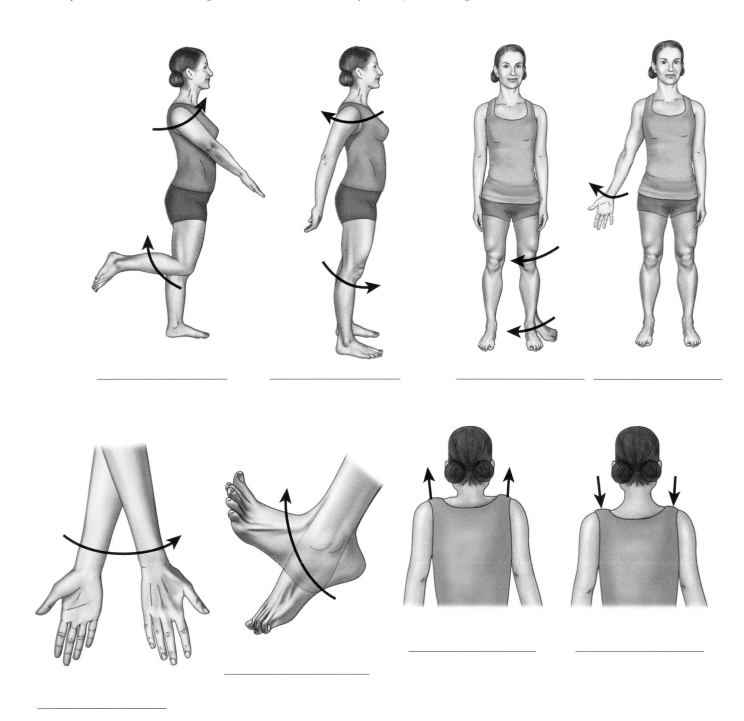

FIGURE **6.15** Motions of synovial joints.

OBSERVING

Complete the following hands-on laboratory activities to apply your knowledge of arthrology.

MATERIALS
Obtain the following items before beginning the laboratory activities:

❏ Skeletal model

ACTIVITY 1 Joint Anatomy

1 Identify the joints labeled in Figures 6.16 through 6.20 on a skeletal model.

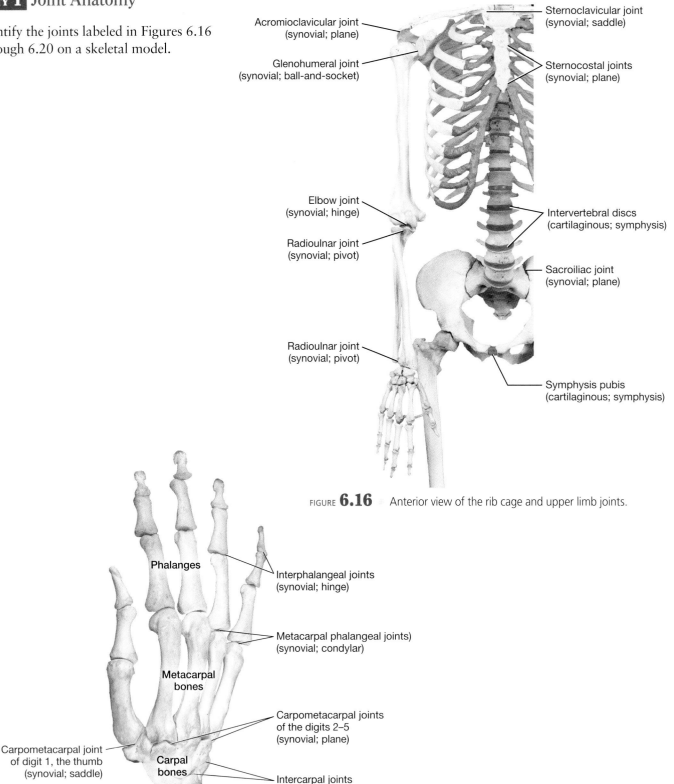

FIGURE **6.16** Anterior view of the rib cage and upper limb joints.

FIGURE **6.17** Joints of the hand, posterior view.

110 ■ *Discovering Anatomy: A Guided Examination of the Cadaver*

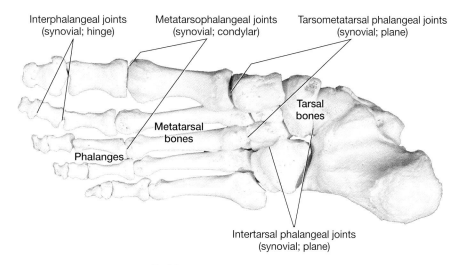

FIGURE **6.18** Joints of the foot, plantar view.

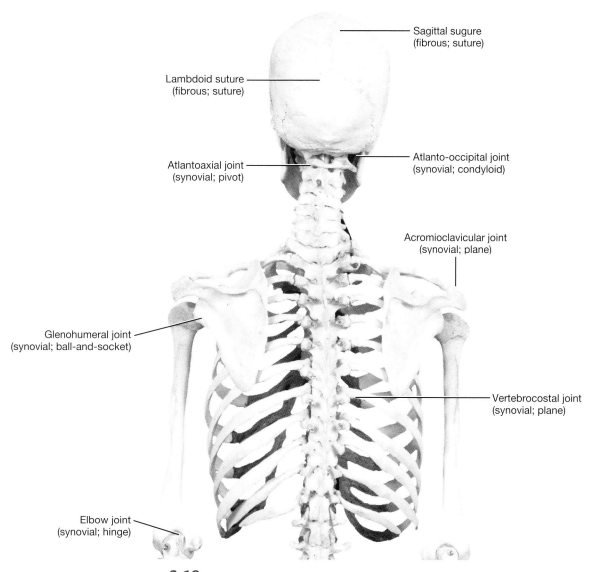

FIGURE **6.19** Posterior view of the rib cage and upper limb joints.

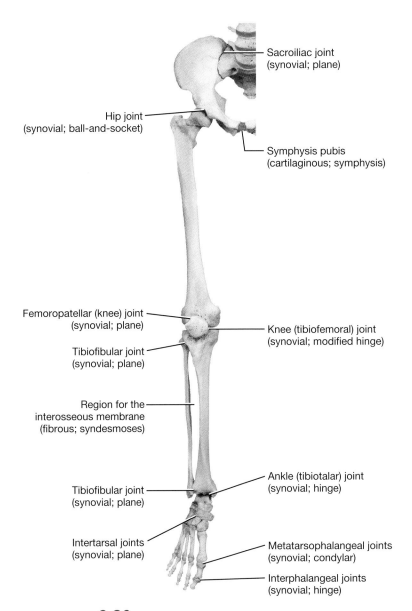

FIGURE **6.20** Anterior view of the joints of the lower limb.

2 Identify the joints labeled in Figures 6.21 and 6.22 on the skull.

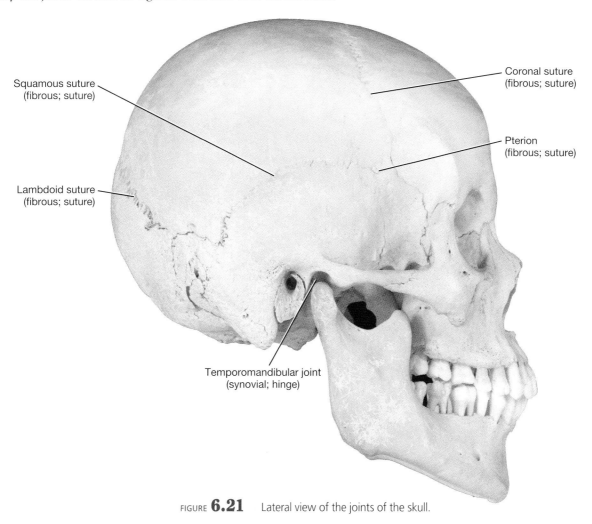

FIGURE **6.21**   Lateral view of the joints of the skull.

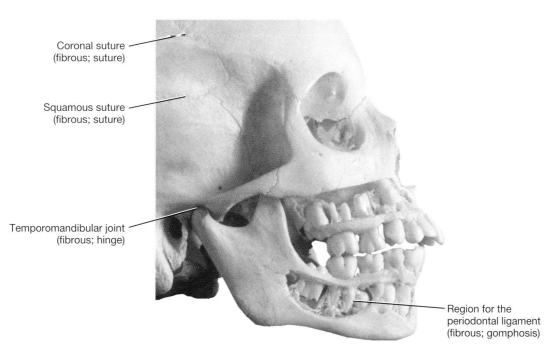

FIGURE **6.22**   Lateral view of the joints of the skull after a step dissection through the outer layer of compact bone in the maxilla and mandible.

UNIT 2: Moving Forward   CHAPTER 6: Arthrology   ■   113

ACTIVITY 2 Synovial Joints

1 For each of the following examples, do the following:
   a. Get into pairs.
   b. Have one partner stand in the anatomical position.
   c. The other partner reads off each of the actions one at a time.
   d. The partner standing in anatomical position moves his or her body after each instruction is given.
   e. Have partners switch places after completing each example.

Example 1
- ☐ Scapula elevation
- ☐ Glenohumeral internal (medial) rotation
- ☐ Glenohumeral abduction
- ☐ Elbow flexion
- ☐ Pronation
- ☐ Wrist extension
- ☐ Finger flexion
- ☐ Thumb extension

Example 2
- ☐ Hip extension
- ☐ Hip lateral rotation
- ☐ Hip adduction
- ☐ Knee flexion
- ☐ Plantar flexion
- ☐ Toe extension

Example 3
- ☐ Scapula elevation
- ☐ Glenohumeral external (lateral) rotation
- ☐ Glenohumeral adduction
- ☐ Elbow flexion
- ☐ Supination
- ☐ Wrist flexion
- ☐ Finger extension
- ☐ Thumb extension

Example 4
- ☐ Hip flexion
- ☐ Hip lateral rotation
- ☐ Hip adduction
- ☐ Knee extension
- ☐ Dorsiflexion
- ☐ Toe flexion

Example 5
- ☐ Cervical lateral flexion
- ☐ Head flexion
- ☐ Mandibular depression
- ☐ Mandibular retraction

Example 6
- ☐ Vertebral column flexion
- ☐ Head extension
- ☐ Glenohumeral joint abduction (right side)
- ☐ Glenohumeral joint adduction (left side)
- ☐ Elbow flexion (right side)
- ☐ Elbow extension (left side)

WRAPPING UP
Complete the following additional activities to help retain your knowledge of arthrology.

Name _____

Date _____ Section _____

1 Describe the movements occurring in the synovial joints in Figure 6.23.

   a. Right glenohumeral joint

   b. Right elbow joint

   c. Left glenohumeral joint

   d. Left elbow joint

   e. Left wrist joint

   f. Head

   g. Spine

FIGURE **6.23** Woman playing tennis.

2 Describe the movements occurring in the synovial joints in Figure 6.24.

   a. Glenohumeral joints

   b. Elbow joints

   c. Wrist joints

   d. Spine

   e. Hip joints

   f. Knee joints

   g. Ankle joints

FIGURE **6.24** Woman lifting weights.

3 Describe the movements occurring in the synovial joints in Figure 6.25.

   a. Glenohumeral joints

   b. Elbow joints

   c. Wrist joints

   d. Right hip joint

   e. Right knee joint

   f. Left hip joint

   g. Left knee joint

   h. Left ankle

FIGURE **6.25** Woman in stretching pose 1.

4 Describe the movements occurring in the synovial joints in Figure 6.26.

   a. Glenohumeral joints

   b. Elbow joints

   c. Wrist joints

   d. Head and neck

   e. Spine

   f. Hip joints

   g. Knee joints

FIGURE **6.26** Woman in stretching pose 2.

WRAPPING UP
(Continued)

Name _____

Date _____ Section _____

5 Describe the movements occurring in the synovial joints in Figure 6.27.
   a. Elbow joints

   b. Wrist joints

   c. Head and neck

   d. Spine

   e. Hip joints

   f. Knee joints

   g. Ankle joints

FIGURE **6.27** Woman in stretching pose 3.

6 Describe the movements occurring in the synovial joints in Figure 6.28.
   a. Right glenohumeral joint

   b. Right elbow joint

   c. Right wrist joint

   d. Left glenohumeral joint

   e. Left elbow joint

   f. Head

   g. Spine

   h. Hip joints

   i. Knee joints

   j. Ankle joints

FIGURE **6.28** Woman in stretching pose 4.

7 Match the type of joint in the right-hand column with its associated structure in the left-hand column.

_____ Shoulder joint

_____ Elbow joint (flexion and extension)

_____ Joint between parietal bones

_____ Joint between parietal bones

_____ Intervertebral disc

_____ Hip joint

_____ Between trapezium and trapezoid

_____ Interosseous membrane between radius and ulna

_____ Tibiotalar joint

_____ Third metacarpophalangeal joint

_____ Knee joint

_____ Trapezium and first metacarpal joint

a. Fibrous; suture

b. Fibrous; syndesmosis

c. Fibrous; gomphosis

d. Cartilaginous; symphysis

e. Synovial; ball-and-socket

f. Synovial; hinge

g. Synovial; plane

h. Synovial; condylar

i. Synovial; saddle

CHAPTER 7

Head and Trunk Muscles

Once you have completed this laboratory session, you should be able to do the following:

1 Identify primary muscles of the head and trunk.
2 Describe the origins and insertions of the head and trunk muscles.
3 Describe the primary movements of head and trunk muscles.

The approach used to learn the primary skeletal muscles of the body is to first cover muscles of the head and trunk, followed by the muscles of the upper limb and lower limb.

Before we begin, we will discuss a couple of foundational points on skeletal muscles. Firstly, a skeletal muscle contracts; it does not push. When a muscle contracts, there are three possible outcomes:

1. Concentric contraction (the muscle shortens during contraction; e.g., curling a dumbbell).
2. Eccentric contraction (the muscle lengthens during contraction; e.g., lowering the dumbbell after it was curled).
3. Isometric contraction (the muscle does not shorten or lengthen during contraction; e.g., gripping the barbell without curling it).

A skeletal muscle usually attaches to at least two different bones and acts upon the synovial joints it crosses. The bony attachments are identified as the origin (the bone that stays stationary) and the insertion (the bone that moves). Some muscles, however, insert into structures other than bones. For example, the zygomaticus major muscle originates on the zygomatic bone and inserts into the orbicularis oris muscle.

These lab activities will help you learn the muscles, their main attachments (origins and insertions), and their primary actions. The muscles for this chapter are divided into head and trunk (neck, thorax, and abdominopelvic) muscles.

GETTING ACQUAINTED
Complete the following activities to become familiar with the head and trunk muscles.

Muscles of the Head
Many of the muscles located in the head are superficially located in the face. Unlike most other muscles of the human body, the muscles of the face do not move any bones. Instead, these muscles move the skin of the face to form communicative expressions, including opening and closing of the eyelids and mouth. Some of the primary muscles of the face include the following:

1. **Frontalis muscle.** Wrinkles the forehead.
2. **Occipitalis muscle.** Moves the scalp.
   a. **Galea aponeurotica.** A tough band of connective tissue that connects the frontalis and occipitalis muscles.
3. **Buccinator muscle.** Draws the cheeks into the oral cavity.
4. **Corrugator supercilli muscle.** Depresses the eyebrows medially (the "angry" muscle).
5. **Depressor anguli oris muscle.** Depresses the corners of the mouth laterally (the "frowning" muscle).
6. **Depressor labii inferioris muscle.** Depresses the lower lip laterally.
7. **Levator anguli oris muscle.** Elevates the corners of the mouth.

MATERIALS
Obtain the following items before beginning the laboratory activities:

❏ Textbook or access to Internet resources
❏ Colored pencils

8. **Levator labii superioris muscle.** Elevates the upper lip.
9. **Mentalis muscle.** Elevates the skin of the chin and protrudes the lower lip (the "pouting" muscle).
10. **Orbicularis oculi muscle.** Closes the eyelids (the "sleeping" muscle).
11. **Orbicularis oris muscle.** Closes and protrudes the lips (the "kissing" muscle).
12. **Platysma muscle.** Tenses the skin of the neck (the "shaving" muscle).
13. **Zygomaticus major muscle.** Elevates the corner of the mouth (the "smiling" muscle).
14. **Risorius muscle.** Retracts the corner of the mouth.
15. **Masseter muscle.** Muscle of mastication responsible for elevating the mandible with power (an "eating" muscle).
16. **Temporalis muscle.** Muscle of mastication that elevates the mandible with power (another "eating" muscle).

Clinical Application

Botox Injections
A common method to treat forehead wrinkles is through Botox injections. Botox is an abbreviation for the botulism toxin, and basically consists of a protein produced by the *Clostridium botulinum* bacteria. When Botox is injected deep to the skin, it temporarily paralyzes the muscles in the area. As such, if a patient wants to reduce the wrinkles in his or her forehead, the provider injects Botox into the forehead where the frontalis and corrugator supercilli muscles are paralyzed. With the forehead muscles relaxed, the skin of the forehead is tighter and firmer, which subsequently reduces the visibility of wrinkles in the area of the injection.

Want a Bite?
If there were a competition based upon what muscle in the body exerts the most force, the gold medal would be awarded to the soleus. The soleus is the strongest body muscle and its contraction propels us along the way as we walk, run, skip, and dance. However, if the medal were to go to the muscle that exerts the most pressure, the masseter muscle would receive the award. Pressure is different from force and takes into account the area over which a force is generated. The average bite strength is between 117 and 265 pounds. One world record states that a man achieved a bite strength of 975 pounds—that is a lot of dumbbells! The masseter achieves its strength due to the biomechanical advantage created by the relationship between the masseter muscle (and its attachments) and the temporomandibular joint.

Muscles of the Neck

The two most superficial muscles in the neck are as follows:

1. **Sternocleidomastoid muscle.** The sternocleidomastoid rotates the head contralaterally and flexes the neck when the muscles on both sides act together bilaterally.
2. **Trapezius muscle.** The trapezius extends the head and elevates the scapula.

Muscles of the Thorax

Muscles in the thoracic region of the trunk can be divided into two groups. There are those that act and arise from the thoracic skeleton but insert and act on the upper limb. There are also those deeper ones that are responsible for moving the trunk. The muscles of the thoracic region are as follows:

1. **Superficial thoracic muscles.** The following muscles originate on the rib cage but insert on the skeleton of the upper limb.
    a. **Pectoralis major muscle.** Adducts, flexes, and medially rotates the humerus at the glenohumeral joint.
    b. **Pectoralis minor muscle.** Stabilizes and protracts the scapula.
    c. **Serratus anterior muscle.** Stabilizes and protracts the scapula.
2. **Deep thoracic muscles.** The area between the ribs is known as the intercostal space. These spaces contain the following three muscles, which assist in respiration:
    a. **External intercostal muscle.** The most superficial intercostal muscle; contributes to inhaling.
    b. **Internal intercostal muscle.** The intermediate intercostal muscle; contributes to exhaling.
    c. **Innermost intercostal muscle.** The deepest intercostal muscle.

Muscles of the Abdomen

There are five paired muscles that form the anterolateral abdominal wall. Contraction of these muscles compress the abdominal contents, which is helpful during forced exhalation, defecation, parturition (childbirth), and vomiting. The muscles also protect vital organs like the intestines and flex and rotate the trunk (e.g., sit-ups and crunches). The muscles of the abdomen are as follows:

1. **External oblique muscle.** Located anterolaterally and is the most superficial abdominal muscle.
2. **Internal oblique muscle.** Located anterolaterally and is the intermediate abdominal muscle.
3. **Transverse abdominis muscle.** Located anterolaterally and is the deepest abdominal muscle.
4. **Rectus abdominis muscle.** Located anteriorly and courses vertically, helping to flex the trunk. It is also known as the "six-pack" muscle.

Clinical Application

TRAM Flap Procedure

The term TRAM is an acronym, which stands for **T**ransverse **R**ectus **A**bdominis **M**yocutaneous. A TRAM flap procedure removes a portion of the rectus abdominis muscle, its vascular supply, and overlying adipose tissue and skin and transplants it onto another site of the body. It is a common procedure used for breast reconstruction surgery following the removal of cancerous and surrounding tissue of the breast. This provides breast cancer patients with the ability to reconstruct their breasts with their own tissue and not foreign transplants.

Muscles of the Back

The muscles in the back can be divided into those that act on the head and upper limb and those that primarily act on the vertebral column. The muscles of the back are as follows:

1. **Superficial back muscles.** Superficial or extrinsic back muscles are involved with movements of the upper limb and head. The superficial back muscles are as follows:
   a. **Trapezius muscle.** Elevates, retracts, and rotates the scapula.
   b. **Levator scapulae muscle.** Elevates the scapula.
   c. **Rhomboid major muscle.** Retracts the scapula.
   d. **Rhomboid minor muscle.** Retracts the scapula.
   e. **Latissimus dorsi muscle.** Adducts, extends, and medially rotates the humerus at the glenohumeral joint.
2. **Deep back muscles.** Deep or intrinsic back mucles are involved in extension, lateral flexion, and rotation of the vertebral column.
   a. **Erector spinae muscles.** A strap of muscles that course from the lower vertebral column to the skull with various attachments along the way. The erector spinae consist of three muscles: the iliocostalis, longissimus, and spinalis muscles.
   b. **Transversospinalis muscles.** A deep group of muscles that course between transverse and spinous processes of the vertebral column (hence the name). The transversospinalis muscle group consists of three muscles: the semispinalis, multifidus, and rotatores muscles.

ACTIVITY 1 Head and Trunk Muscle Attachments and Actions

Before you get started, might we make a suggestion? Muscle attachments are often written in textbooks with extensive origins and insertions. The power of learning attachments is to help students see how the muscle crosses a joint(s) and thus the movement created when the muscle contracts. Sometimes, however, students get so lost in the wordy description that they cannot deduce the action. We suggest writing down the bone and perhaps a major bony landmark for each origin and insertion. In this way, it will become easier to deduce the action. See the following examples:

- Too much detail: The rectus abdominis muscle originates on the crest of the pubis, pubic tubercle, and medial aspect of the pectineal line and inserts along the costal cartilages of ribs 5, 6, and 7 as well as the xiphoid process and body of the sternum.
- The right amount of detail: The rectus abdominis; origin = pubic bone; insertion = rib cage.

1 Provide the origin, insertion, and action for each of the muscles listed in Table 7.1.

TABLE **7.1** Head and Trunk Muscles

Muscle	Origin	Insertion	Action
Muscles of the Head			
Frontalis m.			
Orbicularis oculi m.			
Orbicularis oris m.			
Sternocleidomastoid m.			
Muscles of the Anterior Chest Wall			
Pectoralis major m.			
Pectoralis minor m.			
Subclavius m.			
Muscles of the Trunk			
External intercostal m.			
Internal intercostal m.			
Innermost intercostal m.			
Serratus anterior m.			
External oblique m.			
Internal oblique m.			
Transverse abdominis m.			
Rectus abdominis m.			
Muscles of the Back			
Trapezius m.			
Latissimus dorsi m.			
Levator scapulae m.			
Rhomboid major m.			
Rhomboid minor m.			
Splenius capitis m.			
Erector spinae mm.			

ACTIVITY 2 Identification of Head and Trunk Muscles

1 Identify, label, and color the following structures on Figure 7.1.
 - Orbicularis oculi m.
 - Nasalis m.
 - Levator labii superioris m.
 - Zygomaticus major m.
 - Orbicularis oris m.
 - Buccinator m.
 - Mentalis m.
 - Depressor anguli oris m.
 - Risorius m.
 - Masseter m.
 - Frontalis m.
 - Galea aponeurotica
 - Occipitalis m.
 - Temporalis m.
 - Sternocleidomastoid m.
 - Trapezius m.

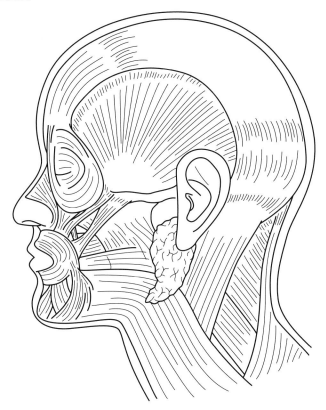

FIGURE **7.1** Lateral view of muscles of the head and neck.

2 Identify, label, and color the following structures on Figure 7.2.
 - Serratus anterior m.
 - External intercostal m.
 - Internal intercostal m.
 - Innermost intercostal m.
 - Ribs
 - Intercostal vein
 - Intercostal artery
 - Intercostal nerve
 - Lung

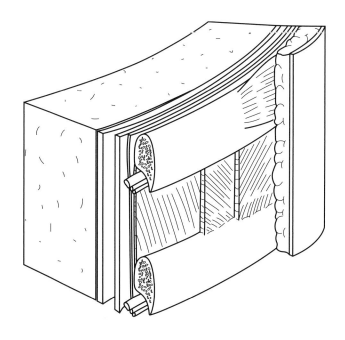

FIGURE **7.2** Layers of muscle.

3 Identify, label, and color the following structures on Figure 7.3.
- ☐ Pectoralis major m.
- ☐ Serratus anterior m.
- ☐ External oblique m.
- ☐ Umbilicus
- ☐ Internal oblique m.
- ☐ Transverse abdominis m.
- ☐ Rectus abdominis m.

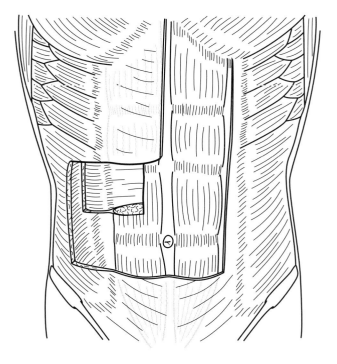

FIGURE **7.3** Anterior view of the trunk muscles.

4 Identify, label, and color the following structures on Figure 7.4.
- ☐ Trapezius m.
- ☐ Latissimus dorsi m.
- ☐ Rhomboid major m.
- ☐ Rhomboid minor m.
- ☐ Deltoid m.
- ☐ Splenius capitis m.
- ☐ Erector spinae mm.
  - Spinalis m.
  - Longissimus m.
  - Iliocostalis m.

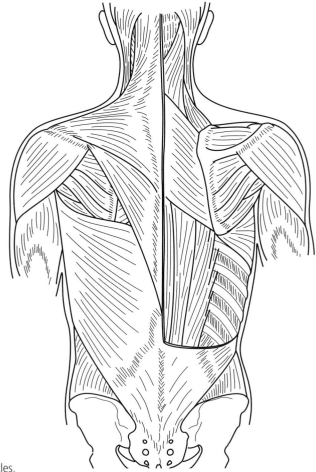

FIGURE **7.4** Posterior view of the trunk muscles.

👁 OBSERVING

Complete the following hands-on laboratory activities to apply your knowledge of head and trunk muscles.

MATERIALS
Obtain the following items before beginning the laboratory activities:
- ❏ Cadaver
- ❏ Gloves
- ❏ Probe
- ❏ Protective gear (lab coat, scrubs, or apron)

ACTIVITY 1 — Head and Trunk Muscle Anatomy

1 Identify the muscles labeled in Figure 7.5 on the cadaver.

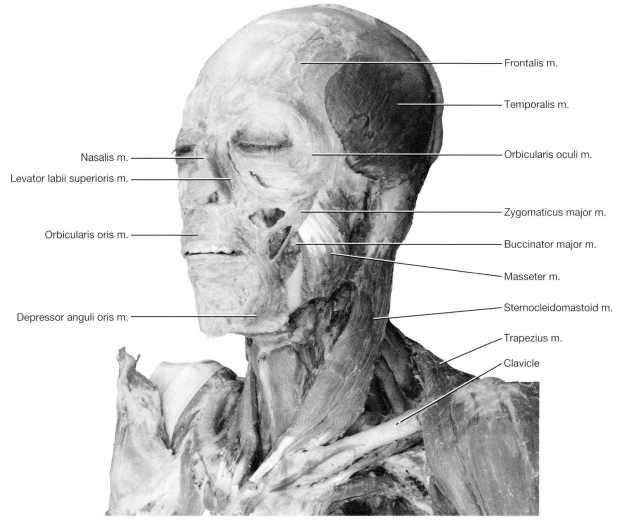

FIGURE **7.5** Anterior view of the face and neck muscles.

UNIT 2: Moving Forward

CHAPTER 7: Head and Trunk Muscles ■ **125**

2 Identify the muscles labeled in Figures 7.6 on the cadaver.

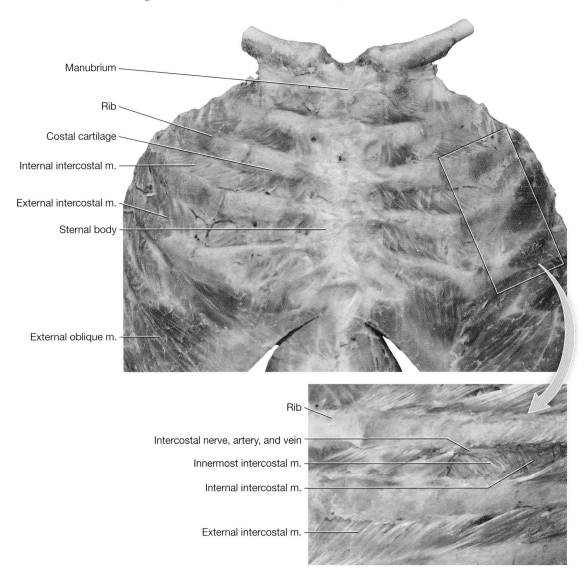

FIGURE **7.6** Anterolateral view of the thoracic muscles.

3 Identify the muscles labeled in Figure 7.7 on the cadaver.

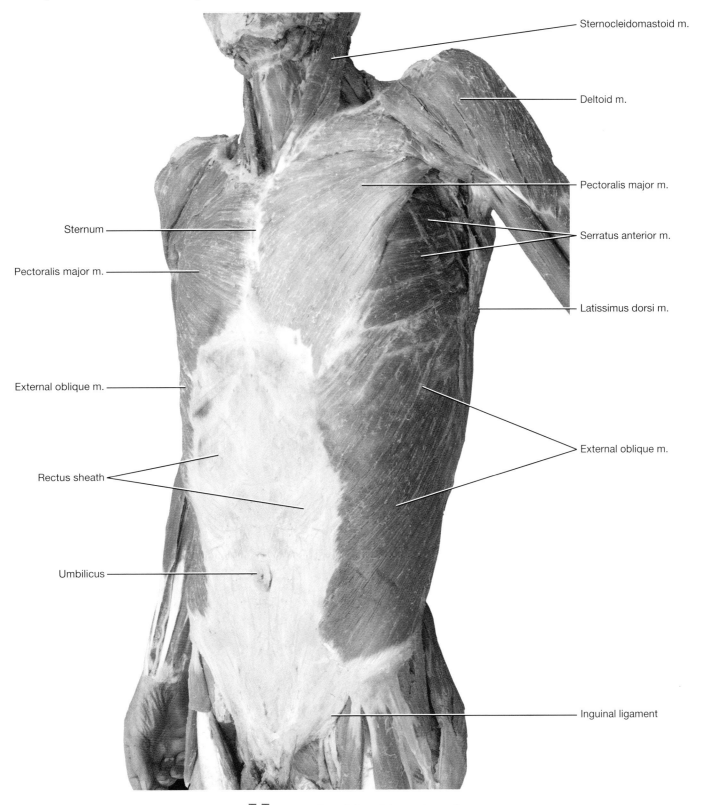

FIGURE **7.7** Anterolateral view of the trunk muscles.

4 Identify the muscles labeled in Figure 7.8 on the cadaver.

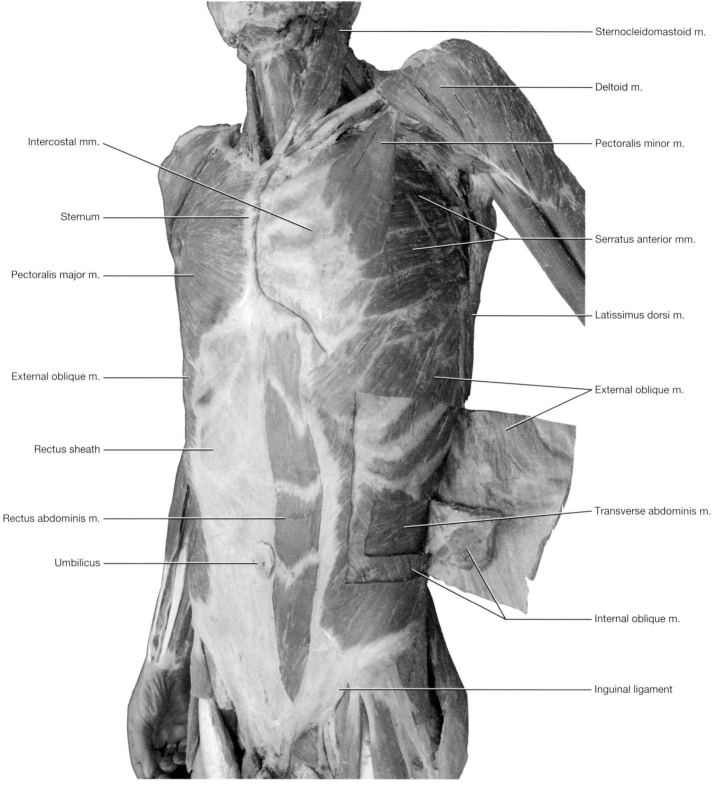

FIGURE **7.8** Anterolateral view of the trunk muscles after a step dissection.

5 Identify the muscles labeled in Figure 7.9 on the cadaver.

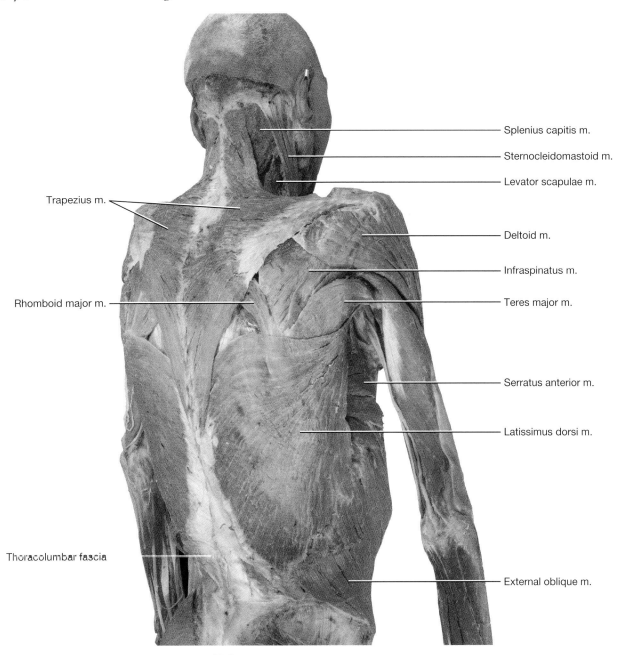

FIGURE **7.9** Posterolateral view of the trunk muscles.

6 Identify the muscles labeled in Figure 7.10 on the cadaver.

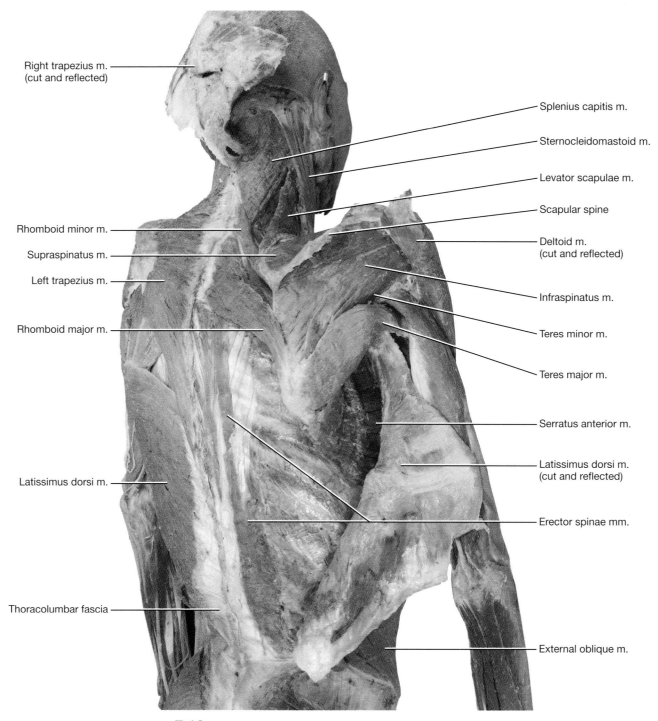

FIGURE **7.10** Posterolateral view of the trunk muscles after a step dissection.

130 ■ *Discovering Anatomy: A Guided Examination of the Cadaver*

WRAPPING UP

Complete the following additional activities to help retain your knowledge of head and trunk muscles.

Name _____

Date _____ Section _____

1 Identify the muscles and structures indicated in Figure 7.11.

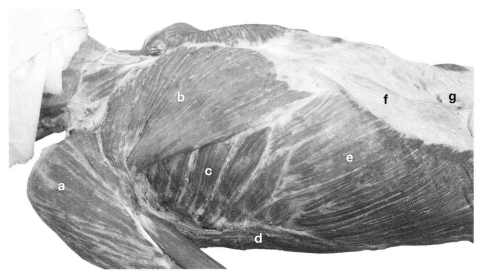

FIGURE **7.11** Anterolateral view of the trunk muscles.

a. _____ b. _____ c. _____ d. _____

e. _____ f. _____ g. _____

2 Identify the muscles and structures indicated in Figure 7.12.

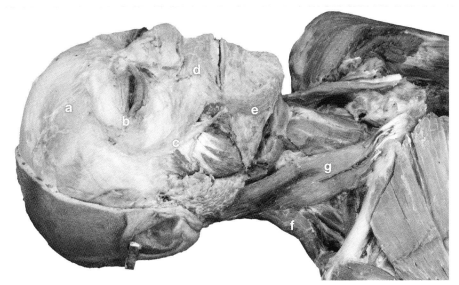

FIGURE **7.12** Anterolateral view of the head and neck muscles.

a. _____ b. _____ c. _____ d. _____

e. _____ f. _____ g. _____

UNIT 2: Moving Forward CHAPTER 7: Head and Trunk Muscles ■ **131**

3 The three different types of muscle contractions are demonstrated in Figure 7.13. In Table 7.2, identify which contraction is shown in each example, and describe what action the muscle is undergoing in each type of contraction.

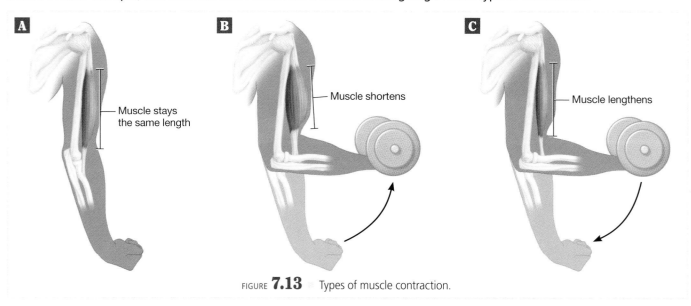

FIGURE 7.13 Types of muscle contraction.

TABLE 7.2 Three Types of Muscle Contractions

	Type of Contraction	Muscle Action
A		
B		
C		

4 Identify the muscles and structures indicated in Figure 7.14.

a. _____

b. _____

c. _____

d. _____

e. _____

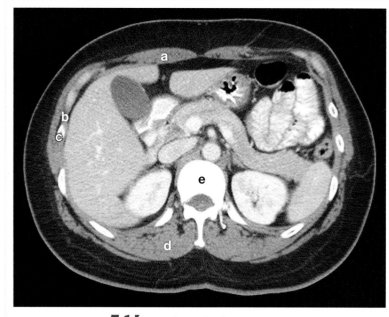

FIGURE 7.14 Radiograph of a transverse section of the human torso at the L2 vertebral level.

132 ■ *Discovering Anatomy: A Guided Examination of the Cadaver*

WRAPPING UP
(Continued)

Name _____

Date _____ Section _____

5 Identify the movement and muscle(s) associated with that movement for each letter in Figure 7.15. Then, complete Table 7.3 with the movements and muscles involved in that movement.

FIGURE **7.15** Movements of various joints of the human body.

TABLE **7.3** Muscle Movements

	Movement	Muscle(s)
A		
B		
C		
D		
E		
F		

Upper Limb Muscles

CHAPTER 8

At the completion of this laboratory session, you should be able to do the following:

1 Name and identify the primary muscles and actions that act on the following joints:
   a. Scapulothoracic joint.
   b. Glenohumeral joint.
   c. Elbow joint.
   d. Wrist joint.
   e. Hand (fingers and thumb).

The upper limb is suspended from the trunk by muscles and the sternoclavicular joint, a small skeletal articulation between the clavicle and sternum. The major joints of the upper limb are the scapulothoracic, glenohumeral, elbow, wrist, and finger and thumb joints. Our approach to learning the muscles of the upper limb is to study the major muscles that act on each of these joints.

 ## GETTING ACQUAINTED

Complete the following activities to become familiar with the muscles of the upper limb.

MATERIALS
Obtain the following items before beginning the laboratory activities:
❏ Textbook or access to Internet resources
❏ Colored pencils

ACTIVITY 1 Muscles of the Scapulothoracic Joint

The scapula possesses only one bony attachment—the clavicle. The rest of its stability is formed via a sling of muscles that move and support the scapula. The scapula moves along the posterolateral surface of the rib cage. Therefore, anatomists termed "scapulothoracic joint" as the movement of the scapula on the rib cage.

The scapulothoracic joint lies between the subscapularis muscle and the serratus anterior muscle. Unlike all other diarthrotic joints in the body, the scapulothoracic joint lacks a bony articulation. In other words, there is no articulation between the scapula and the thoracic cage. However, the movement of the scapula along the rib cage functions as a joint. The following muscles support and move the scapulothoracic joint:

1. **Trapezius muscle.** The trapezius elevates, retracts, and rotates the scapula.
2. **Levator scapulae muscle.** The levator scapulae elevates the scapula.
3. **Rhomboid major and minor muscles.** The rhomboid major and minor retract the scapula.
4. **Serratus anterior muscle.** The serratus anterior protracts the scapula and stabilizes the scapulothoracic joint by keeping the scapula pressed against the rib cage.
5. **Pectoralis minor muscle.** The pectoralis minor protracts and stabilizes the scapula.

Note: Before you get started, please review the text from Chapter 7, Activity 1 (p. 121). You will find it helpful in this activity as well.

1 Using your textbook or online resources, provide the origin, insertion, and action for each of the muscles listed in Table 8.1.

TABLE **8.1** Muscles of the Scapulothoracic Joint

Muscle	Origin	Insertion	Action
Trapezius m.			
Levator scapulae m.			
Rhomboid major m.			
Rhomboid minor m.			
Serratus anterior m.			
Pectoralis minor m.			

Clinical Application

Winged Scapula

One surgical treatment used for women with breast cancer is a radical mastectomy, which involves the removal of breast tissue, pectoralis major and minor muscles, and skin of the breast, as well as full axillary lymph node dissection. While removing the axillary lymph nodes, an injury to the nerve innervating the serratus anterior muscle (long thoracic nerve) may occur. When this occurs, the serratus anterior muscle is denervated (no longer contracts) and thus removes the biomechanical function of this muscle, which is to hold the scapula against the back of the rib cage. Thus, a "winged scapula" may occur as a result of this injury.

2 Describe the movement of the scapulothoracic joint shown in Figures 8.1 through 8.3 and write down the muscle or muscles that are responsible for that movement

FIGURE **8.1** Example 1.

Scapulothoracic joint movement:

Muscle(s):

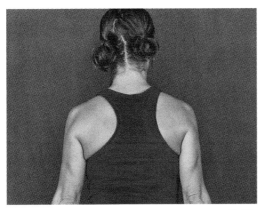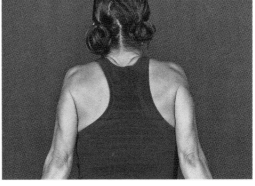

FIGURE **8.2** Example 2.

Scapulothoracic joint movement:

Muscle(s):

FIGURE **8.3** Example 3.

Scapulothoracic joint movement:

Muscle(s):

3 Identify, label, and color the following structures on Figure 8.4.
- ☐ Trapezius m.
- ☐ Levator scapulae m.
- ☐ Rhomboid minor m.
- ☐ Rhomboid major m.
- ☐ Serratus anterior m.

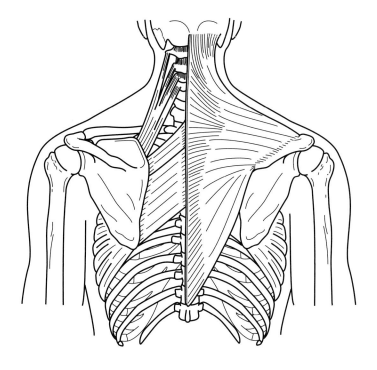

FIGURE **8.4** Scapulothoracic joint muscles: posterior view after a step dissection. The right side of the image shows superficial muscles; the left side of the image shows deeper muscles.

ACTIVITY 2 Muscles of the Glenohumeral Joint

The glenohumeral joint is a ball-and-socket joint composed of the glenoid fossa of the scapula and head of the humerus. Due to a loose joint capsule and a limited interface between the humeral head and scapula, it is the most flexible and mobile synovial joint in the body. The primary muscles that act on the glenohumeral joint are as follows:

1. **Deltoid muscle.** The deltoid flexes, abducts, and extends the glenohumeral joint.
2. **Rotator cuff muscles.** The rotator cuff is a collection of muscles that function together to stabilize the head of the humerus in the glenoid fossa. In addition, each muscle has the following actions:
   a. **Supraspinatus muscle.** Abducts the glenohumeral joint in the initial 15 degrees.
   b. **Infraspinatus muscle.** Externally rotates the glenohumeral joint.
   c. **Teres minor muscle.** Externally rotates the glenohumeral joint.
   d. **Subscapularis muscle.** Internally rotates the glenohumeral joint.
3. **Pectoralis major muscle.** The pectoralis major flexes, adducts, and internally rotates the glenohumeral joint.
4. **Teres major muscle.** The teres major adducts, extends, and internally rotates the glenohumeral joint.
5. **Latissimus dorsi muscle.** The latissimus dorsi adducts, extends, and internally rotates the glenohumeral joint.
6. **Coracobrachialis muscle.** The coracobrachialis flexes and adducts the glenohumeral joint.

Generally speaking, muscles located anteriorly to the glenohumeral joint, including the pectoralis major, coracobrachialis, and anterior deltoid, produce flexion.

Muscles located posteriorly to the glenohumeral produce extension, including the latissimus dorsi, teres major, and posterior deltoid. The middle region of the deltoid muscle is the principle abductor of the glenohumeral joint.

1 Using your textbook or online resources, provide the origin, insertion, and action for each of the muscles listed in Table 8.2.

TABLE **8.2** Muscles of the Glenohumeral Joint

Muscle	Origin	Insertion	Action
Deltoid m.			
Pectoralis major m.			
Latissimus dorsi m.			
Teres major m.			
Coracobrachialis m.			
Rotator cuff mm.			
Subscapularis m.			
Supraspinatus m.			
Infraspinatus m.			
Teres minor m.			

Have a Heart

Dynamic cardiomyoplasty is a technique for assisting heart contractions that involves the use of an electrically stimulated skeletal muscle (often the latissimus dorsi muscle) wrapped around region of the heart that has suffered myocardial damage. During this procedure the latissimus dorsi muscle is:

1. Conditioned to become less like skeletal muscle and more like cardiac muscle through pulsation with electrical impulses.
2. Detached from its attachments to the vertebral column.
3. Transplanted into the pericardical sac along with its nerve (thoracodorsal nerve) and blood supply (thoracodorsal artery and vein).
4. Wrapped like a towel around the region of the heart that is damaged.

A pacemaker is implanted in order to provide the rhythmic contractions necessary to continuously pump blood throughout the body.

Clinical Application

Rotator Cuff

Four of the nine muscles that cross the glenohumeral joint are collectively referred to as the rotator cuff muscles. These muscles include the supraspinatus, infraspinatus, teres minor, and subscapularis. Each of the four tendons blend with and reinforce the fibrous capsule of the glenohumeral joint en route to their insertion on the greater and lesser tubercles of the humerus. This structural arrangement plays a major role in stabilizing the shoulder joint. Rotator cuff injuries are common in baseball players because the action of moving the arm overhead, as is required in throwing a baseball, requires complete abduction of the shoulder followed by a rapid and forceful rotation and flexion of the shoulder. This movement causes a great deal of strain and may lead to tearing of the rotator cuff.

Baseball pitcher.

2 Describe the movements of the glenohumeral joint shown in Figures 8.5 through 8.9 and write down the muscle or muscles that are responsible for that movement.

FIGURE **8.5** Example 1.

Glenohumeral joint movement:

Muscle(s):

140 ■ *Discovering Anatomy: A Guided Examination of the Cadaver*

FIGURE **8.6** Example 2.

Glenohumeral joint movement:

Muscle(s):

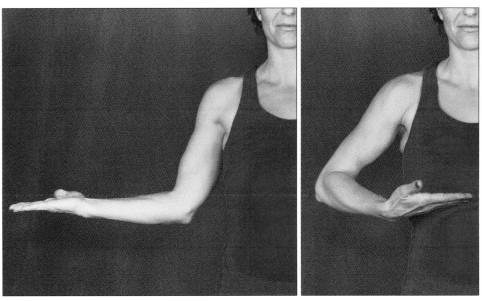

FIGURE **8.7** Example 3.

Glenohumeral joint movement:

Muscle(s):

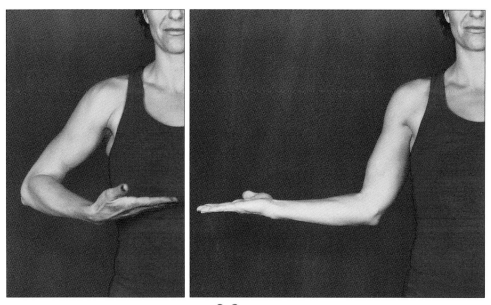

FIGURE **8.8** Example 4.

Glenohumeral joint movement:

Muscle(s):

FIGURE **8.9** Example 5.

Glenohumeral joint movement:

Muscle(s):

3 Identify, label, and color the following structures on Figure 8.10.
- ☐ Deltoid m.
- ☐ Latissimus dorsi m.
- ☐ Pectoralis major m.
- ☐ Teres major m.
- ☐ Rotator cuff
  - Supraspinatus m.
  - Infraspinatus m.
  - Teres minor m.

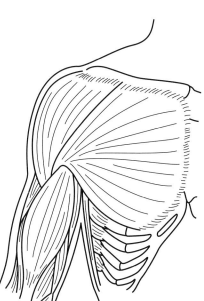

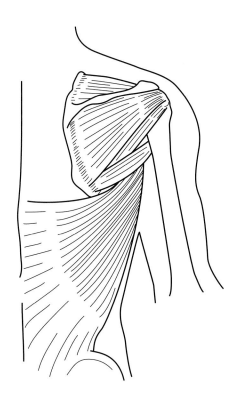

FIGURE **8.10**   Glenohumeral joint muscles: (**A**) anterior (superficial); (**B**) posterior (the trapezius and latissimus dorsi muscles have been cut and reflected).

ACTIVITY 3 Muscles of the Elbow Joint

The elbow joint is formed by the articulation between the humerus, ulna, and radius. The humerus and ulna form a synovial hinge joint, which produces elbow flexion and extension. The humerus, ulna, and radius form a synovial pivot joint, which produces elbow pronation and supination. The primary muscles that act on the elbow joint are as follows:

1. **Biceps brachii muscle.** The biceps brachii flexes and supinates the elbow joint.
2. **Brachialis muscle.** The brachialis flexes the elbow joint.
3. **Triceps brachii muscle.** The triceps brachii extends the elbow joint.

The muscles that primarily act on the elbow joint are located in the arm and are divided into two anterior muscles and one posterior muscle. The two anterior muscles are the biceps brachii and brachialis, and because they cross vertically and anterior to the elbow they produce flexion at the hinge joint. In addition to elbow flexion, the biceps brachii is the prime supinator of the forearm. The one posterior muscle is the triceps brachii. Because the triceps brachii crosses vertically and posterior to the elbow, it produces extension at the hinge joint.

1 Using your textbook or online resources, provide the origin, insertion, and action for each of the muscles listed in Table 8.3.

TABLE **8.3** Muscles of the Elbow Joint

Muscle	Origin	Insertion	Action
Biceps brachii m.			
Brachialis m.			
Triceps brachii m.			
Brachioradialis m.			

2 Describe the movements of the elbow joint shown in Figures 8.11 through 8.14 and write down the muscle or muscles that are responsible for that movement.

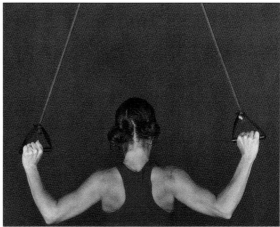

FIGURE **8.11** Example 1.

Elbow joint movements (2):

Muscle(s):

FIGURE **8.12** Example 2.

Elbow joint movements (2):

Muscle(s):

FIGURE **8.13** Example 3.

Elbow joint movements (2):

Muscle(s):

FIGURE **8.14** Example 4.

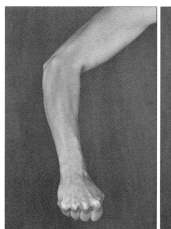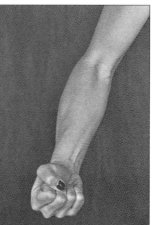

Elbow joint movements (2):

Muscle(s):

3 Identify, label, and color the following structures on Figure 8.15.
- ☐ Biceps brachii m.
- ☐ Brachialis m.
- ☐ Triceps brachii m.

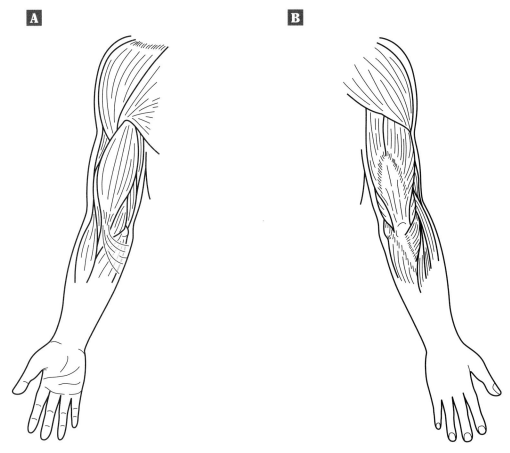

FIGURE **8.15** Upper limb muscles: (**A**) anterior; (**B**) posterior.

ACTIVITY 4 Muscles of the Wrist (Anterior Forearm)

The muscles that primarily act on the wrist, digits, and thumb joints are located in the forearm and are divided into eight anterior muscles and eleven posterior muscles. The majority of muscles that move the fingers and thumb are located in the forearm and operate via their long tendons, akin to operating a puppet by strings. The anterior forearm muscles are primarily flexors of the wrist and digits.

The wrist joint is comprised of the articulation between the radius and the proximal row of carpal bones. The wrist moves in flexion, extension, ulnar deviation (adduction), and radial deviation (abduction). The muscles that act on the wrist are as follows:

1. **Anterior forearm muscles.** The muscles that flex the wrist and digits are located in the anterior region of the forearm.
   a. **Pronator teres muscle.** Pronates the forearm.
   b. **Flexor carpi radialis muscle.** Flexion and radial deviation of the wrist joint.
   c. **Palmaris longus muscle.** Flexion of the wrist joint.
   d. **Flexor carpi ulnaris muscle.** Flexion and ulnar deviation of the wrist joint.
   e. **Flexor digitorum superficialis muscle.** Flexion of the wrist joint and digits.
   f. **Flexor digitorum profundus muscle.** Flexion of the wrist joint and digits.
   g. **Flexor pollicis longus muscle.** Flexion of the thumb.
   h. **Pronator quadratus muscle.** Pronation of the forearm.

1 Using your textbook or online resources, provide the origin, insertion, and action for each of the muscles listed in Table 8.4.

TABLE **8.4** Muscles of the Anterior Forearm

Muscle	Origin	Insertion	Action
Pronator teres m.			
Flexor carpi radialis m.			
Palmaris longus m.			
Flexor carpi ulnaris m.			
Flexor digitorum superficialis m.			
Flexor digitorum profundus m.			
Flexor pollicis longus m.			
Pronator quadratus m.			

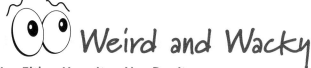

You Either Have It or You Don't

The palmaris longus muscle is one of the most variable muscles in the body. It is totally absent in approximately 1 out of 10 people, and it is absent in one or the other forearm in approximately 1 out of 20 people. Furthermore, it is absent more often in females than males, and absent in the left forearm more than the right in both sexes. Due to its superficial position, the palmaris longus muscle can be readily detected as present or absent on oneself. To check, tightly clench your left and right hands into a fist and examine the proximal part of your wrist for the tendon. Do you have one?

That's Shocking!

Why is it that when a person grabs an electrical source, such as a live wire, he or she may be unable to let go? Alternating electric current is hazardous because the current repetitively stimulates the nerves and muscles, resulting in a sustained contraction, or tetanus, as long as the contact with the wire is continued. This means that both the anterior forearm muscles, which flex the digits, and posterior forearm muscles, which extend the digits, are affected. Given that the current flow in the forearm stimulates both the muscles of flexion and extension, it seems surprising that one cannot let go. However, forearm flexor muscles, which wrap around the wire, are stronger than forearm extensor muscles. In other words, all upper limb muscles are stimulated to contract, but the forearm flexor muscles, being stronger, cause the hands to close tightly.

2 Describe the movements of the wrist and hand joints shown in Figures 8.16 through 8.19 and write down the muscle or muscles that are responsible for that movement.

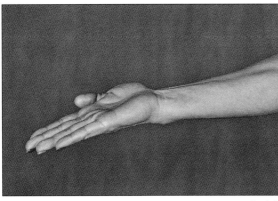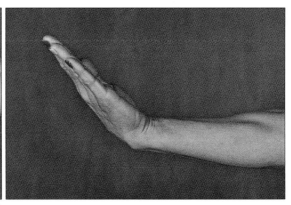

FIGURE **8.16** Example 1.

Wrist movement:

4 muscles:

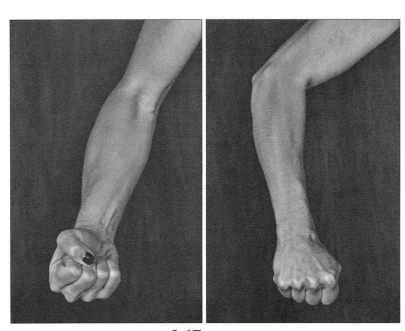

FIGURE **8.17** Example 2.

Elbow and wrist movement:

1 elbow muscle:

1 wrist muscle:

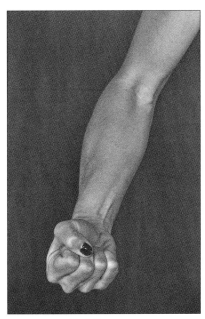

FIGURE **8.18** Example 3.

Fingers and thumb position:

2 finger muscles:

1 thumb muscle:

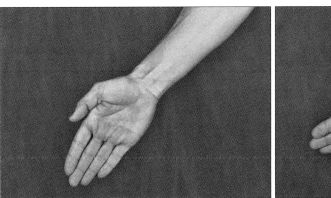 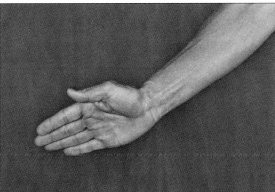

FIGURE **8.19** Example 4.

Wrist and/or hand joint movement:

2 muscles:

3 Identify, label, and color the following structures on Figure 8.20.

- ☐ Pronator teres m.
- ☐ Flexor carpi radialis m.
- ☐ Palmaris longus m.
- ☐ Palmar aponeurosis
- ☐ Flexor carpi ulnaris m.
- ☐ Flexor digitorum superficialis m.
- ☐ Flexor pollicis longus m.
- ☐ Flexor digitorum profundus m.
- ☐ Pronator quadratus m.

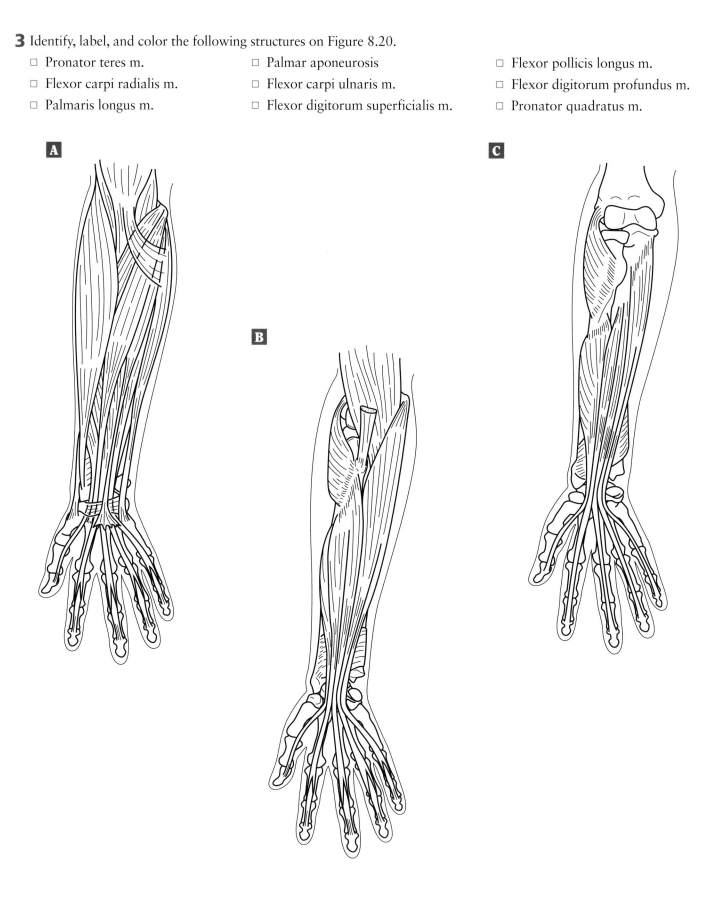

FIGURE **8.20** Anterior view of the right forearm muscles: (**A**) superficial dissection (skin removed); (**B**) intermediate dissection (pronator teres, flexor carpi radialis, palmaris longus, and flexor carpi ulnaris mm. removed); (**C**) deep dissection (flexor digitorum superficialis m. removed).

ACTIVITY 5 Muscles of the Wrist (Posterior Forearm)

The posterior muscles of the forearm are divided into two distinct muscle groups: a lateral group of eight muscles that arise from the lateral epicondyle, and a radial group of four muscles that descend along the distal end of the radius to attach to the index finger and thumb. With a couple of exceptions, all of the posterior muscles are extensors of the elbow, wrist, or fingers.

1. **Posterior forearm muscles.** The muscles that extend the wrist and digits are located in the posterior region of the forearm.
   a. **Brachioradialis muscle.** Flexes the elbow joint.
   b. **Extensor carpi radialis longus muscle.** Extension and radial deviation (abduction) of wrist joint.
   c. **Extensor carpi radialis brevis muscle.** Extension and radial deviation of wrist joint.
   d. **Extensor digitorum communis muscle.** Extension of wrist and digits.
   e. **Extensor carpi ulnaris muscle.** Extension and ulnar deviation of wrist joint.
   f. **Extensor indicis muscle.** Extension of index finger.
   g. **Extensor digiti minimi muscle.** Extension of digit five (pinkie).
   h. **Abductor pollicis longus muscle.** Abduction of the thumb.
   i. **Extensor pollicis longus muscle.** Extension of the thumb.
   j. **Extensor pollicis brevis muscle.** Extension of the thumb.
   k. **Supinator muscle.** Supination of the forearm.
   l. **Anconeus muscle.** Extension of the elbow.

1 Provide the origin, insertion, and action for each of the muscles listed in Table 8.5.

TABLE **8.5** Muscles of the Posterior Forearm

Muscle	Origin	Insertion	Action
Brachioradialis m.			
Extensor carpi radialis longus m.			
Extensor carpi radialis brevis m.			
Extensor digitorum communis m.			
Extensor carpi ulnaris m.			
Extensor indicis m.			
Extensor digiti minimi m.			
Abductor pollicis longus m.			
Extensor pollicis longus m.			
Extensor pollicis brevis m.			
Supinator m.			
Anconeus m.			

Weird and Wacky

Take a Whiff From Where?
In the 17th century, "snuff," or powdered tobacco, became quite popular. A person would take a pinch of snuff, place it on the back of his wrist, and snort it up into his nose. The region on the back of the wrist was then called, and is still called to this day, the "anatomical snuffbox." The tendons of the extensor pollicis longus, extensor pollicis brevis, and abductor pollicis longus make this depression where snuff was placed. In contemporary medical standards, the anatomical snuffbox has the following clinical relations: the scaphoid bone forms the floor, the radial artery traverses deep within the anatomical snuffbox, and the superficial radial nerve courses over it.

2 Describe the movements of the wrist and hand joints shown in Figures 8.21 through 8.23 and write down the muscle or muscles that are responsible for that movement.

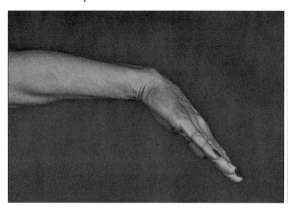 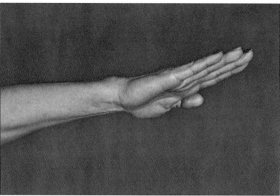

FIGURE **8.21** Example 1.

Wrist movement:

4 muscles:

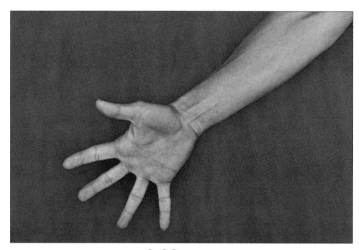

FIGURE **8.22** Example 2.

Digit and thumb movement:

1 digit muscle:

1 thumb muscle:

152 ■ *Discovering Anatomy: A Guided Examination of the Cadaver*

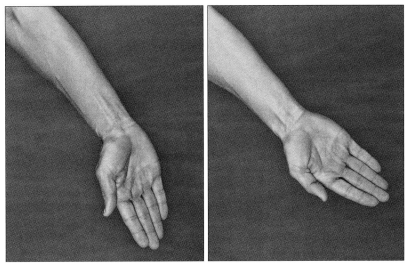

FIGURE **8.23** Example 3.

Wrist movement:

2 muscles:

3 Identify, label, and color the following structures on Figure 8.24.
- ☐ Brachioradialis m.
- ☐ Extensor carpi radialis longus m.
- ☐ Extensor carpi radialis brevis m.
- ☐ Extensor digitorum communis m.
- ☐ Extensor carpi ulnaris m.
- ☐ Extensor digiti minimi m.
- ☐ Abductor pollicis longus m.
- ☐ Extensor pollicis brevis m.
- ☐ Extensor pollicis longus m.
- ☐ Extensor indicis m.
- ☐ Supinator m.
- ☐ Anconeus m.

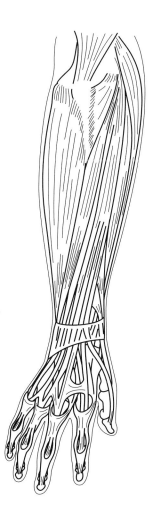
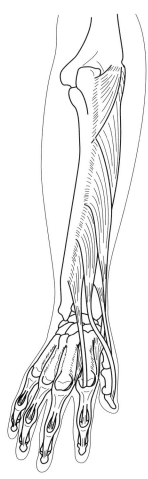

FIGURE **8.24** Posterior view of the right forearm muscles: (**A**) superficial dissection; (**B**) deep dissection.

OBSERVING

Complete the following hands-on laboratory activities to apply your knowledge of the upper limb muscles.

ACTIVITY 1 Upper Limb Muscle Anatomy

1 Identify the muscles labeled in Figure 8.25 on the cadaver.

2 Identify the muscles labeled in Figure 8.26 on the cadaver.

MATERIALS

Obtain the following items before beginning the laboratory activities:

- ❏ Cadaver
- ❏ Gloves
- ❏ Probe
- ❏ Protective gear (lab coat, scrubs, or apron)

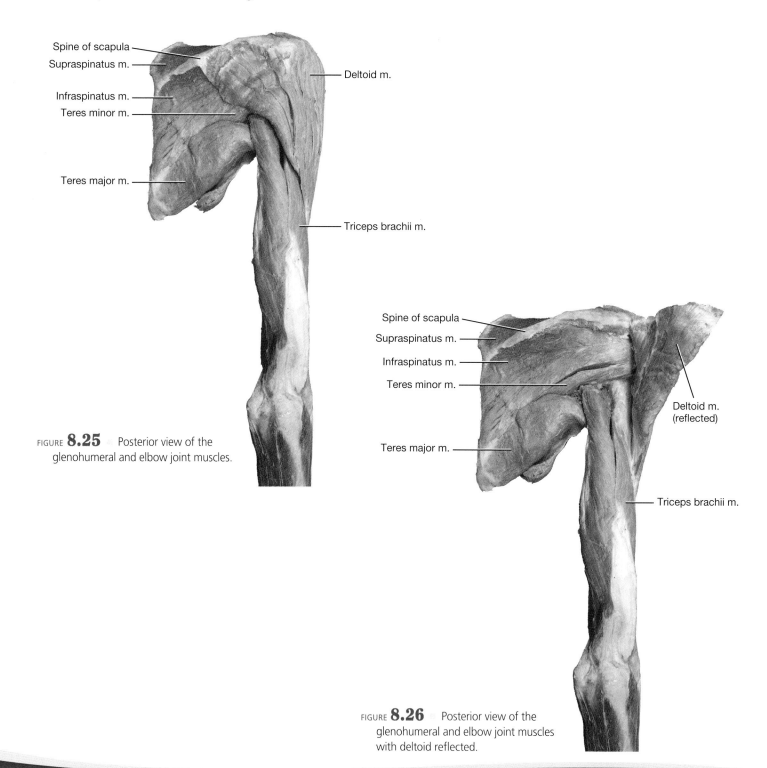

FIGURE **8.25** Posterior view of the glenohumeral and elbow joint muscles.

FIGURE **8.26** Posterior view of the glenohumeral and elbow joint muscles with deltoid reflected.

154 ■ *Discovering Anatomy: A Guided Examination of the Cadaver*

3 Identify the muscles labeled in Figure 8.27 on the cadaver.

4 Identify the muscles labeled in Figure 8.28 on the cadaver.

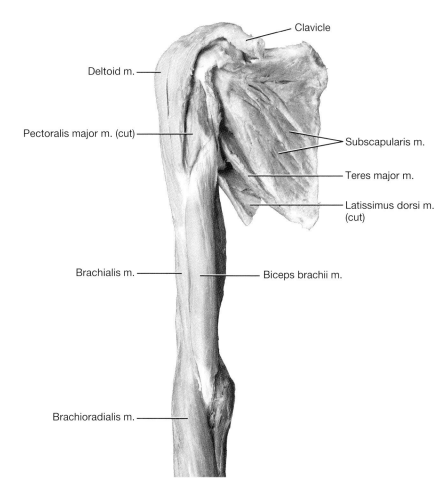

FIGURE **8.27** Anterior view of the right glenohumeral and elbow joint muscles.

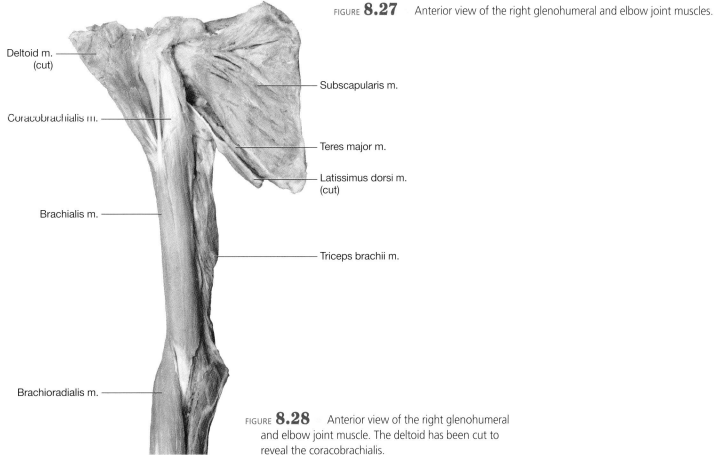

FIGURE **8.28** Anterior view of the right glenohumeral and elbow joint muscle. The deltoid has been cut to reveal the coracobrachialis.

5 Identify the muscles labeled in Figure 8.29 on the cadaver.

6 Identify the muscles labeled in Figure 8.30 on the cadaver.

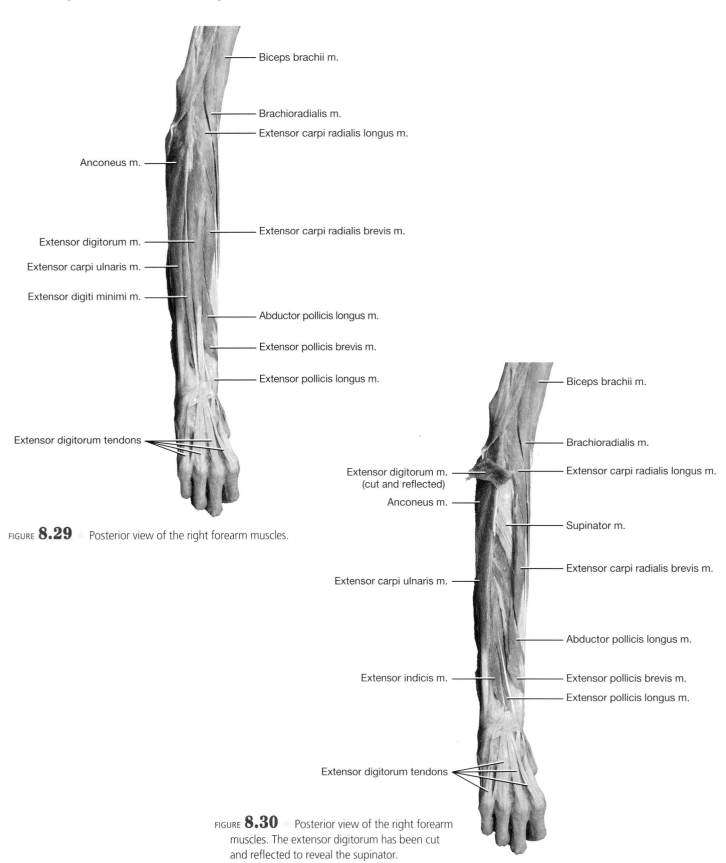

FIGURE **8.29** Posterior view of the right forearm muscles.

FIGURE **8.30** Posterior view of the right forearm muscles. The extensor digitorum has been cut and reflected to reveal the supinator.

7 Identify the muscles labeled in Figure 8.31 on the cadaver.

8 Identify the muscles labeled in Figure 8.32 on the cadaver.

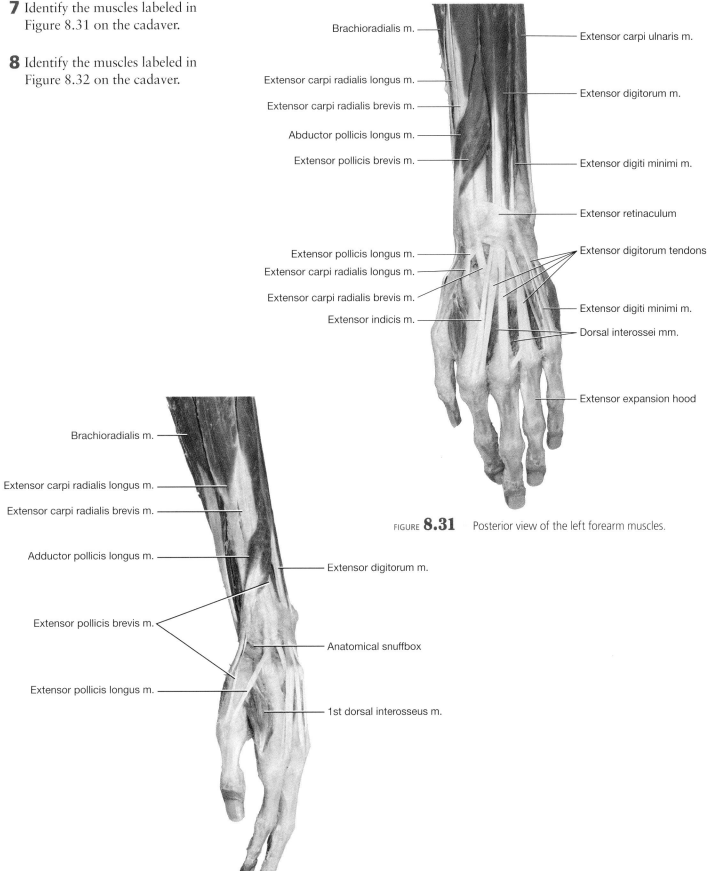

FIGURE **8.31** Posterior view of the left forearm muscles.

FIGURE **8.32** Medial view of the left forearm muscles.

UNIT 2: Moving Forward

CHAPTER 8: Upper Limb Muscles ■ 157

9 Identify the muscles labeled in Figure 8.33 on the cadaver.

10 Identify the muscles labeled in Figure 8.34 on the cadaver.

11 Identify the muscles labeled in Figure 8.35 on the cadaver.

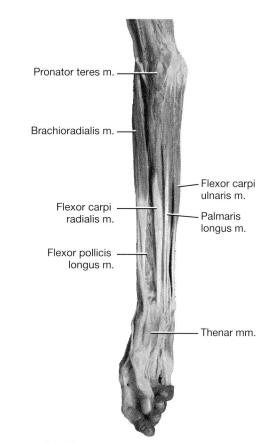

FIGURE **8.33** Medial view of the right forearm muscles.

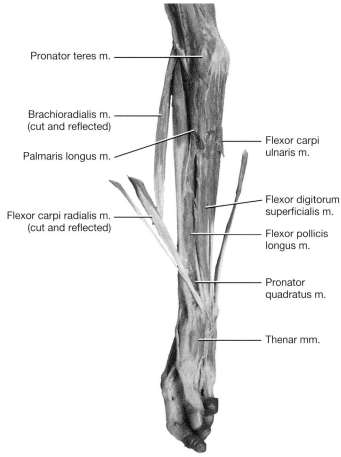

FIGURE **8.34** Medial view of the right forearm muscles after a step dissection.

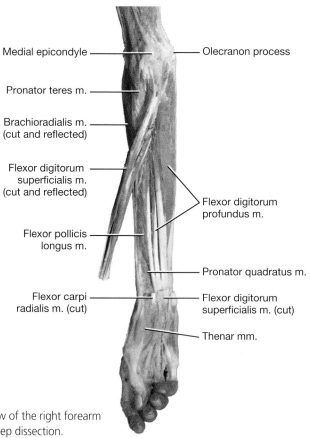

FIGURE **8.35** Medial view of the right forearm muscles after a second step dissection.

158 ■ *Discovering Anatomy: A Guided Examination of the Cadaver*

WRAPPING UP
Complete the following additional activities to help retain your knowledge of the upper limb muscles.

Name _____

Date _____ Section _____

1. Identify the muscles and structures indicated in Figure 8.36.

   a. _____
   b. _____
   c. _____
   d. _____
   e. _____
   f. _____
   g. _____
   h. _____
   i. _____

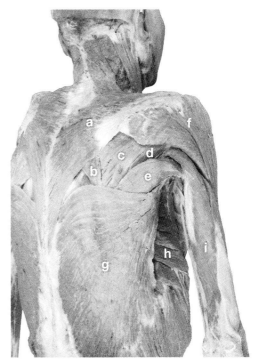

FIGURE **8.36** Posterolateral view of the trunk muscles.

2. Identify the muscles and structures indicated in Figure 8.37.

   a. _____
   b. _____
   c. _____
   d. _____
   e. _____
   f. _____
   g. _____
   h. _____
   i. _____
   j. _____
   k. _____

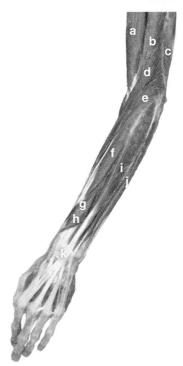

FIGURE **8.37** Posterior view of the forearm muscles.

UNIT 2: Moving Forward   CHAPTER 8: Upper Limb Muscles ■ **159**

3 Identify the muscles and structures indicated in Figure 8.38.

a. _____
b. _____
c. _____
d. _____
e. _____
f. _____
g. _____
h. _____

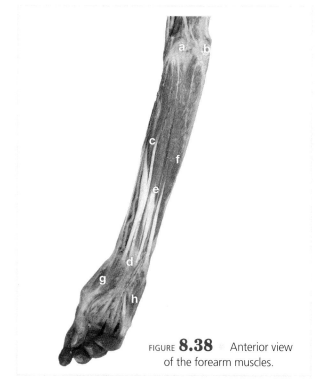

FIGURE **8.38** Anterior view of the forearm muscles.

4 In Figures 8.39 and 8.40, identify the movement and muscle(s) associated with that movement.

FIGURE **8.39** Woman in yoga pose.

FIGURE **8.40** Woman stretching arms outward.

a. Movement: _____
   Muscle(s): _____

b. Movement: _____
   Muscle(s): _____

c. Movement: _____
   Muscle(s): _____

d. Movement: _____
   Muscle(s): _____

e. Movement: _____
   Muscle(s): _____

f. Movement: _____
   Muscle(s): _____

g. Movement: _____
   Muscle(s): _____

h. Movement: _____
   Muscle(s): _____

Lower Limb Muscles

CHAPTER 9

At the completion of this laboratory session, you should be able to do the following:

1 Name and identify the primary muscles and actions that act on the following joints:
   a. Hip joint.
   b. Knee joint.
   c. Ankle joint.
   d. Foot.

In humans, the lower limb is the primary method of propulsion during walking and running and must bear the vast majority of body weight. We will study the lower limb by focusing on the muscles that act upon the hip, knee, ankle, and foot.

 GETTING ACQUAINTED

Complete the following activities to become familiar with muscles of the lower limb.

MATERIALS

Obtain the following items before beginning the laboratory activities:

❏ Textbook or access to Internet resources
❏ Colored pencils

ACTIVITY 1   Muscles of the Hip Joint

The muscles that act on the hip joint produce a combination of flexion, extension, external rotation, internal rotation, abduction, and adduction. The hip muscles responsible for these actions are the primary flexors (iliacus and psoas major muscles), abductors (gluteal muscles), and adductors (adductor brevis, longus, and magnus) of the thigh at the hip joint.

The hip joint is a synovial ball-and-socket joint, consisting of the head of the femur (ball) and the acetabulum (socket). The hip joint allows for a great deal of freedom, including flexion and extension, abduction and adduction, as well as internal and external rotation. However, the hip joint is not as flexible as the glenohumeral joint is because the acetabulum is a much deeper socket than the glenoid cavity. This deeper socket is what provides greater support and stability for the weight of the body. The primary muscles that act on the hip joint include the gluteal, iliopsoas, and medial thigh muscles.

1. **Gluteal muscles.** The gluteal muscles form the soft pad of tissue in the posterior region of the pelvis (rump). The muscles cross the hip joint from a variety of positions, enabling these muscles to collectively extend, abduct, and medially and laterally rotate the thigh at the hip joint.
   a. **Gluteus maximus muscle.** Extends the hip from a flexed position (as in the action of climbing stairs) and externally rotates the hip.
   b. **Gluteus medius muscle.** Abducts the hip joint (as in the action of spreading legs apart) and internally rotates the hip.
   c. **Gluteus minimus muscle.** Abducts the hip joint (as in the action of spreading legs apart) and internally rotates the hip.
   d. **Tensor fascia latae muscle.** Abducts and flexes the hip joint (as in the actions of spreading legs apart and lifting a knee off the ground).
2. **Iliopsoas muscle.** The primary flexors of the hip joint are the iliacus and psoas major muscles. The iliacus muscle arises from the iliac fossa and the psoas major muscle originates from the lumbar vertebrae. When the tendons from these two muscles travel deep to the inguinal ligament and enter the thigh, they have a common attachment to the lesser trochanter of the femur. As a result of their common attachment, their collective name in the thigh is the iliopsoas muscle. Due to the vertical course of these muscles anterior to the hip joint, they produce hip flexion when contracted. The iliopsoas muscle group primarily flexes the femur at the hip joint.

161

3. **Medial thigh muscles.** The deep fascia of the lower limb encircles the thigh and sends three distinct septae (walls, partitions) to knit with the periosteum of the femur. This forms three thigh compartments, one of which is the medial compartment. The muscles of the medial thigh compartment follow a vertical course medial to the hip joint, thus resulting in hip adduction when contracted (e.g., keeps thighs together while riding a horse). The muscles included in this compartment are the following:
   a. **Adductor brevis muscle.**
   b. **Adductor longus muscle.**
   c. **Adductor magnus muscle.**
   d. **Gracilis muscle.**
   e. **Pectineus muscle.**

1 Complete Table 9.1 for muscles of the hip joint.

TABLE **9.1** Muscles of the Hip Joint

Muscle	Origin	Insertion	Action
Gluteus maximus m.			
Gluteus medius m.			
Gluteus minimus m.			
Tensor fasciae latae m.			
Iliopsoas m.			
Pectineus m.			
Adductor longus m.			
Adductor brevis m.			
Adductor magnus m.			
Gracilis m.			

Weird and Wacky

It's All About that Base

One of the biggest muscles in the body is the gluteus maximus. Its biomechanical advantage (when it functions best) is contracting when the hip is already flexed (e.g., climbing stairs). The gluteus maximus is located in our "rump" and, in fact, the term "gluteus" is Greek for "rump." As such, the literal meaning for the gluteus maximus is the "biggest rump."

Clinical Application

That's a Pain in the Derrière

The gluteal region is a common site for an intramuscular (IM) injection of a medicine. These injections penetrate skin, fat, and gluteal muscles. The gluteal region is ideal because the muscles provide a substantial area for absorption of the injected medicine via the gluteal veins. Care must be taken with gluteal IM injections because the sciatic nerve lies deep within this region. Injections into the buttock are safest in the superior and lateral quadrant of the gluteal region, or superior to a line extending from the posterior superior iliac spine (PSIS) to the greater trochanter.

2 Complete the following for Figures 9.1 through 9.4:
- Describe the movement of the hip joint shown in the photo.
- Write down the muscle or muscles that are responsible for that movement.

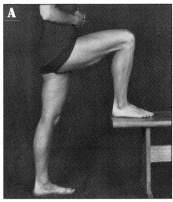 

FIGURE **9.1** Example 1.

Hip joint movement:

Muscle(s):

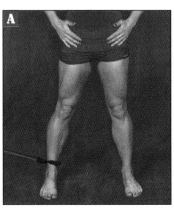 

FIGURE **9.2** Example 2.

Hip joint movement:

Muscle(s):

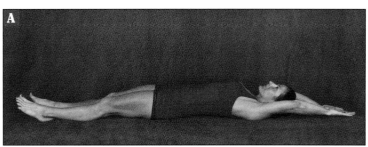

FIGURE **9.3** Example 3.

Hip joint movement:

Muscle(s):

FIGURE **9.4** Example 4.

Hip joint movement:

Muscle(s):

3 Identify, label, and color the following structures in Figures 9.5 and 9.6.
- ☐ Adductor longus m.
- ☐ Adductor magnus m.
- ☐ Gracilis m.
- ☐ Iliopsoas m.
- ☐ Tensor fasciae latae m.
- ☐ Iliotibial band
- ☐ Pectineus m.
- ☐ Gluteus maximus m.
- ☐ Gluteus medius m.
- ☐ Gluteus minimus m.

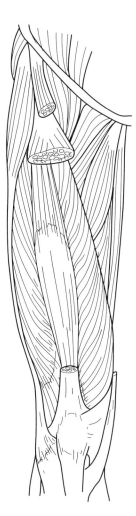

FIGURE **9.5** Anterior view of the right lower limb muscles.

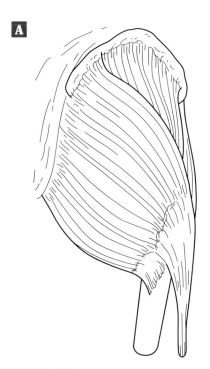

FIGURE **9.6** Posterior view: (**A**) gluteal region; (**B**) gluteus maximus and medius mm. (cut and reflected).

ACTIVITY 2 | Muscles of the Knee Joint

The primary muscles that act on the knee joint are located in the thigh and consist of anterior thigh muscles (knee extensors) and posterior thigh muscles (knee flexors). The anterior thigh muscles are known as the quadriceps femoris muscles and the posterior thigh muscles are known as the hamstrings.

The knee joint consists of two synovial joints: (1) between the condyles of the femur and tibia, and (2) between the patellar surface of the femur and the patella. The knee joint plays a critical role in stability and mobility of the lower limb. The following muscles support and move the knee joint:

1. **Anterior thigh muscles.**
    a. **Quadriceps femoris muscle.** Primarily extends the knee joint.
        i. **Rectus femoris muscle.**
        ii. **Vastus medialis muscle.**
        iii. **Vastus lateralis muscle.**
        iv. **Vastus intermedius muscle.**
    b. **Sartorius muscle.** Flexes, abducts, and externally rotates the hip joint.
2. **Posterior thigh muscles.** The posterior thigh muscles primarily flex the knee joint and assist in hip extension.
    a. **Biceps femoris muscle.**
    b. **Semitendinosus muscle.**
    c. **Semimembranosus muscle.**

1 List the origin, insertion, and action associated with each muscle listed in Table 9.2.

TABLE **9.2** Muscles of the Knee Joint

Muscle	Origin	Insertion	Action
Sartorius m.			
Quadriceps femoris muscle group			
Rectus femoris m.			
Vastus lateralis m.			
Vastus medialis m.			
Vastus intermedius m.			
Hamstrings			
Biceps femoris m.			
Semitendinosus m.			
Semimembranosus m.			

 **Weird and Wacky**

You Are Such a Ham!
The term "ham" comes from Old English and originally referred to the region on the posterior aspect of the knee. If you stand and bend your knees, you can feel two hard tendons coursing vertically on either side. They were literally called the "strings on the back of the knee" or, simply put, "hamstrings." Later on the term hamstrings came to denote the muscles in the posterior region of the thigh that give rise to the tendons in the "ham." This old English terminology helps us understand the group of muscles on the posterior region of the thigh, but not so much when explaining why someone who tells corny jokes is "such a ham."

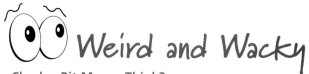 **Weird and Wacky**

Charley Bit My . . . Thigh?
A charley horse is a muscle spasm or cramp. This injury often occurs in the quadriceps muscles during sporting events like football, soccer, and baseball. Some say the term came from the lame horse, named Charley, that pulled the roller in the ballpark for the Chicago White Sox in the late 1800s. Others say that it originated from the pitcher Charley Radbourne, nicknamed Old Hoss, who suffered a leg injury during a baseball game in the same time period. Evidently his real name "Charley" was combined with his nickname "Hoss" to form the term "Charley Hoss" to describe this leg injury. However, both of these stories are like Swiss cheese—they have lots of holes in them (but they sure are fun to retell!).

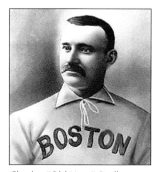
Charley "Old Hoss" Radbourne.

 Clinical Application

Avulsion Injuries
Hamstring injuries are one of the most common sports-related injuries. Hamstring avulsion injuries are particularly difficult. They result from forceful eccentric contraction (i.e., the muscle contracts while its belly is lengthening) and may lead to the avulsion of the proximal tendons from the ischial tuberosity. Hamstring avulsion may occur in water-skiing injuries (resulting from a rough jump start), football (caused by a direct blow to the posterior thigh), and repetitive eccentric loading (as in track and field). Trainers suggest that engaging in eccentric strengthening programs can prevent avulsion injuries.

2 Describe the movement of the knee joint shown in Figures 9.7 and 9.8 and write down the muscle or muscles that are responsible for that movement.

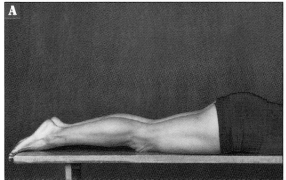

FIGURE **9.7** Example 1.

Knee joint movement:

Muscle(s):

FIGURE **9.8** Example 2.

Knee joint movement:

Muscle(s):

3 Identify, label, and color the following structures in Figure 9.9.
- ☐ Sartorius m.
- ☐ Rectus femoris m.
- ☐ Vastus lateralis m.
- ☐ Vastus medialis m.
- ☐ Patella
- ☐ Biceps femoris m.
- ☐ Semitendinosus m.
- ☐ Semimembranosus m.

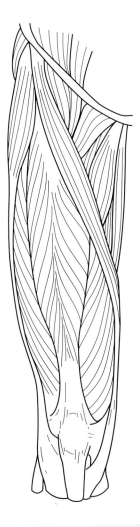

FIGURE **9.9** Muscles of the right lower limb: (**A**) anterior; (**B**) posterior.

ACTIVITY 3 Muscles of the Ankle (Anterior and Lateral Leg)

The ankle joint consists of two primary synovial joints: (1) between the tibia and talus (dorsiflexion and plantar flexion movements), and (2) between the tarsal bones below the talus (inversion and eversion movements). The ankle joint plays a critical role in stability and mobility of the lower limb.

The muscles of the leg that support and move the ankle joint are organized into anterior, lateral, and posterior. The following muscles are associated with the anterior and lateral leg compartments:

1. **Anterior leg muscles.** Anterior leg muscles are primarily involved in dorsiflexion and inversion. These muscles cross the ankle joint anteriorly, thus producing dorsiflexion at the ankle joint along with more specialized functions, such as extension of the digits and inversion of the ankle. The muscles of the anterior compartment include the following:
   a. **Tibialis anterior muscle.**
   b. **Extensor hallucis longus muscle.** Extends great toe.
   c. **Extensor digitorum longus muscle.** Extends lesser toes.
   d. **Peroneus (fibularis) tertius.** Dorsiflexion.

2. **Lateral leg muscles.** Lateral leg muscles are primarily involved in eversion and plantar flexion. These muscles cross the ankle joint laterally and posteriorly to the ankle joint thus producing eversion and weak plantar flexion at the ankle joints. The muscles of the lateral compartment include the following:
   a. **Peroneus (fibularis) longus muscle.**
   b. **Peroneus (fibularis) brevis muscle.**

1 List the origin, insertion, and action associated with each muscle listed in Table 9.3.

TABLE **9.3** Muscles of the Ankle (Anterior and Lateral Leg)

Muscle	Origin	Insertion	Action
Anterior compartment			
Tibialis anterior m.			
Extensor hallucis longus m.			
Extensor digitorum longus m.			
Peroneus (fibularis) tertius m.			
Lateral compartment			
Peroneus (fibularis) longus m.			
Peroneus (fibularis) brevis m.			

2 Describe the movement(s) of the ankle joint shown in Figures 9.10 through 9.12 and write down the muscle or muscles that are responsible for that movement.

FIGURE **9.10** Example 1.

Right ankle joint movement(s):

Right ankle muscle(s):

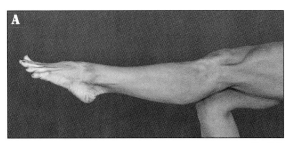 

FIGURE **9.11** Example 2.

Ankle joint movement(s):

Muscle(s):

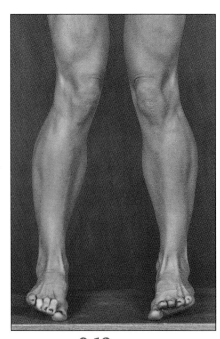

Ankle joint movement(s):

Muscle(s):

FIGURE **9.12** Example 3.

3 Identify, label, and color the following structures in Figure 9.13.

☐ Anterior compartment
- Tibialis anterior m.
- Extensor hallucis longus m.
- Extensor digitorum longus m.
- Peroneus tertius m.

☐ Lateral compartment
- Peroneus (fibularis) longus m.
- Peroneus (fibularis) brevis m.

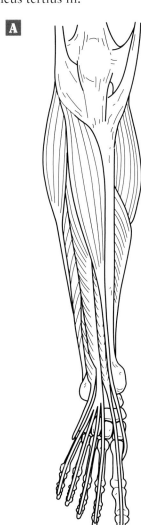
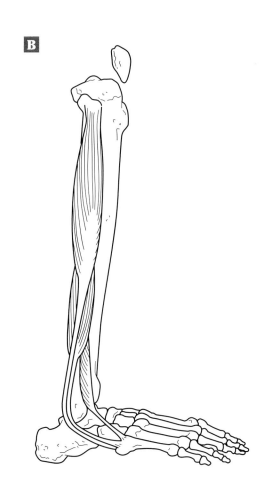

FIGURE **9.13** Right leg muscle compartments: (**A**) anterior; (**B**) lateral.

ACTIVITY 4 Muscles of the Ankle (Posterior Leg)

The fascia within the posterior compartment further divides the muscles into superficial and deep groups. The posterior leg muscles are primarily involved in plantar flexion and include the following:

1. **Superficial group.**
    a. **Gastrocnemius muscle.**
    b. **Soleus muscle.**
    c. **Plantaris muscle.**
2. **Deep group.**
    a. **Popliteus muscle.**
    b. **Tibialis posterior muscle.**
    c. **Flexor digitorum longus muscle.** Flexes lesser toes.
    d. **Flexor hallucis longus muscle.** Flexes great toe.

The superficial and deep leg muscles cross the ankle joint posteriorly and result in plantar flexion. Additionally, two of the muscles continue to the toes and produce flexion of the digits and inversion. The tibial nerve innervates these muscles.

1 List the origin, insertion, and action associated with each muscle listed in Table 9.4.

TABLE **9.4** Muscles of the Ankle (Posterior Compartment)

Muscle	Origin	Insertion	Action
Gastrocnemius m.			
Soleus m.			
Plantaris m.			
Tibialis posterior m.			
Flexor digitorum longus m.			
Flexor hallucis longus m.			
Popliteus m.			

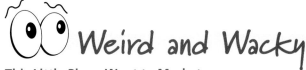

This Little Piggy Went to Market
The big toe (often referred to as the "great" toe) is called "hallux." However, this term is a corruption of the Latin *allex*, which is what REALLY means "great toe." The rest of the toes (digits 2 through 5) are referred to as "lesser" toes. That has to be a hit to their egos.

Don't Be a Heel
The calcaneal tendon (also known as the Achilles tendon) attaches the gastrocnemius and soleus muscles to the calcaneus. The tendon received its name in reference to the one vulnerable spot of the Greek hero Achilles. Supposedly his mother held him by his heel and dipped him in the River Styx to render him invulnerable. As such, his entire body was protected with the exception of the region behind his heel (where the calcaneal tendon is located), hence the name "Achilles tendon." In earlier anatomy the term "Achilles sinew" was used, which was coined by the Flemish physician Philip Verheyen. Throughout his life Verheyen vacillated between religion and medicine. At one point an illness resulted in the amputation of his left leg. Some historians say he decided to dissect his amputed leg (why not?) while others claim this event is fictional. Regardless, it seems that somewhere between reading the Good Book and dissecting his own leg he became a physician.

The Wellington Monument depicting Achilles in Hyde Park, London.

2 Describe the movement(s) of the ankle joint shown in Figures 9.14 through 9.16 and write down the muscle or muscles that are responsible for that movement.

FIGURE **9.14**   Example 1.

Left ankle joint movement(s):

Right ankle joint movement(s):

Left ankle muscle(s):

Right ankle muscle(s):

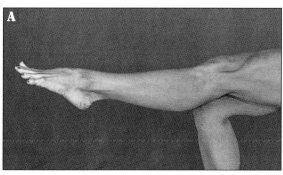 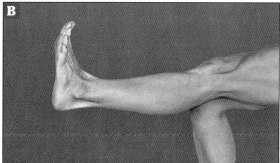

FIGURE **9.15**   Example 2.

Ankle joint movement(s):

Muscle(s):

Toe joint(s) movement(s):

Muscle(s):

FIGURE **9.16** Example 3.

3 Label and color the following structures in Figure 9.17.
- ☐ Gastrocnemius m.
- ☐ Plantaris m.
- ☐ Soleus m.
- ☐ Achilles tendon
- ☐ Popliteus m.
- ☐ Tibialis posterior m.
- ☐ Flexor hallucis longus m.
- ☐ Flexor digitorum longus m.

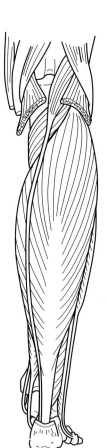

FIGURE **9.17** Leg muscles (posterior compartment): (**A**) superficial dissection; (**B**) intermediate dissection; (**C**) deep dissection.

OBSERVING

Complete the following hands-on laboratory activities to apply your knowledge of lower limb muscles.

MATERIALS

Obtain the following items before beginning the laboratory activities:
- ❏ Cadaver
- ❏ Gloves
- ❏ Probe
- ❏ Protective gear (lab coat, scrubs, or apron)

ACTIVITY 1 Lower Limb Muscle Anatomy

1 Identify the muscles labeled in Figure 9.18 on the cadaver.

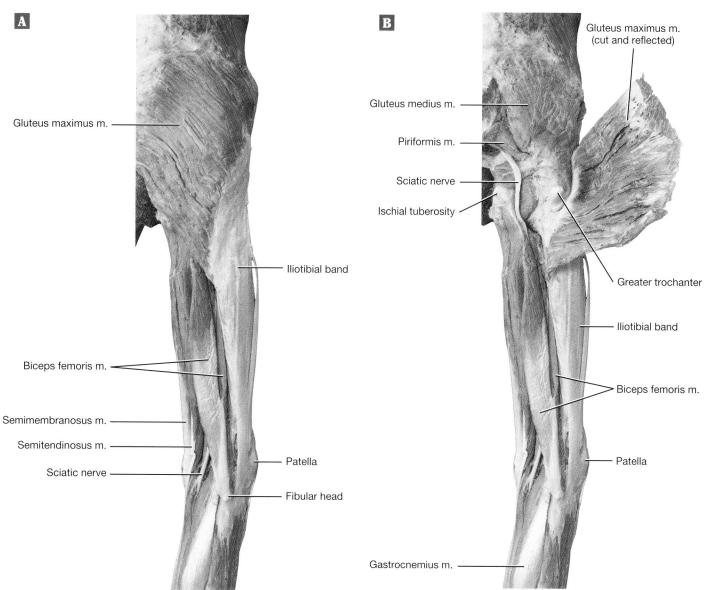

FIGURE **9.18** Hip joint muscles (posterior view of right hip): (**A**) superficial dissection; (**B**) intermediate dissection (gluteus maximus cut and reflected).

2 Identify the muscles labeled in Figure 9.19 on the cadaver.

3 Identify the muscles labeled in Figure 9.20 on the cadaver.

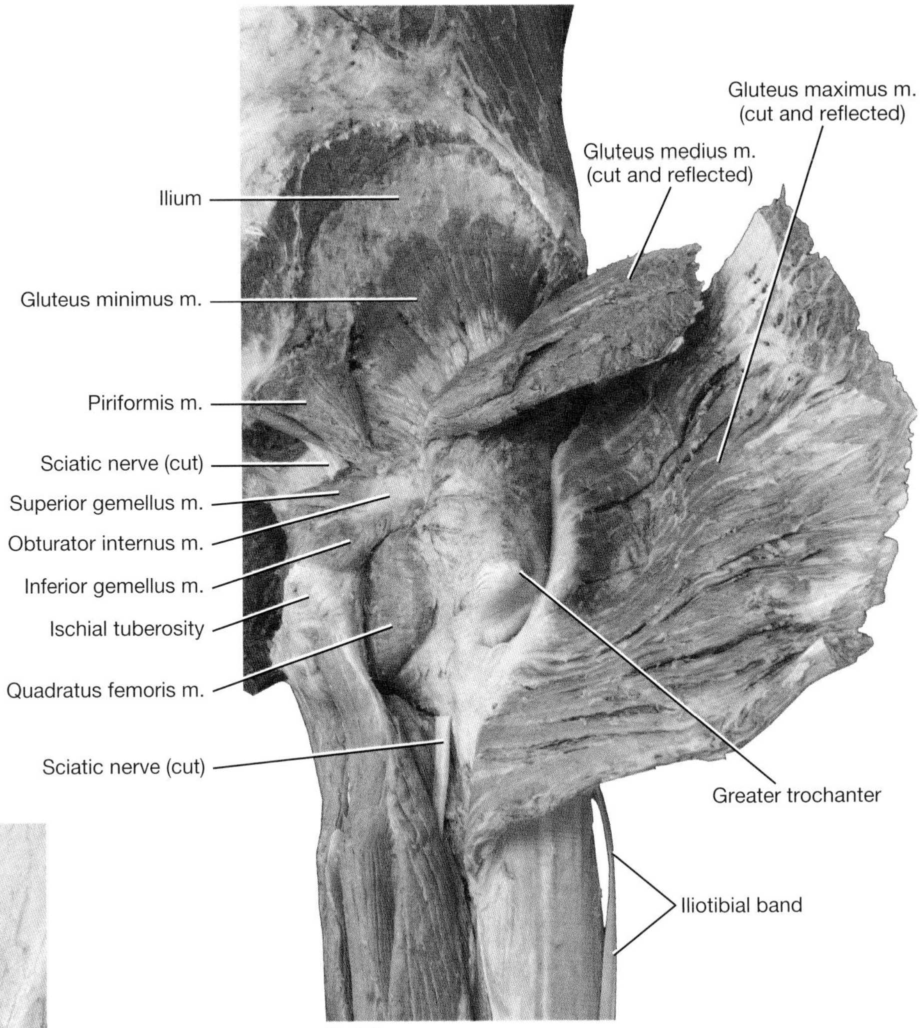

FIGURE **9.19** Hip and knee muscles: deep dissection (gluteus maximus and medius mm. are cut and reflected).

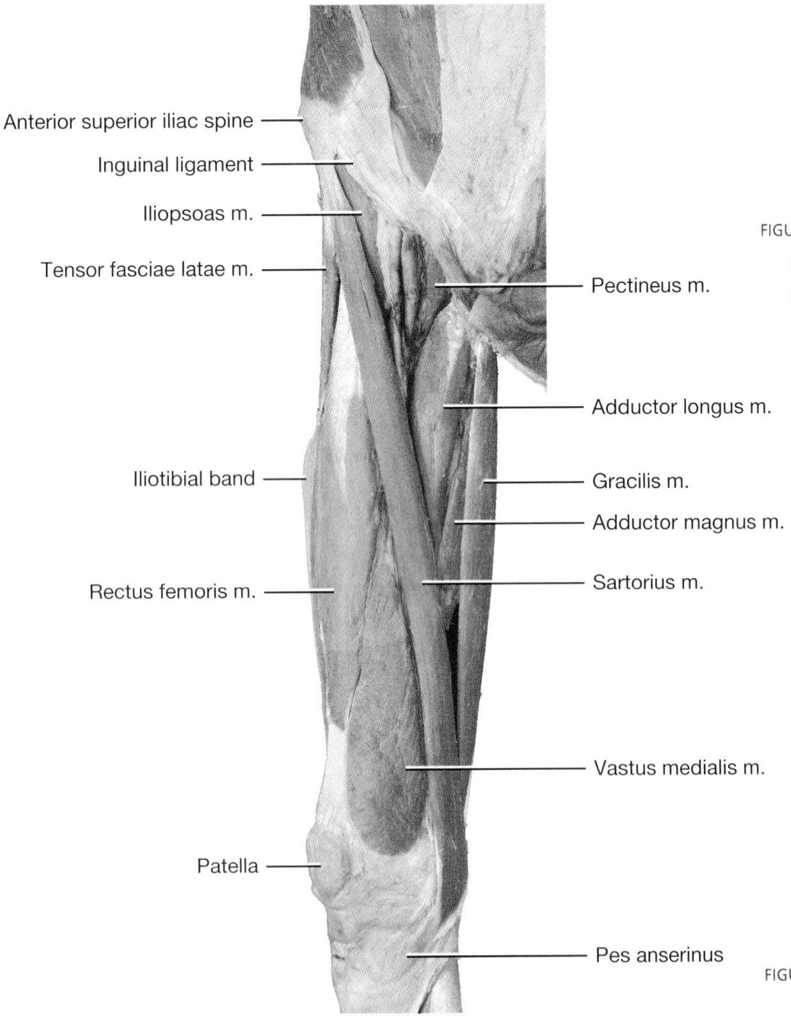

FIGURE **9.20** Hip and knee muscles: superficial dissection (anteromedial view; right limb).

4 Identify the following muscles in Figure 9.21 on the cadaver.

5 Identify the following muscles in Figure 9.22 on the cadaver.

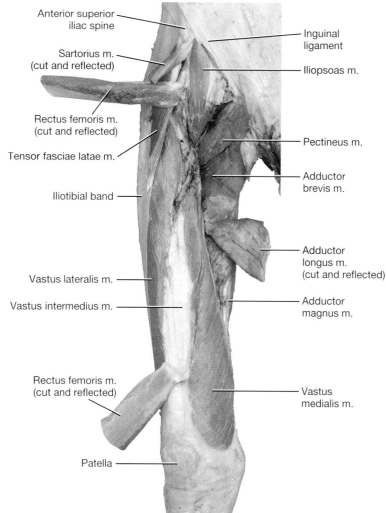

FIGURE **9.21** Hip and knee muscles: intermediate dissection (anterior view; right limb).

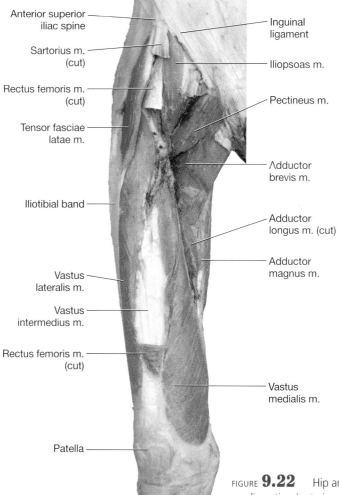

FIGURE **9.22** Hip and knee muscles: deep dissection (anterior view; right limb).

6 Identify the following muscles in Figure 9.23 on the cadaver.

7 Identify the following muscles in Figure 9.24 on the cadaver.

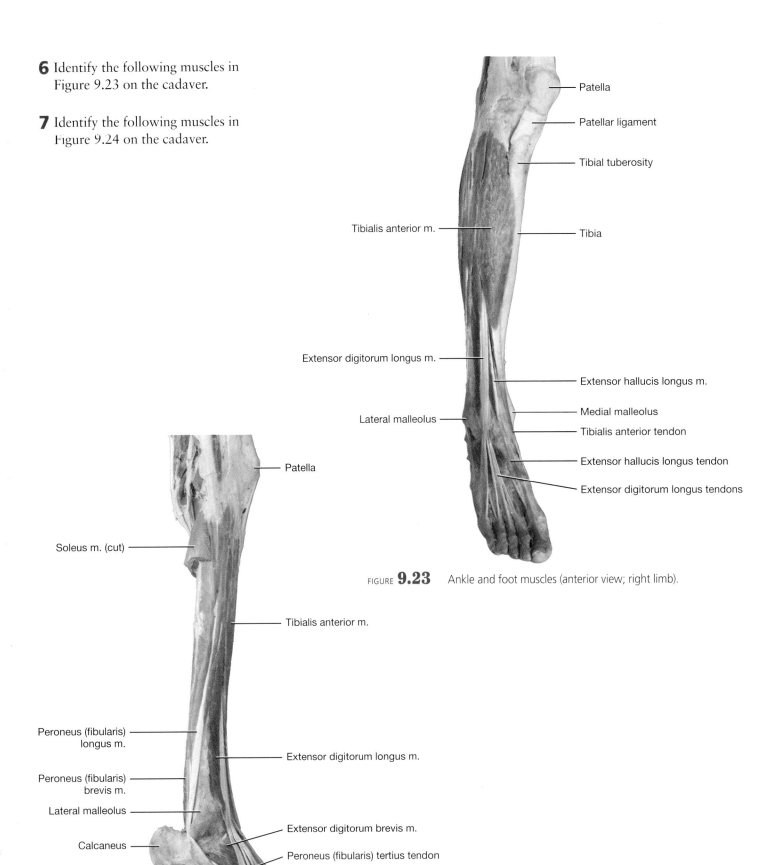

FIGURE **9.23** Ankle and foot muscles (anterior view; right limb).

FIGURE **9.24** Ankle and foot muscles (lateral view; right limb).

8 Identify the following muscles in Figure 9.25 on the cadaver.

9 Identify the following muscles in Figure 9.26 on the cadaver.

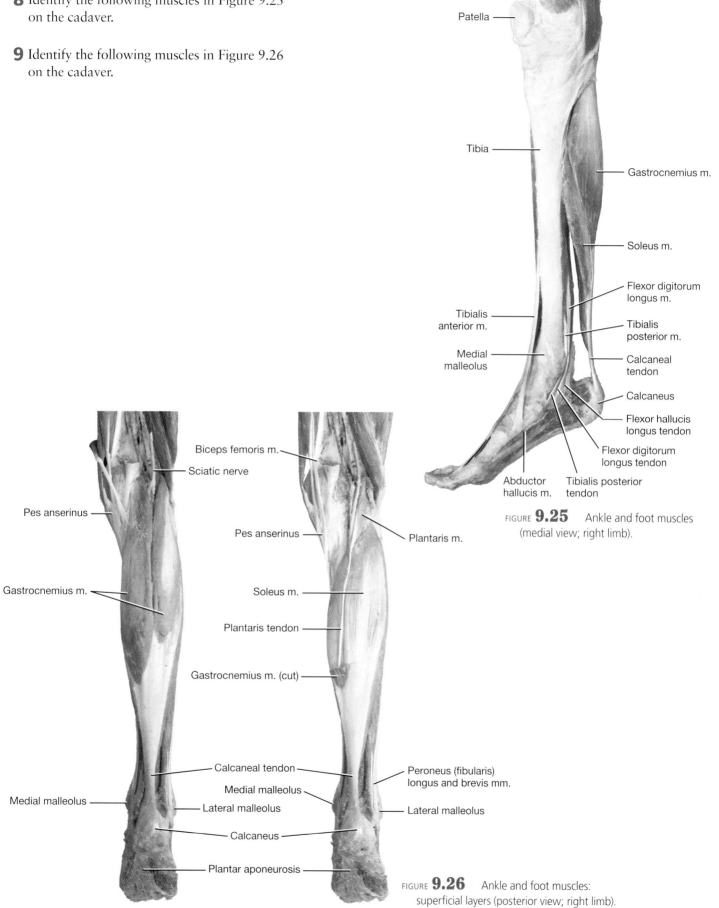

FIGURE **9.25** Ankle and foot muscles (medial view; right limb).

FIGURE **9.26** Ankle and foot muscles: superficial layers (posterior view; right limb).

UNIT 2: Moving Forward

CHAPTER 9: Lower Limb Muscles ■ **179**

10 Identify the following muscles in Figure 9.27 on the cadaver.

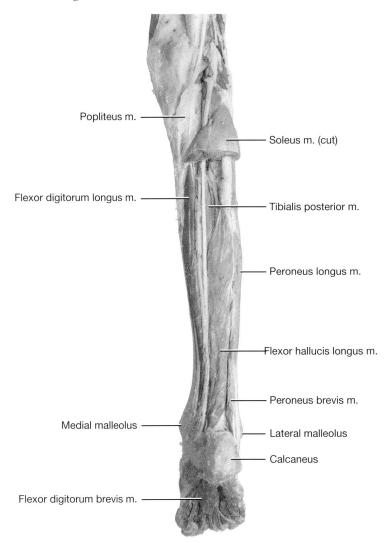

FIGURE **9.27** Ankle and foot muscles: deep layer (posterior view; right limb).

WRAPPING UP

Complete the following additional activities to help retain your knowledge of lower limb muscles.

Name _____

Date _____ Section _____

1. Identify the muscles and structures indicated in Figure 9.28.

   a. _____
   b. _____
   c. _____
   d. _____
   e. _____
   f. _____
   g. _____
   h. _____

2. Identify the muscles and structures indicated in Figure 9.29.

   a. _____
   b. _____
   c. _____
   d. _____
   e. _____
   f. _____
   g. _____
   h. _____
   i. _____
   j. _____
   k. _____
   l. _____

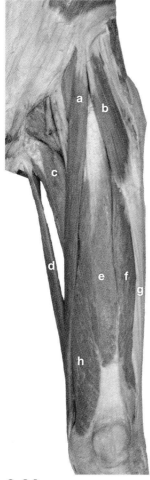

FIGURE **9.28** Anterior view of left thigh.

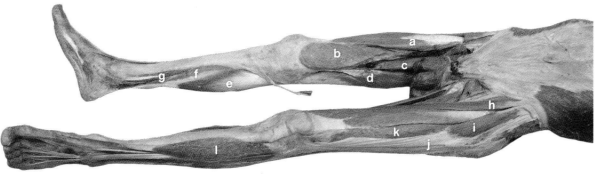

FIGURE **9.29** Anterolateral view of lower limb muscles.

UNIT 2: Moving Forward   CHAPTER 9: Lower Limb Muscles ■ **181**

3 Identify the muscles and structures indicated in Figure 9.30.

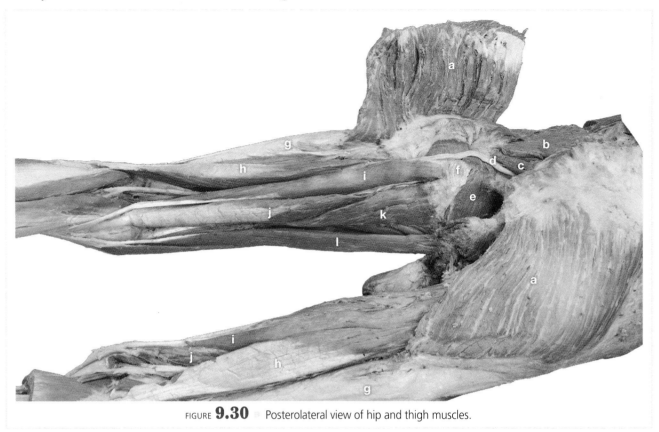

FIGURE 9.30   Posterolateral view of hip and thigh muscles.

a. _____
b. _____
c. _____
d. _____
e. _____
f. _____

g. _____
h. _____
i. _____
j. _____
k. _____
l. _____

WRAPPING UP
(Continued)

Name _____

Date _____ Section _____

4 Identify the movement indicated by the letter in Figures 9.31 and the muscle(s) resulting in that movement.

FIGURE **9.31** Limb movements: (**A**) anterior view; (**B**) lateral view.

a. Movement: _____

   Muscle(s): _____

b. Movement: _____

   Muscle(s): _____

c. Movement: _____

   Muscle(s): _____

d. Movement: _____

   Muscle(s): _____

e. Movement: _____

   Muscle(s): _____

f. Movement: _____

   Muscle(s): _____

g. Movement: _____

   Muscle(s): _____

h. Movement: _____

   Muscle(s): _____

i. Movement: _____

   Muscle(s): _____

j. Movement: _____

   Muscle(s): _____

3

Coming to Your Senses

CHAPTER 10 Central Nervous System **187**

CHAPTER 11 Peripheral and Autonomic
Nervous Systems **203**

CHAPTER 12 Special Senses **227**

Central Nervous System

CHAPTER 10

At the completion of this laboratory session, you should be able to do the following:

1 Identify and describe the different parts of the brain.
2 Describe the ventricular system of the brain including the flow of CSF.
3 Identify the meninges protecting the brain.
4 Identify key structures of the spinal cord.

The central nervous system (CNS) consists of two organs: the **brain** and the **spinal cord**. The brain and spinal cord govern sensory input, integration of motor output, homeostatic functions, and higher cognitive functions.

 GETTING ACQUAINTED
Complete the following activities to become familiar with the central nervous system.

MATERIALS
Obtain the following items before beginning the laboratory activities:
- ❏ Textbook or access to Internet resources
- ❏ Colored pencils

ACTIVITY 1 The Brain

The brain consists of the ventricular system, cerebrum, diencephalon, brainstem, and cerebellum.

1. **Ventricular system.** The ventricular system consists of four interconnected hollow spaces in the brain. Each ventricle contains **choroid plexuses**, which produce a fluid similar to plasma called **cerebrospinal fluid (CSF)**. CSF is located inside the brain (ventricular system) as well as outside of the brain and spinal cord (subarachnoid space). As a result, the brain and spinal cord float in a fluid. The four ventricles are as follows:
    a. **Lateral ventricles.** The largest of the ventricles and located within the right and left cerebral hemispheres. The paired lateral ventricles resemble the horns of a ram when viewed from the anterior side.
    b. **Third ventricle.** Located between the two diencephalons and is continuous with the lateral ventricles, via the **interventricular foramina (of Monro)**, and the fourth ventricle, via the **cerebral aqueduct (of Sylvius)**.
    c. **Fourth ventricle.** A diamond-shaped chamber within the pons and medulla that is continuous with the third ventricle via the cerebral aqueduct, the spinal cord via the central canal, and the subarachnoid space via the paired lateral apertures (of Luschka) and median aperture (of Magendie).

 Clinical Application

Hydrocephalus
The relationship between the cerebrospinal fluid (CSF) and the ventricular system is much like that of a water tap that is constantly running and a sink. The tap lets water flow into the sink and the sink lets the water out through the drain. However, if the drain is stopped, then water accumulates in the sink. In the brain, CSF is produced by choroid plexus ("tap") and enables CSF to flow through the ventricular system and subarachnoid space ("sink"). The CSF is removed from the subarachnoid space by way of arachnoid granulations ("drain"). Now, what would happen if the drain clogged somewhere along the way? It would cause CSF to build up in the sink. This is what is known as hydrocephalus. In children, hydrocephalus often results in the enlargement of the head because the skull bones have not yet fused (remember the fontanelles from Chapter 4?). In adults, hydrocephalus is a different challenge because the skull is rigid (remember the sutures?). As such, the accumulating CSF compresses the brain tissue and can result in severe damage or even death. In most cases, inserting a shunt into the ventricles and draining the excess fluid into the internal jugular vein or abdominal cavity is the treatment for hydrocephalus.

2. **Cerebrum.** The cerebrum is the largest and most developed region of the brain. It consists of left and right cerebral hemispheres, which are divided by the **longitudinal cerebral fissure**. Real brain tissue has the consistency of a gelatin dessert. The brains used in the lab have been preserved in formaldehyde, which changes the consistency to be more like clay. The primary anatomical parts of the cerebrum are as follows:
    a. **Lobes.** The cerebral cortex consists of lobes, which are named according to the overlying skull bones (**frontal lobe, parietal lobe, temporal lobe, occipital lobe,** and the deeper **insular lobe**). Large sulci divide the cerebral hemispheres. The central sulcus separates the frontal from the parietal lobe and the lateral sulcus separates the temporal lobe from the parietal and frontal lobes.
    b. **Cerebral cortex.** The outer 2 to 4 millimeters of the cerebrum. It functions as the "executive suite" because it enables us to communicate, initiate muscle movement, receive sensory input, store memories, and understand qualities associated with conscious behavior. The cerebral cortex consists mainly of neuronal cell bodies, dendrites, and unmyelinated axons (which cause its gray appearance). It also possesses **gyri** (domed ridges) and **sulci** (shallow grooves), which increase its surface area.
    c. **Cerebral white matter.** Provides communication between cerebral areas and between the cerebral cortex and lower CNS centers such as the spinal cord. Located deep to the gray matter of the cerebral cortex, white matter consists primarily of myelinated fibers bundled into large **tracts**. These fibers and the tracts they form are sorted into the following classifications according to the direction in which they run:
        i. **Commissures.** Axonal tracts that course side-to-side between the two cerebral hemispheres. The largest commissure is the **corpus callosum**, which is positioned superior to the lateral ventricles.
        ii. **Association fibers.** Axonal tracts that course forward and backward within the same cerebral hemisphere.
        iii. **Projection fibers.** Projection axonal fibers course up and down either from the cerebral cortex down to lower CNS areas (e.g., spinal cord) or from lower CNS areas (e.g., thalamus) up to the cerebral hemispheres.
3. **Diencephalon.** The diencephalon, positioned deep to the cerebral hemispheres in the central core of the brain, consists of the following parts:
    a. **Thalamus.** The "gateway" for the cerebrum. The thalamus integrates, relays, and sorts incoming afferent information from peripheral tissues to the cerebrum.
    b. **Hypothalamus.** Directs many homeostatic functions such as the endocrine and autonomic nervous systems, hunger and thirst, the sleep-wake cycle, and body temperature. The **infundibulum** connects the **pituitary gland** to the hypothalamus.
    c. **Epithalamus.** Helps to regulate the sleep cycle by way of the **pineal gland** (an endocrine gland), which secretes the hormone **melatonin**.

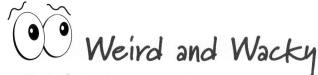

My Brain Hurts
The brain consists of millions and millions of neurons and is responsible for integrating all incoming sensory information from body tissues from head to toe. Whenever you stub your toe, or touch an ice cube, your brain integrates this information so you consciously know that your toe hurts or the ice is cold. However, even with these millions of neurons, the brain cannot feel anything. At least, not like your finger or toes can. The brain does not have any sensory neurons and as such does not feel pain, temperature, vibration, or touch. In fact, during brain surgery sometimes the patient is still conscious and when the cut in the brain is made the doctor can still communicate with the patient. But what about people who have headaches and migraines? Are they not feeling pain? Yes, but headaches and migraines are often associated with the meninges that surround the brain.

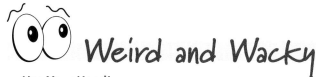

Use Your Head!
Here are a few more interesting things about the brain:
1. **That is deep.** The brain is the only part of the human body that is capable of contemplating itself. In other words, it can be aware that it is being aware of being aware of itself!
2. **Need a light?** The brain is comprised of millions of neurons, which communicate with each other via electrochemical impulses. When all of the current coursing through all of the neurons is combined, it is estimated that a brain produces enough power to light a 25-watt bulb!
3. **I'm all in.** There is some misinformation floating around that we only use 10% of our brains. This is not true; we actually use 100% of our brains.

4. **Brainstem.** From superior to inferior, the brainstem regions are the following:
   a. **Midbrain.** Located between the diencephalon and pons; contains the cerebral aqueduct.
   b. **Pons.** Located between the midbrain and medulla oblongata; contains the upper part of the fourth ventricle.
   c. **Medulla oblongata.** Located between the pons and spinal cord; contains the caudal part of the fourth ventricle.
5. **Cerebellum.** The cerebellum is located in the posterior cranial fossa and filters afferent information pertaining to muscle movement and proprioception at a subconscious level. The cerebellum has two lobes connected by a middle region called the **vermis**.

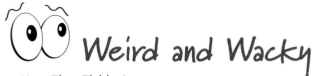

Hey, That Tickles!

Have you ever tried tickling yourself? If you have, you already know that it's impossible. The reason is not fully known; however, neuroimaging studies have provided a rough explanation. Areas of the brain that coordinate sensation with muscle response are activated less when self-produced tactile stimulus are used (tickling oneself) in contrast to an externally produced stimulus (your friend tickles you). In other words, your brain can predict when you tickle yourself and prevent the motor response, but cannot often predict when someone else tickles you. Fingers seem to point to the cerebellum for preventing this self-tickling void because it is responsible for coordinating controlled skeletal muscle contractions.

1 Identify, label, and color the following structures on Figure 10.1.

☐ Lateral ventricles   ☐ Third ventricle   ☐ Fourth ventricle

☐ Interventricular foramen   ☐ Cerebral aqueduct

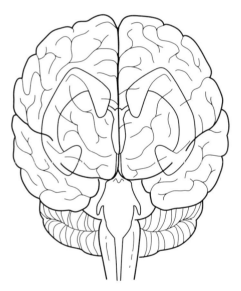
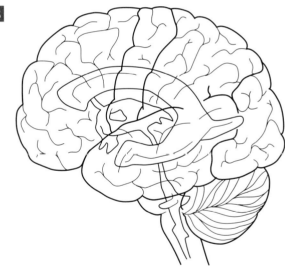

FIGURE **10.1**   Ventricular system: (**A**) anterior; (**B**) left lateral.

2 Identify, label, and color the following structures on Figure 10.2.

- ☐ Frontal lobe
- ☐ Parietal lobe
- ☐ Occipital lobe
- ☐ Temporal lobe
- ☐ Central sulcus
- ☐ Lateral sulcus
- ☐ Cerebellum
- ☐ Pons
- ☐ Medulla
- ☐ Spinal cord

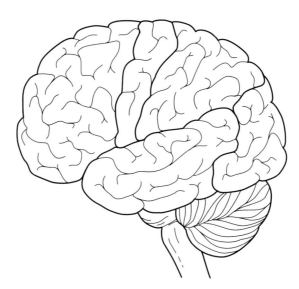

FIGURE **10.2** Lateral view of the brain.

3 Identify, label, and color the following structures on Figure 10.3.

- ☐ Frontal lobe
- ☐ Temporal lobe
- ☐ Cerebellum
- ☐ Occipital lobe
- ☐ Pons
- ☐ Medulla
- ☐ Pituitary gland
- ☐ Spinal cord gray matter
- ☐ Spinal cord
- ☐ White matter

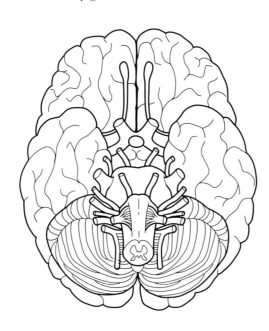

FIGURE **10.3** Inferior view of the brain.

Discovering Anatomy: A Guided Examination of the Cadaver

4 Identify, label, and color the following structures in Figure 10.4.

- ☐ Frontal lobe
- ☐ Parietal lobe
- ☐ Occipital lobe
- ☐ Cerebellum
- ☐ Corpus callosum
- ☐ Thalamus
- ☐ Hypothalamus
- ☐ Epithalamus
- ☐ Pituitary gland
- ☐ Region of the third ventricle
- ☐ Midbrain
- ☐ Pons
- ☐ Medulla
- ☐ Spinal cord
- ☐ Cerebral aqueduct
- ☐ Fourth ventricle

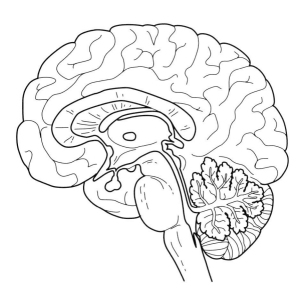

FIGURE **10.4** Medial view of the midsagittal section of the brain.

ACTIVITY 2 Meninges

Nerves and nervous tissue are very delicate and are injured by even the slightest pressure. Therefore, a set of connective tissue membranes, called the meninges (singular: meninx), surrounds the brain. The meninges are as follows:

1. **Dura mater.** The dura mater is a thick, double-layered meninx. It is the most superficial meninx. The superficial **periosteal dural layer** adheres to the internal surface of the skull, and the deeper **meningeal dural layer** becomes the dura mater of the spinal cord. The two layers of the dura are fused, but in some regions the layers separate to form the following structures:
   a. **Falx cerebri.** Forms a partition between the right and left cerebral hemispheres.
   b. **Falx cerebelli.** Courses along the vermis to separate the cerebellar hemispheres.
   c. **Tentorium cerebelli.** Separates the cerebrum from the cerebellum.
   d. **Dural venous sinuses.** The dural venous sinuses are spaces containing deoxygenated blood from the brain, which primarily drain into the internal jugular vein exiting at the base of the skull.

2. **Arachnoid mater.** Arachnoid mater is the middle meninx. It contains **arachnoid granulations,** which project into the dural venous sinuses and enable CSF to reenter the systemic circulation. The subarachnoid space (region between arachnoid and pia mater) contains CSF.

3. **Pia mater.** The thinnest and most internal of the meninges is the pia mater. It is intimately attached to the cerebral hemispheres, and courses along in the gyri and sulci.

1 Identify, label, and color the following structures in Figure 10.5.
- ☐ Skin
- ☐ Periosteum
- ☐ Skull bone
- ☐ Dura mater (periosteal layer)
- ☐ Dura mater (meningeal layer)
- ☐ Dural venous sinus
- ☐ Arachnoid mater
- ☐ Subarachnoid space
- ☐ Arachnoid granulations
- ☐ Pia mater
- ☐ Cerebrum (gray matter)
- ☐ Cerebrum (white matter)

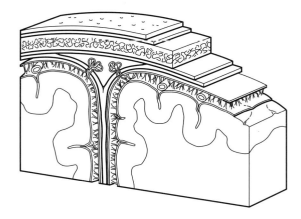

FIGURE **10.5** Step dissection of brain, skull, and meninges (coronal section).

ACTIVITY 3 Spinal Cord

The spinal cord is the continuation of the medulla oblongata. The spinal cord ends at the L1–L2 vertebral level due to the unequal growth of the vertebral column and spinal cord in the developing fetus. The caudal end of the spinal cord is called the **conus medullaris**, which gives rise to nerve roots called the **cauda equina** (meaning "horse's tail").

1. **Spinal meninges.** The spinal meninges are arranged the same way as those that surround the brain:
   a. **Dura mater.** The superficial meningeal layer that forms a tough connective tissue sheath that surrounds and protects the spinal cord.
   b. **Arachnoid matter.** The intermediate meningeal layer that forms the subarachnoid space, which is filled with cerebrospinal fluid (CSF).
   c. **Pia mater.** Intimately associated with the external surface of the spinal cord. Additionally, pia mater also tethers the spinal cord laterally within the vertebral canal via small extensions called **denticulate ligaments** and inferiorly to the coccygeal bone via the **filum terminale**.

2. **Gray matter.** Gray matter is "H"-shaped in appearance and subdivided into the following horns:
   a. **Anterior (ventral) horns.** Contain the cell bodies of motor neurons.
   b. **Posterior (dorsal) horns.** Receive sensory information from sensory neurons in the posterior (or dorsal) roots.
   c. **Lateral (intermediate) horns.** The T1–L2 and S2–S4 spinal cord levels possess a lateral horn, which contains the cell bodies of preganglionic sympathetic and parasympathetic neurons respectively.

3. **White matter.** White matter surrounds the spinal cord gray matter and contains myelinated axons grouped into vertical bundles called tracts. These tracts either course upward with sensory information from tissues to the brain, or downward with motor information from the brain to muscles or organs.

Clinical Application

Lumbar Puncture

The conus medullaris (inferior region of the spinal cord) is located at the L1–L2 vertebral level. Cerebrospinal fluid (CSF) flows in the subarachnoid space all around the spinal cord, including the space below the spinal cord. A lumbar puncture (spinal tap) is a diagnostic procedure used to obtain samples of CSF in order to help diagnose diseases of the CNS such as meningitis or subarachnoid hemorrhages. In this procedure the space between the L4 and L5 vertebrae is identified and a needle is inserted into the subarachnoid space. This procedure is done with very little risk of damaging the spinal cord because its inferior end is located two to three vertebral segments superior to the needle insertion. A patient undergoing a lumbar puncture is typically asked to lie on their side or sit in a sitting position in order to flex the vertebral column and increase the space between adjacent vertebrae.

1 Identify, label, and color the following structures in Figure 10.6.
- ☐ Cerebrum
- ☐ Cerebellum
- ☐ Spinal cord
- ☐ Dura mater
- ☐ Spinal nerves
- ☐ Vertebrae

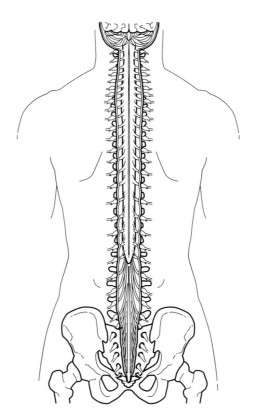

FIGURE **10.6**   Coronal section through the brain and spinal cord, posterior view.

2 Identify, label, and color the following structures in Figure 10.7.
- ☐ Dura mater
- ☐ Arachnoid mater
- ☐ Spinal cord (gray matter)
- ☐ Spinal cord (white matter)
- ☐ Spinal nerves
- ☐ Pia mater

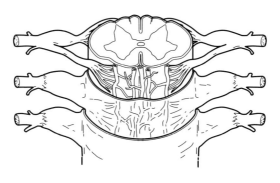

FIGURE **10.7**   Step dissection of the spinal cord, cross section.

OBSERVING

Complete the following hands-on laboratory activities to apply your knowledge of the central nervous system.

MATERIALS

Obtain the following items before beginning the laboratory activities:
- ❏ Cadaver
- ❏ Gloves
- ❏ Probe
- ❏ Protective gear (lab coat, scrubs, or apron)

ACTIVITY 1 Central Nervous System Anatomy

1 Identify the structures of the central nervous system labeled in Figure 10.8.

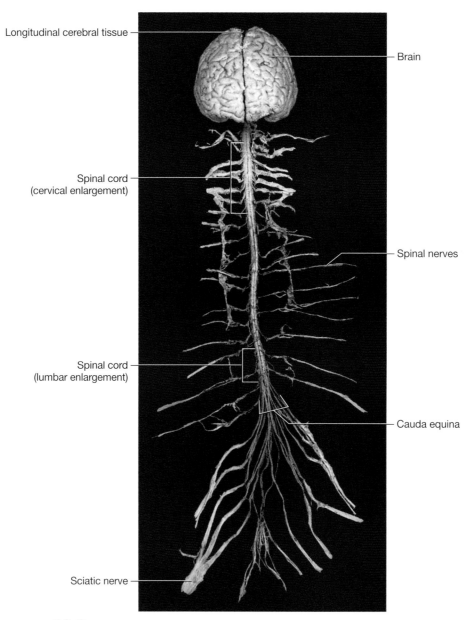

FIGURE **10.8** Central nervous system removed in entirety from a cadaver, anterior view.

2 Identify the structures labeled in Figure 10.9 on the brain.

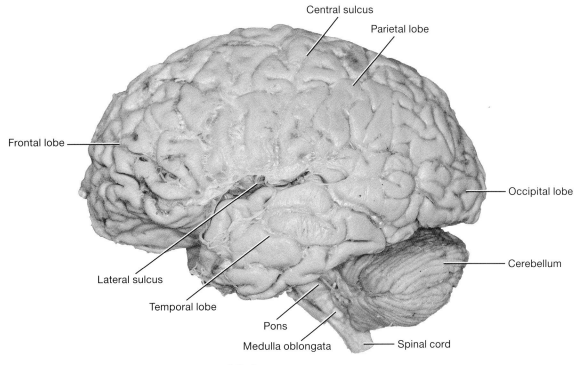

FIGURE **10.9** Lateral view of the brain.

3 Identify the structures labeled in Figure 10.10 on the brain.

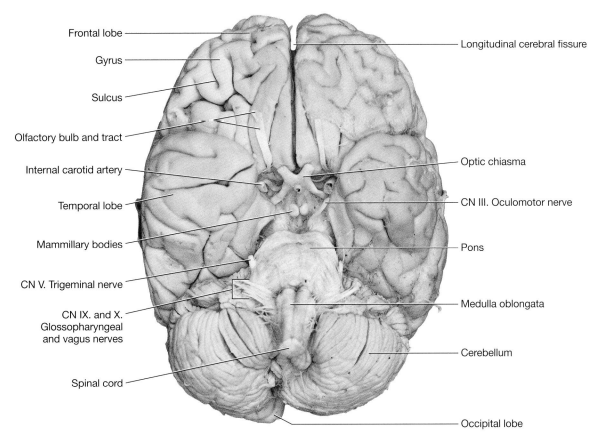

FIGURE **10.10** Inferior view of the brain.

4 Identify the structures labeled in Figure 10.11 on the brain.

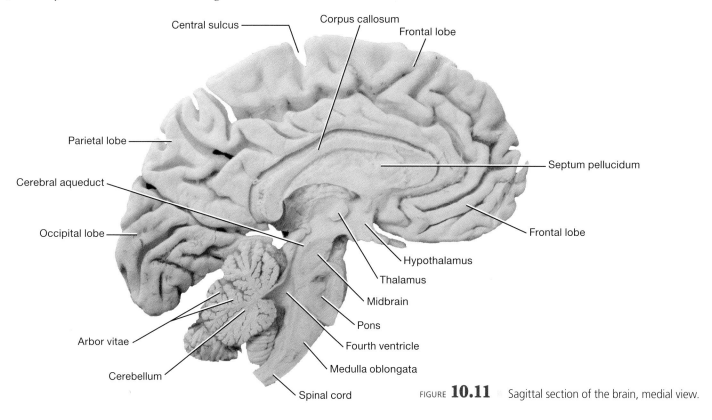

FIGURE **10.11** Sagittal section of the brain, medial view.

5 Identify the structures labeled in Figure 10.12 on the cadaver.

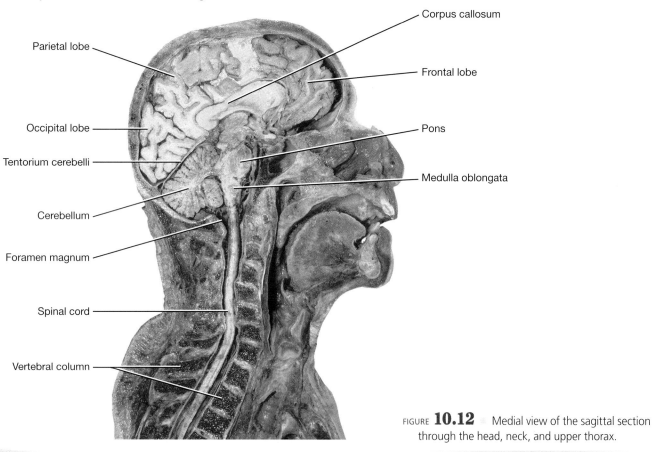

FIGURE **10.12** Medial view of the sagittal section through the head, neck, and upper thorax.

6 Identify the structures labeled in Figure 10.13 on the brain.

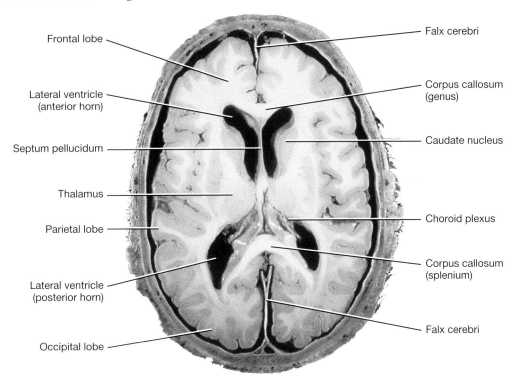

FIGURE **10.13** Axial (transverse) section through the head, inferior view.

7 Identify the structures labeled in Figure 10.14 on the vertebral column step dissection.

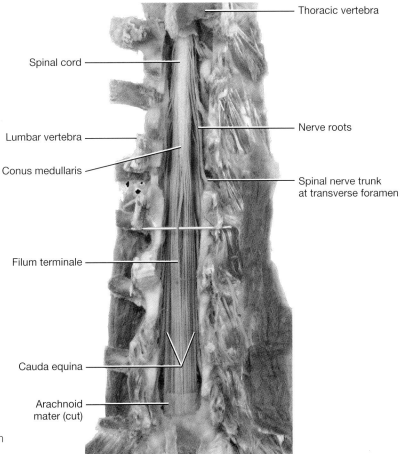

FIGURE **10.14** Coronal section through the vertebral column, posterior view.

UNIT **3:** Coming to Your Senses

CHAPTER **10:** Central Nervous System ■ **197**

8 Identify the structures labeled in Figure 10.15 on the spinal cord.

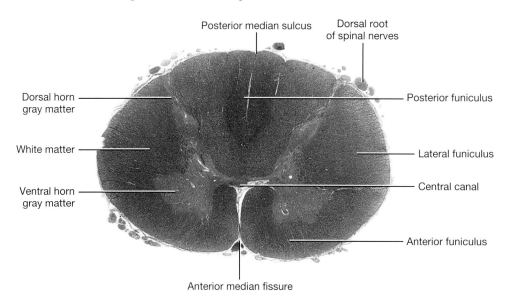

FIGURE **10.15**   Axial (transverse) section through the spinal cord.

WRAPPING UP

Complete the following additional activities to help retain your knowledge of the central nervous sytem.

Name _____

Date _____ Section _____

1 Identify the structures indicated in Figure 10.16.

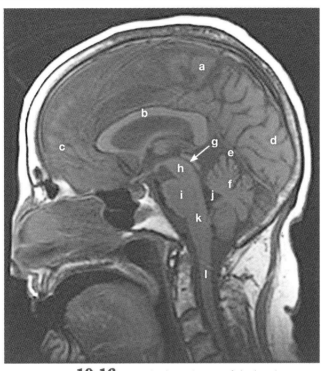

FIGURE **10.16** Sagittal MRI image of the head.

a. _____

b. _____

c. _____

d. _____

e. _____

f. _____

g. _____

h. _____

i. _____

j. _____

k. _____

l. _____

UNIT 3: Coming to Your Senses

2 Identify the structures indicated in Figure 10.17.

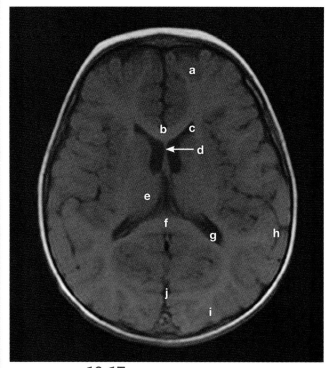

FIGURE **10.17** Axial MRI image of the head.

a.

b.

c.

d.

e.

f.

g.

h.

i.

j.

WRAPPING UP
(Continued)

Name _____

Date _____ Section _____

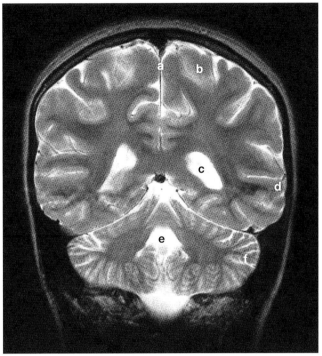

FIGURE **10.18**   Axial MRI of the head.

3 Identify the structures indicated in Figure 10.18.

a. _____

b. _____

c. _____

d. _____

e. _____

f. _____

UNIT **3:** Coming to Your Senses   CHAPTER **10:** Central Nervous System   **201**

Peripheral and Autonomic Nervous Systems

CHAPTER 11

At the completion of this laboratory session, you should be able to do the following:

1. Compare and contrast the somatic and autonomic nervous systems.
2. Outline the connective tissues that encompass a peripheral nerve.
3. Describe each of the 12 cranial nerves using the associated Roman numeral, cranial nerve name, and function.
4. Compare and contrast roots and rami.
5. Describe the spinal nerve levels contributing to the cervical plexus and their overall function.
6. Describe the brachial plexus including its roots, trunks, divisions, cords, and terminal branches; name and describe the function of the primary terminal branches.
7. Describe the intercostal nerves, lumbar plexus, and sacral plexus including major nerve branches and functions.
8. Describe the components of a spinal reflex.
9. Compare and contrast the sympathetic and parasympathetic nervous systems.

The peripheral nervous system (PNS) provides communication for the CNS with the rest of the tissues of the body. It is organized into a somatic division (somatic nervous system) and autonomic division (autonomic nervous system). The CNS is protected from mechanical damage by way of the skull and vertebral column and chemical damage via the blood brain barrier. In contrast, the PNS lacks both protective mechanisms and, therefore, is more susceptible to injury.

The peripheral nervous system consists of the following two components:

1. **Somatic nervous system.** A primary function of the somatic division of the PNS is transmitting sensory information from the skin (pain, temperature, touch) and tendons (proprioception) to the CNS and voluntary motor innervation from the CNS to skeletal muscles. These sensory and motor pathways are transported by way of 12 pairs of cranial nerves and 31 pairs of spinal nerves, which arise segmentally from the top of the brain to the bottom of the spinal cord.

2. **Autonomic nervous system.** A primary function of the autonomic nervous system (ANS) is the involuntary and unconscious control of cardiac muscle, smooth muscle, and glands. It consists of the sympathetic and parasympathetic nervous systems. As a general rule, the sympathetic and parasympathetic divisions create responses that counter each other. The sympathetic nervous system is involved in the "fight-or-flight" responses of the body, while the parasympathetic nervous system initiates the "rest-and-digest" responses.

 GETTING ACQUAINTED

Complete the following activities to become familiar with the peripheral and autonomic nervous systems.

MATERIALS

Obtain the following items before beginning the laboratory activities:

- ❑ Textbook or access to Internet resources
- ❑ Colored pencils

ACTIVITY 1 Tissue Structure of Nerves

The peripheral nervous system (PNS) is classified by location as either cranial nerves or spinal nerves. Each of these nerves houses sensory and motor neurons. Sensory neurons convey sensory information from peripheral tissues to the CNS. Motor neurons convey motor information from the CNS to a peripheral tissue. The motor and sensory divisions may be further subdivided based upon the structures served:

1. **Somatic sensory.** The somatic sensory innervates skin (pain, temperature, vibration, and touch) and tendons (proprioception).
2. **Visceral sensory.** The visceral sensory provides sensory information from organs and glands (baroreceptors, chemoreceptors, etc.).
3. **Somatic motor.** The somatic motor innervates skeletal muscle (biceps or calf muscles, etc.).
4. **Visceral motor.** The visceral motor innervates organs and glands (heart, lungs, or lacrimal glands, etc.).

A nerve is a collection of numerous neurons organized by the following three connective tissue sheaths:

1. **Endoneurium.** The endoneurium is loose connective tissue that surrounds the cell membrane (axolemma) of individual neurons.
2. **Perineurium.** The perineurium is dense collagenous connective tissue that surrounds a fascicle (bundle) of axons.
3. **Epineurium.** The epineurium is dense collagenous connective tissue that surrounds a group of fascicles, including blood and lymphatic vessels.

1 Identify, label, and color the following parts of a nerve on Figure 11.1.
- ❑ Epineurium
- ❑ Perineurium
- ❑ Nerve fascicle
- ❑ Endoneurium
- ❑ Schwann cell
- ❑ Axon

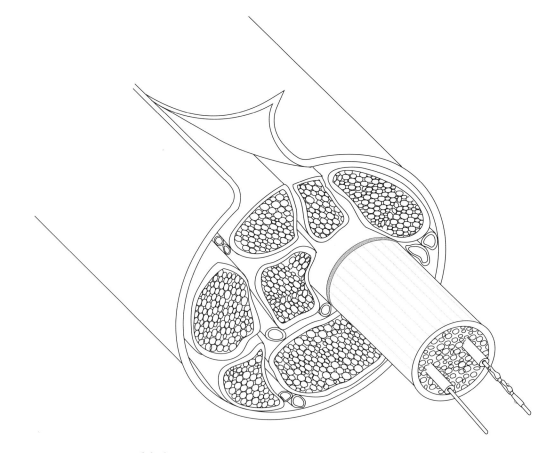

FIGURE **11.1** Tissue structure of a nerve.

ACTIVITY 2 Cranial Nerves

Each of the 12 pairs of cranial nerves (CN) arises from the brain and is named in two ways:

1. **Roman numeral.** A Roman numeral based on its consecutive origin from the brain (e.g., CN III).
2. **Descriptive name.** A name based upon the nerve's location or function (e.g., oculomotor nerve).

For the most part cranial nerves innervate structures of the head and neck. The neurons within cranial nerves are either sensory, motor, or a combination of both. In other words, some cranial nerves contain only sensory neurons (CN I, CN II, CN VIII), some only contain motor neurons (CN III, CN IV, CN VI, CN XI, CN XII), and the rest have a combination of sensory and motor neurons (CN V, CN VII, CN IX, CN X).

The following provides a brief overview of the Roman numeral, descriptive name, and function for each of the cranial nerves:

1. **CN I: Olfactory nerve (sensory).** The olfactory nerve provides innervation of olfactory mucosa in the nasal cavity for the sense of olfaction (smell).
2. **CN II: Optic nerve (sensory).** The optic nerve provides innervation of the retina for vision. The optic nerve meets at the **optic chiasma**, where axonal fibers either stay on the same side, or decussate (cross) to the opposite side to form the **optic tracts**, which course primarily to the occipital lobe.
3. **CN III: Oculomotor nerve (motor).** The oculomotor nerve provides innervation to the muscle that elevates the eyelid (levator palpebrae superioris muscle) and four of the six muscles that move the eyeball (superior rectus, medial rectus, inferior rectus, and inferior oblique muscles). In addition, CN III is responsible for parasympathetic innervation to the sphincter pupillae muscle, which constricts the pupil, and ciliary muscles, which change the shape of the lens for near vision.
4. **CN IV: Trochlear nerve (motor).** The trochlear nerve provides innervation for one of the six muscles that move the eyeball (superior oblique muscle).
5. **CN V: Trigeminal nerve (mixed).** The trigeminal nerve provides sensory innervation to the face and motor innervation to the muscles of mastication (chewing) via three branches, named for the region of the face supplied: ophthalmic nerve, maxillary nerve, and mandibular nerve.
6. **CN VI: Abducens nerve (motor).** The abducens nerve provides innervation for the lateral rectus muscle, one of the six muscles that move the eyeball.
7. **CN VII: Facial nerve (mixed).** The facial nerve innervates the following:
    a. Muscles of facial expression.
    b. Lacrimal glands (produce tears).
    c. Submandibular and sublingual salivary glands (produce saliva).
    d. Nasopalatine glands (produce mucous lining the nasal and palatal region).
    e. Tongue (taste to the anterior two-thirds of the tongue).
    f. External ear (general sensory).
8. **CN VIII: Vestibulocochlear nerve (sensory).** Provides innervation to structures of the inner ear responsible for hearing (cochlea) as well as equilibrium and balance (vestibular apparatus).
9. **CN IX: Glossopharyngeal nerve (mixed).** Provides innervation to the following structures:
    a. Stylopharyngeus muscle.
    b. Parotid gland (produces saliva).
    c. Tongue (taste to the posterior one-third of the tongue).
    d. Auditory tube (general sensory).
    e. Carotid sinus (baroreceptor) and carotid body (chemoreceptor).
10. **CN X: Vagus nerve (mixed).** Provides innervation to the following structures:
    a. Palatal muscles (swallowing).
    b. Pharyngeal muscles (swallowing).
    c. Laryngeal muscles (speaking).
    d. Digestive and respiratory organs (parasympathetic innervation to heart, lungs, stomach, liver, small intestines,

and part of the large intestine).
   e. Pharynx and larynx (general sensory).
   f. Carotid body (chemoreceptor).
   g. Thoracic and abdominal viscera (visceral sensation).
11. **CN XI: Spinal accessory nerve (motor).** Provides innervation to the trapezius and sternocleidomastoid muscles.
12. **CN XII: Hypoglossal nerve (motor).** Provides innervation to muscles that move the tongue.

 Weird and Wacky

Mnemonics

A mnemonic is a memorization device, such as patterns of letters that help you remember something. For example, a mnemonic device could be a silly phrase where the first letter of each word helps you remember the order of a number of words. For generations, anatomy students have used mnemonic devices to help them remember the order in which the cranial nerves emerge from the brain. One mnemonic is as follows: "On Old Olympus' Towering Top, A Fine Vested German Viewed Sea Hawks." In other words, it looks like this:

On	**O**ptic	**F**ine	**F**acial
Old	**O**lfactory	**V**ested	**V**estibulocochlear
Olympus	**O**culomotor	**G**erman	**G**lossopharyngeal
Towering	**T**rochlear	**V**iewed	**V**agus
Top	**T**rigeminal	**S**ea	**S**pinal accessory
A	**A**bducens	**H**awks	**H**ypoglossal

This is one of dozens of mnemonics that students have used over the years. Feel free to use this one or create one of your own.

1 Identify, label, and color the following nerves of the brain on Figure 11.2.
- ☐ CN I: Olfactory n.
- ☐ CN II: Optic n.
- ☐ CN III: Oculomotor n.
- ☐ CN IV: Trochlear n.
- ☐ CN V: Trigeminal n.
- ☐ CN VI: Abducens n.
- ☐ CN VII: Facial n.
- ☐ CN VIII: Vestibulocochlear n.
- ☐ CN IX: Glossopharyngeal n.
- ☐ CN X: Vagus n.
- ☐ CN IX: Spinal accessory n.
- ☐ CN XII: Hypoglossal n.

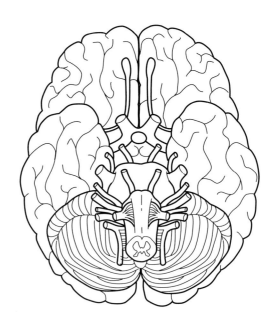

FIGURE **11.2** Inferior view of the brain.

ACTIVITY 3 Spinal Nerves

The spinal cord bilaterally gives rise to ventral and dorsal spinal roots, which join to form spinal nerves that further branch into ventral and dorsal rami. The following provides a brief overview of each of these structures:

1. **Spinal roots.** Spinal roots are "one-way streets" in that each root contains only one type of neuron. **Dorsal roots** transport sensory neurons (afferent information) and **ventral roots** transport motor neurons (efferent information).
2. **Spinal nerves.** The union of a dorsal root and a ventral root forms spinal nerves. Therefore, spinal nerves are "two-way streets," mixing sensory and motor neurons.
3. **Spinal rami.** The rami are also "two-way streets." **Dorsal rami** provide motor innervation to deep back muscles (e.g., erector spinae muscles) and sensory innervation to the skin of the back between the scapulae. **Ventral rami** provide motor and sensory innervation to the body wall (e.g., intercostal muscles) and limb muscles (e.g., triceps or quadriceps) and skin.

Ventral rami arise bilaterally from each segmental level of the spinal cord and in some regions the ventral rami form plexuses. A brief description of the ventral rami is listed here:

1. **C1–C4 ventral rami: Cervical plexus.** The C1–C4 ventral rami (plus a small contribution from C5 ventral ramus) form a plexus of nerves known as the cervical plexus, which supplies the skin of the neck, part of the scalp, and body wall neck muscles. Its major branch is the phrenic nerve (C3–C5), which innervates the diaphragm.
2. **C5–T1 ventral rami: Brachial plexus.** The C5–T1 ventral rami form a plexus of nerves known as the brachial plexus, which supplies the skin and muscles of the upper limb. Anatomists refer to the ventral rami that form the brachial plexus as the "roots" of the brachial plexus (not to be confused with the spinal roots). The C5 and C6 roots combine to form the superior trunk, the C7 root forms the middle trunk, and the C8 and T1 roots combine to form the inferior trunk. Each trunk splits into an anterior division and a posterior division to innervated flexor and extensor muscles. The anterior division of the inferior trunk forms the medial cord, which descends in the medial arm. The anterior divisions of the superior and middle trunks unite to form the lateral cord, which descends in the lateral arm. The posterior divisions of each trunk combine to form the posterior cord, which lies in the posterior arm. The following terminal branches arise from the brachial plexus cords:
    a. **Radial nerve.** Innervates the triceps, forearm extensors, and skin on the posterior arm and hand.
    b. **Axillary nerve.** Innervates the deltoid, teres minor muscles, and skin on the lateral shoulder.
    c. **Musculocutaneous nerve.** Innervates the muscles in the anterior region of the arm (i.e., coracobrachialis, biceps brachii, and brachialis muscles) and skin of the lateral forearm.
    d. **Ulnar nerve.** Innervates two forearm flexor muscles and many intrinsic hand muscles and skin over the medial hand.
    e. **Median nerve.** Innervates most of the forearm flexors and some intrinsic hand muscles and skin over the anterior and lateral hand. As the median nerve enters the wrist, it travels under the band of connective tissue called the flexor retinaculum. Occasionally the median nerve becomes compressed under the flexor retinaculum resulting in carpal tunnel syndrome.
3. **T1–T12 ventral rami: Intercostal nerves.** The ventral rami in the thoracic region course between the intercostal muscles, thus changing their name to intercostal nerves. Intercostal nerves innervate intercostal muscles, abdominal muscles, and the skin of the chest and abdomen.
4. **Lumbar plexus (L1–L4).** The L1–L4 ventral rami form a plexus of nerves known as the lumbar plexus. The femoral nerve is the largest branch from the lumbar plexus. It provides motor innervation to the quadriceps femoris muscles and sensory innervation to the thigh and leg. The obturator nerve innervates the medial compartment thigh muscles, which adduct the hip. The lateral femoral cutaneous nerve provides sensation to the lateral side of the thigh.
5. **Sacral plexus (L4–S3).** The L4–S3 ventral rami form a plexus of nerves known as the sacral plexus. The largest nerve in the body and sacral plexus is known as the sciatic nerve, which courses down the back of the thigh and gives rise to the **tibial nerve** and the **common peroneal (fibular) nerve**. The tibial nerve provides motor and sensory innervation to the posterior thigh, posterior leg, and foot. The common peroneal nerve provides motor and sensory innervation to the anterolateral leg and dorsum of the foot.

Clinical Application

The Sciatic Nerve

The sciatic nerve is the largest and longest peripheral nerve in the body. It arises from the L4–S3 spinal nerve levels in the lower back and courses through the buttock, down the lower limb to the plantar surface of the foot. This nerve provides much of the sensory innervation for the lower limb as well as motor to all muscles with the exception of the quadriceps and adductors. The sciatic nerve is really two different nerves: the tibial nerve and the common peroneal nerve. Compression and irritation of the sciatic nerve results in a condition known as "sciatica." This condition may result from herniation from an intervertebral disc, lumbar spinal stenosis, pregnancy, or muscle impingement.

Posteromedial view of the posterior structures of the thigh region.

Clinical Application

Don't Push Me

A nerve is much like a person—it hates to be pushed. A pinched nerve occurs when some structure is pushing on a nerve. It is referred to by many names: nerve compression, entrapment, or impingement, and if it occurs near the spinal cord, it is called radiculopathy. If the nerve is pinched further along its course, it is called a neuropathy. A common place where nerve compression occurs is where a nerve courses near a bone (e.g., intervertebral foramen). Due to the fact that peripheral nerves conduct sensory and motor impulses, it is possible that both sensation (pain, temperature, touch, and vibration) as well as motor pathways (skeletal muscle movement and reflexes) may be affected. Some common examples include carpal tunnel syndrome, thoracic outlet syndrome, and piriformis syndrome.

1 Identify, color, and label the main components of spinal nerves in Figure 11.3.

- ☐ Spinal cord
- ☐ Dorsal root
- ☐ Ventral root
- ☐ Spinal nerve trunk
- ☐ Sympathetic ganglion
- ☐ Dorsal ramus
  - Deep back muscles
  - Skin of the back
- ☐ Ventral ramus
  - Intercostal muscles
  - Skin of the anterolateral trunk

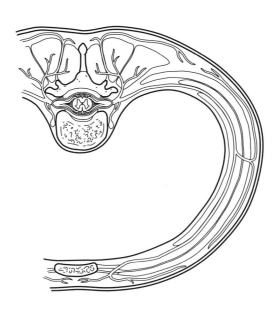

FIGURE **11.3** Cross section of the thorax.

2 Identify, color, and label the following parts of the brachial plexus in Figure 11.4.

- ☐ C5, C6, C7, C8, and T1 roots
- ☐ Superior, middle, and inferior trunks
- ☐ Anterior division
- ☐ Posterior division
- ☐ Medial, lateral, and posterior cords
- ☐ Axillary nerve
- ☐ Radial nerve
- ☐ Musculocutaneous nerve
- ☐ Median nerve
- ☐ Ulnar nerve

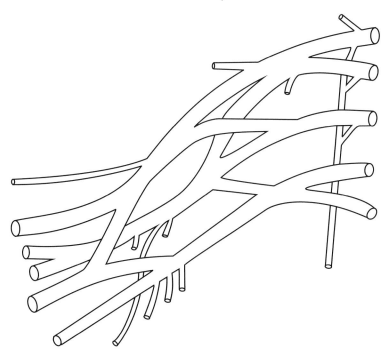

FIGURE **11.4** Brachial plexus.

3 Identify, color, and label the following spinal nerves in Figure 11.5.

Right side:
- ☐ Cervical plexus
- ☐ Brachial plexus
- ☐ Intercostal nerves
- ☐ Lumbar plexus
- ☐ Sacral plexus

Left side:
- ☐ C1–C8 segmental spinal nerve levels
- ☐ T1–T12 segmental spinal nerve levels
- ☐ L1–L5 segmental spinal nerve levels
- ☐ S1–S5 segmental spinal nerve levels

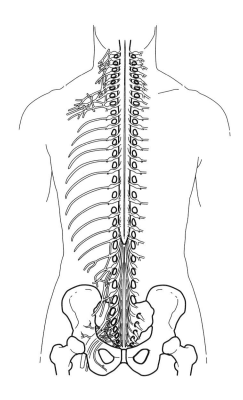

FIGURE **11.5** Segmental spinal nerves.

ACTIVITY 4 Spinal Reflexes

A reflex is an involuntary movement in response to a stimulus (e.g., tapping the patellar tendon causes the knee to jerk). Reflexes are important diagnostically because they provide a consistent, predictable, and dependable way to test sensory pathways, CNS integration levels, and motor pathways.

The human body has many different reflex arcs. The most simple is the stretch reflex. Stretch reflexes are important in maintaining posture and equilibrium and are initiated when a muscle is stretched. The stretch is detected by specialized stretch receptors in the muscles (muscle spindles), whereby sensory neurons conduct this information to the CNS. Sensory neurons act directly on motor neurons resulting in contraction of the same muscle to counter the stretch. For example, when the patellar ligament is tapped, the following events take place:

1. The stimulus is detected by specialized muscle spindles (stretch receptors) in the quadriceps femoris muscles when the patellar ligament is hit.
2. Sensory neurons within the femoral nerve conduct the stimulus from the muscle spindles to the L4 ventral horn of the spinal cord.
3. Synapse occurs between the sensory neurons from the patellar tendon and motor neurons on route to the quadriceps muscles.
4. Motor neurons within the L4 ventral horn of the spinal cord travel in the femoral nerve to the quadriceps muscles.
5. The quadriceps muscles contract, triggering the knee to extend (jerk).

1 Label the following structures on Figure 11.6.
- ☐ Patellar ligament
- ☐ Sensory neuron from the quadriceps
- ☐ Synapse with motor neuron
- ☐ Dorsal horn gray matter
- ☐ Ventral horn gray matter
- ☐ Motor neuron
- ☐ Quadriceps muscle
- ☐ Hamstring muscles

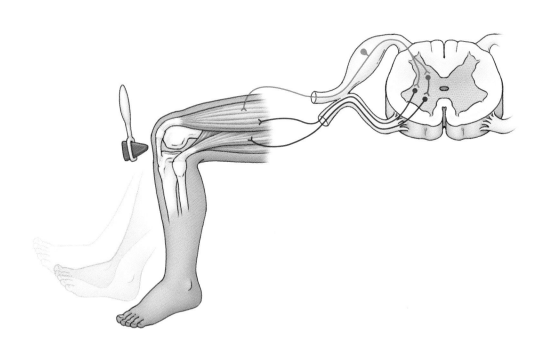

FIGURE **11.6** Reflex arc (patellar tendon reflex example).

ACTIVITY 5 Autonomic Nervous System

The autonomic nervous system (ANS) is the division of the peripheral nervous system responsible for maintaining homeostasis and is, therefore, involuntary. The ANS has two branches—the sympathetic nervous system (SNS) and the parasympathetic nervous system (PNS).

1. **Sympathetic nervous system (SNS).** Functionally, the SNS is described in terms of "fight or flight" because it helps the body respond to emergency or stressful situations (e.g., running from a bear). The cell bodies of the neurons of the SNS are located within the T1–L2 spinal cord levels (lateral horns) and the axons of the SNS release the neurotransmitters epinephrine (i.e., adrenaline) and norepinephrine (i.e., noradrenaline), which result in the following:
    a. Increased heart rate, blood pressure, metabolic rate, blood glucose, and release of fatty acids.
    b. Dilation of the pupils, skeletal muscles, coronary arteries, and airways.
    c. Decreased digestive and urinary functions.
    d. Constriction of peripheral blood vessels serving the abdominal viscera and skin, which also increases blood pressure.
    e. Release of epinephrine from the adrenal medulla.

2. **Parasympathetic nervous system (PNS).** Functionally, the PNS is described in terms of "rest and digest" because it helps slow down heart and respiratory rate. The cell bodies of the neurons of the PNS are located in the brain stem (cranial nerve nuclei) and S2–S4 spinal cord levels. The axons release the neurotransmitter acetylcholine, which results in the following:
    a. Decreased heart rate and blood pressure.
    b. Constriction of airways and pupils.
    c. Increased digestion, secretion of digestive products, urine production, storage of fats, and glucose production.
    d. Adjustment of the lens for near vision.

Note that many of the effects of the SNS and PNS are opposite. Generally, the PNS is subordinate to the SNS (for example, digesting that doughnut is not all that important when you are being chased by a bear). However, neither branch of the ANS is ever completely quiet. The divisions work together constantly to provide balance for any given situation.

1 Follow the given instructions for Figure 11.7.
- ☐ Sympathetic side (left side)
  - Color T1–L2 spinal cord segments
  - Color the sympathetic chain
- ☐ Parasympathetic side (right side)
  - Color the brain stem
  - Color the S2–S4 spinal cord segments

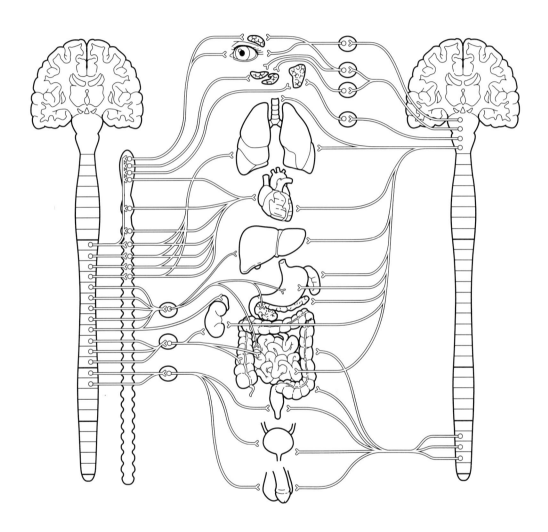

FIGURE **11.7** Schematic of the autonomic nervous system.

👁 OBSERVING

Complete the following hands-on laboratory activities to apply your knowledge of the peripheral and autonomic nervous systems.

MATERIALS
Obtain the following items before beginning the laboratory activities:
- ❏ Cadaver
- ❏ Pen light (the light from a phone will do)
- ❏ Piece of candy or gum
- ❏ Cotton swab

ACTIVITY 1 Identification of Structures on the Cadaver

1 Identify the following structures on the brain in Figure 11.8.

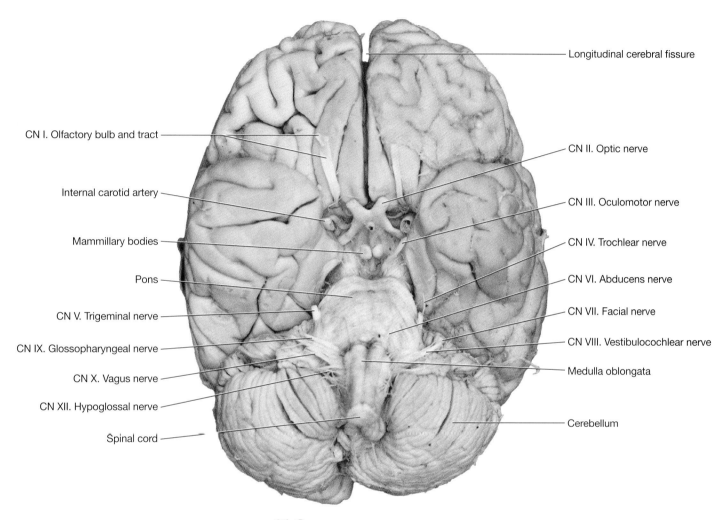

FIGURE **11.8** Inferior view of the cadaver brain.

2 Identify the following components on spinal nerves in Figure 11.9.

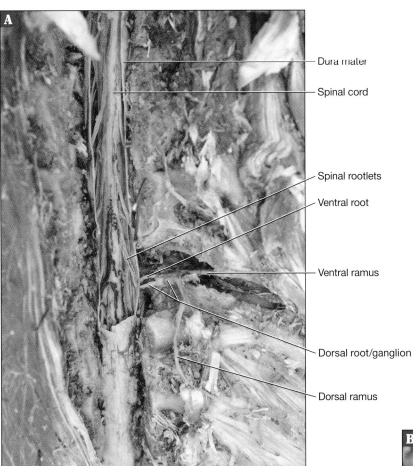

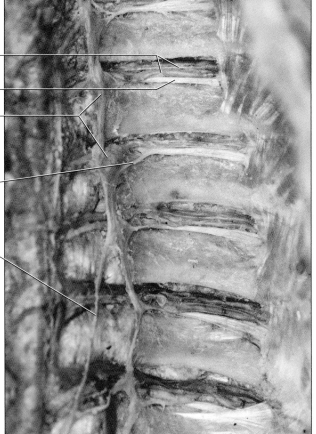

FIGURE **11.9** Spinal nerve dissection: (**A**) posterior view of the vertebral column with laminectomy (dura mater is cut to reveal the spinal cord; step dissection shows the elements of a spinal nerve); (**B**) anterior view of the posterior mediastinum; the sympathetic trunk, giving rise to communicating rami and intercostal nerves as well as a splanchnic nerve.

3 Identify the components of the brachial plexus in Figure 11.10.

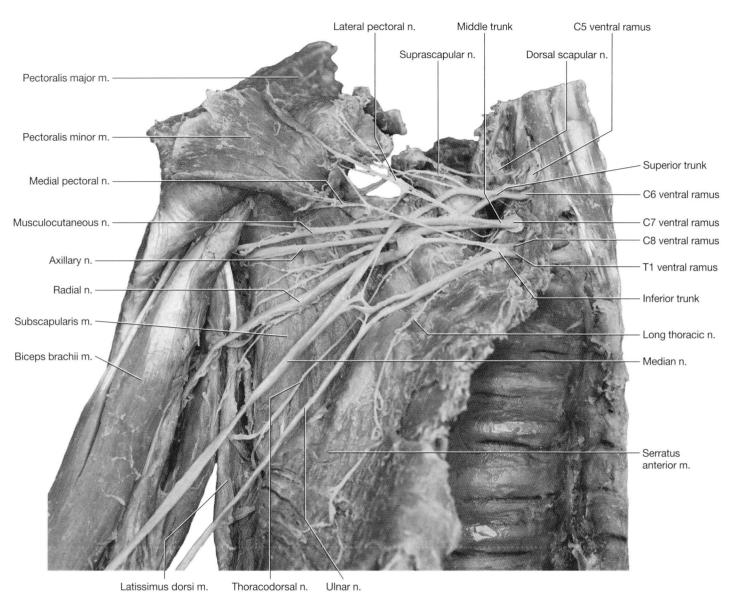

FIGURE **11.10** Brachial plexus dissection.

4 Identify the components of the lumbar plexus in Figure 11.11.

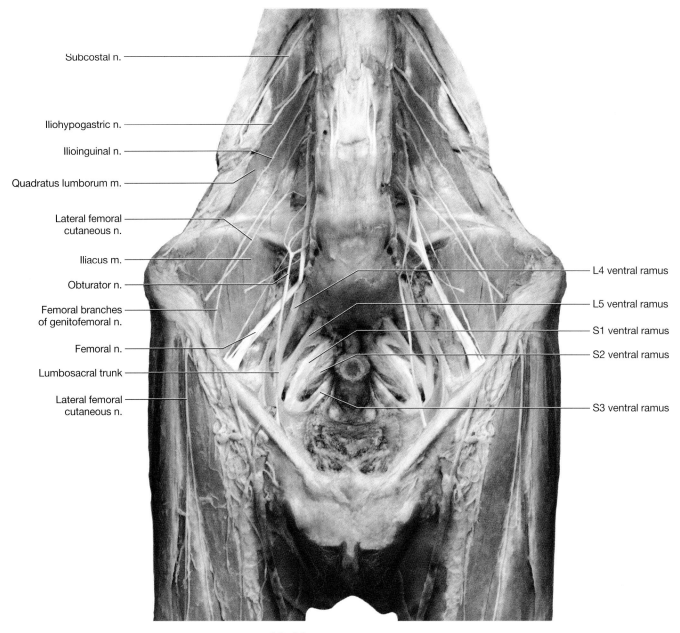

FIGURE **11.11** Lumbar plexus dissection.

5 Identify the components of the sacral plexus in Figure 11.12.

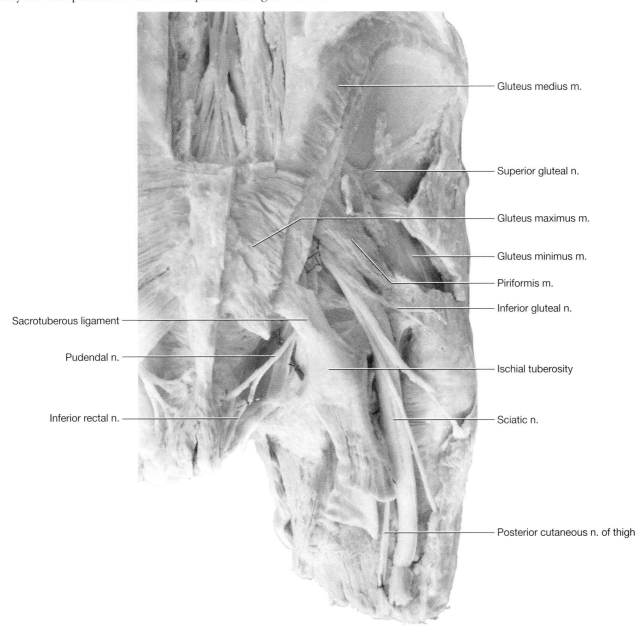

FIGURE **11.12** Sacral plexus dissection.

ACTIVITY 2 Clinical Tests for Peripheral Nerves

A component of a physical exam is to test peripheral nerves. In this way the health and function of tissues of the body, sensory nerves, levels of the CNS, motor nerves, and effectors such as muscles are tested. With one or two of your lab mates perform the following tests on each other and record your results.

Part A: Pupillary Light Reflex

1 Shine a penlight (or light from your phone) into the left eye from an angle.

2 What happened to the pupil in the left eye?

3 The right eye?

4 Move the penlight away from the left eye.

5 What happened to the pupil in the left eye?

6 The right eye?

7 Shine the light into the right eye from an angle.

8 What happened to the pupil in the left eye?

9 The right eye?

10 Move the penlight away from the right eye.

11 What happened to the pupil in the left eye?

12 The right eye?

13 What does a "consensual" reflex mean? *Note:* You may need to reference your textbook or Internet resources to answer this question.

14 What tissue or structure did the light stimulate?

15 What cranial nerve conducted the "light" stimulus from the eye to the brain?

16 What brain stem level was being examined by this test?

17 What cranial nerve conducted the motor output from the brain stem to cause the pupil to constrict?

Part B: Eye Movements

1 Ask your lab partner to look at your finger and follow it with his/her eyes without moving the head.

2 Draw an "H" pattern in the air.

3 Which cranial nerves were being tested by this test?

4 Which brain stem levels (midbrain, pons, or medulla oblongata) were also being examined by this test?

Part C: Tongue

1 Ask your lab partner to sit in a chair and close his/her eyes.

2 Have your lab partner protrude the tongue out of his/her mouth.

3 Place a piece of candy or gum on the tongue and ask your lab partner whether he/she can feel and taste it.

4 Which cranial nerve was responsible for innervating the tongue muscles so your lab partner could stick out his/her tongue?

5 Which cranial nerve was responsible for sensing food on the tongue?

6 Which cranial nerve was responsible for tasting the food on the tongue?

Part D: Ears

1 Have your lab partner sit in a chair facing forward with his/her eyes closed.

2 Stand behind your lab partner and rub your fingers together on one side of their head. Ask whether he/she can hear anything and, if so, which side.

3 Which cranial nerve was being tested?

Part E: Throat

1 Have your lab partner sit in a chair facing forward.

2 Ask your lab partner to swallow his/her saliva and then to hum.

3 Which muscles enabled your lab partner to swallow, and which cranial nerve was being tested?

4 Which muscles enabled your lab partner to hum, and which cranial nerve was being tested?

Part F: Face Sensation

1 Have your lab partner sit in a chair with his/her eyes closed.

2 Using a cotton swab or the tip of a pencil, touch your lab partner's forehead and ask whether he/she can feel it.

3 Repeat with the skin below the bottom eyelid and again on the skin below the bottom lip.

4 Have your lab partner open and close his/her jaw.

5 What cranial nerve was being tested?

6 Why was the test repeated in three different places on the face? Why those three places?

7 How was having your lab partner open and close his/her mouth helping to test this cranial nerve?

Part G: Shoulder and Neck Movements

1 Have your lab partner elevate his/her shoulders against resistance.

2 Have your lab partner turn his/her head to the left and then to the right against resistance.

3 Which muscle was being tested when your lab partner was elevating his/her shoulders?

4 Which muscle was being tested when your lab partner turned his/her head?

5 Which cranial nerve was being examined by this test?

Part H: Facial Muscles
1 Have your lab partner perform the following actions: smile, frown, raise the eyebrows, pucker the lips, close the eyes, and flare the nose.

2 Which cranial nerve was being tested?

3 Which muscles were performing the actions?

Part I: Patellar Ligament Reflex
1 Have your lab partner sit on the edge of a table with legs dangling off the side.

2 Palpate the patellar ligament (between the patella and tibial tuberosity).

3 Strike the patellar tendon with tips of your fingers (or reflex hammer if one is available).

4 What happened?

5 Which muscles, nerve, and spinal nerve levels are tested with this reflex?

Part J: Stretch Reflex
1 Sit down on a chair with legs bent and relaxed.

2 Palpate the hamstrings (muscle along the posterior aspect of your thigh). Do they feel tight or relaxed?

3 Stand up and bend over to touch your toes.

4 Palpate your hamstrings again. Do they feel tight or relaxed?

5 Why were your hamstrings relaxed when you were sitting?

6 Why did your hamstrings become tight when you stood up and bent over to touch your toes?

WRAPPING UP

Complete the following additional activities to help retain your knowledge of the peripheral and autonomic nervous sytems.

Name _____

Date _____ Section _____

1. Label the following diagram of the brachial plexus on Figure 11.13.
   - ☐ C5, C6, C7, C8, and T1 roots
   - ☐ Superior, middle, and inferior trunks
   - ☐ Anterior division
   - ☐ Posterior division
   - ☐ Medial, lateral, and posterior cords
   - ☐ Axillary nerve
   - ☐ Radial nerve
   - ☐ Musculocutaneous nerve
   - ☐ Median nerve
   - ☐ Ulnar nerve

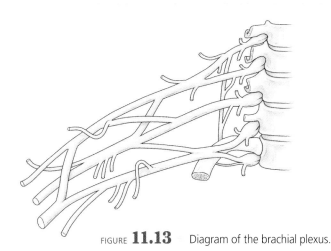

FIGURE **11.13** Diagram of the brachial plexus.

2. Label the following diagram of the lumbosacral plexus on Figures 11.14 and 11.15.
   - ☐ L1 ventral ramus
   - ☐ L4 ventral ramus
   - ☐ Femoral nerve
   - ☐ Lateral femoral cutaneous nerve
   - ☐ Obturator nerve
   - ☐ Saphenous nerve
   - ☐ Superior gluteal nerve
   - ☐ Inferior gluteal nerve
   - ☐ Sciatic nerve
   - ☐ Tibial nerve
   - ☐ Common peroneal nerve

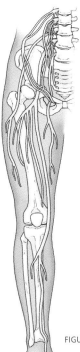

FIGURE **11.14** Anterior view of the leg.

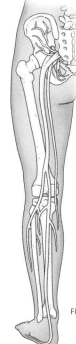

FIGURE **11.15** Posterior view of the leg.

3 Match the cranial nerve in the right-hand column with its function in the left-hand column (some cranial nerves may be used once, more than once, or not at all).

Cranial nerve function

_____ Senses tickling of the nose to wake you up in the morning

_____ Constricts the pupil when your roommate turns on the light in the morning

_____ Enables you to hear your dad's corny joke

_____ Helps to roll your eyes at the corny joke

_____ Enables you to stick your tongue out of your mouth to lick vanilla ice cream

_____ Helps you know that ice cream is cold

_____ Helps you know the flavor of the ice cream is vanilla

_____ Enables you to smile because of how tasty the ice cream is

_____ Enables you to use your vocal folds to say, "I love vanilla ice cream"

_____ Enables you to stand on one leg with eyes closed

Cranial nerve

a. CN I. Olfactory nerve
b. CN II. Optic nerve
c. CN III. Oculomotor nerve
d. CN IV. Trochlear nerve
e. CN V. Trigeminal nerve
f. CN VI. Abducens nerve
g. CN VII. Facial nerve
h. CN VIII. Vestibulocochlear nerve
i. CN IX. Glossopharyngeal nerve
j. CN X. Vagus nerve
k. CN XI. Spinal accessory nerve
l. CN XII. Hypoglossal nerve

4 Match the peripheral nerve in the right-hand column with its function in the left-hand column (some nerves may be used once, more than once, or not at all).

Peripheral nerve function

_____ Keeps your thighs together so you do not fall off the horse while riding

_____ Innervates your diaphragm so you can breathe faster while the horse is galloping out of control

_____ Increases your heart rate because you are stressed

_____ Flexes elbows when curling barbells

_____ Causes you to slam on the brakes when stopping the car quickly

_____ Causes an increase in blood pressure when you slam on the brakes in your car

_____ Lifts foot off the ground

_____ Enables you to straighten your knees when moving to stand from a seated position

_____ Enables you to spread your fingers apart

_____ Is compressed in carpal tunnel syndrome

_____ Causes you to salivate when you smell homemade pizza

_____ Helps a patient in a wheelchair to straighten his elbows and raise himself out of his seat

Peripheral spinal nerve

a. Common fibular nerve
b. Phrenic nerve
c. Radial nerve
d. Musculocutaneous nerve
e. Femoral nerve
f. Ulnar nerve
g. Median nerve
h. Tibial nerve
i. Obturator nerve
j. Sympathetic nervous system (SMS)
k. Parasympathetic nervous system (PMS)

WRAPPING UP
(Continued)

Name _____

Date _____ Section _____

5 Answer the following questions using Figure 11.16 (note that not all letters will be used):

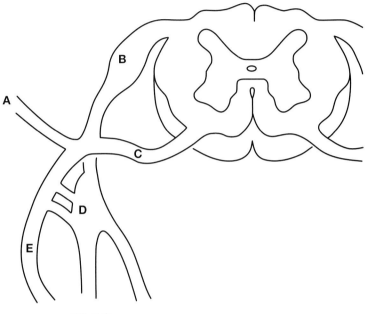

FIGURE **11.16** Schematic of spinal cord and spinal nerves.

a. If injured, I will lose motor function to muscles of the back and body wall. No sensory deficits are noted. What letter am I?

_____

b. If injured, I will lose motor function to the erector spinae muscles and sensation to the skin of the back. Sensation to the upper limb and anterolateral body wall is intact. No motor problems are noted with upper limb muscles and body wall muscles. What letter am I?

_____

c. The Varicella zoster virus (VZV) can often remain dormant for years in the dorsal root ganglion. The virus can re-emerge, causing shingles in a dermatomal pattern. Identify the letter that best indicates the dormancy location for the VZV.

_____

Special Senses

CHAPTER 12

At the completion of this laboratory session, you should be able to do the following:

1. Identify and describe the different special senses.
2. Identify the primary parts of the following:
   a. Eye (responsible for vision).
   b. Nose (responsible for smell).
   c. Ear (responsible for hearing).
   d. Tongue (responsible for taste).
3. Describe how the special senses are a helpful diagnostic tool in health care.

The general sensory component of the peripheral nervous system conveys information from tissues, like the skin and stomach, to the brain. The term "general sensation" is often used because these sensory neurons are located all over the body. For example, sensory neurons that process information about temperature reside in the skin of the back, hand, and foot, as well as in the tongue and nasal cavity. They are general in their distribution.

In contrast, special sensory organs convey information from unique tissues located in the eye, nose, ear, and tongue. For example, if lemon ice cream were placed on your foot, thigh, back, hand, or tongue, it would be sensed as cold by general sensory thermoreceptors. However, the only place you could detect the sweetness and tartness of the lemon ice cream is on the tongue where special sensory taste receptors are located.

GETTING ACQUAINTED

Complete the following activities to become familiar with the special senses.

MATERIALS
Obtain the following items before beginning the laboratory activities:

❏ Textbook or access to Internet resources
❏ Colored pencils

ACTIVITY 1 Key Terms of the Special Senses

Special sensory organs enable us to interact with our external environment through a series of stimuli and impulses. A stimulus excites a sense organ, which in turn transduces that stimulus into an electrochemical impulse. Sensory nerves transmit the impulse or sensation from the sense organ to the brain to be perceived and acted upon. Ultimately, it is the brain that enables us to perceive the sight, smell, taste, and sounds of our external environment. For example, light stimulates the retina, which transduces the photo energy into an electrochemical impulse and propagates along the optic nerve to the occipital lobe.

1 Define the function(s) for each of the structures list in Table 12.1.

TABLE **12.1** ■ Key Terms of the Special Senses

Sensory Organ	Function(s)
Eye	
Conjunctiva	
Sclera	
Cornea	
Iris	
Pupil	
Lens	
Ciliary body	
Choroid	
Retina	
Rods and cones	
Optic disc	
Aqueous humor	
Vitreous humor	
Nose	
Chemoreceptor	
Olfactory nerve	
Cribriform plate	
Ear	
Auricle	
External auditory canal	
Tympanic membrane	
Auditory ossicles	
Auditory tube	
Semicircular canals	
Cochlea	
Tongue	
Taste buds	
Circumvallate papillae	
Chorda tympani nerve	

ACTIVITY 2 Eye Anatomy

Vision is incredible. Our eyes enable us to see objects from far away, or close to our nose, as well as a seemingly endless array of colors, shapes, and movements. The only place in our body where a tissue is able to transduce light waves into a nervous impulse is in the retina in our eyes. The eyeball consists of the following three layers, or tunics:

1. **Fibrous tunic.** The fibrous tunic consists of the sclera and cornea. The sclera forms the white part of the eyeball and is composed of dense collagenous connective tissue, giving the globe its structure. It also provides an area for extraocular muscle attachment. The cornea is a clear, dome-shaped structure that helps refract light and is also where a contact lens is placed.

2. **Vascular tunic.** The vascular tunic consists of the choroid, ciliary body, and iris. The ciliary body consists of smooth muscle that changes the shape of the lens via suspensory ligaments when contracted or relaxed. The iris contains muscles that change the diameter of the pupil through constriction or dilation.

3. **Internal tunic (retina).** The internal tunic consists of photoreceptors called rods and cones. Rods are scattered throughout the retina and assist with peripheral vision, adjusting sight in the dark, following objects that move, and observing shades of black and white. The macula lutea is an oval region at the center of the retina that surrounds the fovea centralis and controls visual acuity. Cones are concentrated at the fovea centralis and assist with sharp vision, color vision, observing still objects, and seeing in daylight. There are no rods or cones in the back of the eye where the optic nerve exits the globe. This area is called the optic disc, but is also referred to as the "blind spot" because no vision occurs at this location.

The eyeball is hollow on the inside and is compartmentalized into the following cavities:

1. **Anterior cavity.** The anterior cavity is located between the lens (structure that focuses light onto the retina) and the cornea. This anterior cavity is further subdivided into the area in front of the iris (anterior chamber) and behind the iris (posterior chamber). Aqueous humor fills both of these chambers.

2. **Posterior cavity (vitreous chamber).** The posterior cavity is located posterior to the lens and retina and contains vitreous humor.

The pupil and lens act in concert to focus the light on the rods and cones within the retina. The rods and cones transduce the light energy into an electrochemical impulse that is conveyed along the optic nerve to the occipital lobe of the cerebrum where visual sensations are perceived.

Clinical Application

Watch Out!

The eye has several different protective mechanisms. The eye resides in sockets completely surrounded by the palatine, frontal, lacrimal, ethmoid, maxilla, zygomatic, and sphenoid bones. The eye sockets and eyelids enable only one-sixth of the eye to be exposed to the external environment. Eyelids provide an additional protective mechanism by closing to wash sterilizing tears over the eye and to prevent irritation from wind or dust. If anything touches the eyelashes or cornea, the corneal reflex is engaged and both eyes automatically close. Additionally, eyebrows are designed to provide shade to the eye to help protect it from the sunlight.

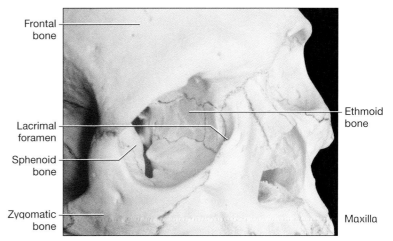

Orbit of the skull.

Clinical Application

Macular Degeneration

Macular degeneration (MD) is deterioration and damage to the macula lutea. As such, a patient loses central vision, which is needed in activities such as driving, reading, watching a computer screen, and recognizing faces. Therefore, a patient with MD is unable to see fine details whether close or at a distance but will most likely retain peripheral (side) vision. For example, if a patient with MD looks directly at a clock on his kitchen wall, he will likely be able to see the outside rim of the clock and perhaps the numbers but will be unable to see the center of the clock with the minute and hour hands. For Americans over 50 years of age, macular degeneration is the leading cause of vision loss. As life expectancy continues to increase, the number of people affected by macular degeneration is predicted to increase in large measure.

1 Identify, label, and color the following structures of the eye on Figure 12.1.

- ☐ Sclera
- ☐ Cornea
- ☐ Anterior chamber
- ☐ Posterior chamber
- ☐ Vitreous chamber
- ☐ Choroid
- ☐ Ciliary body
- ☐ Suspensory ligaments
- ☐ Lens
- ☐ Pupil
- ☐ Iris
- ☐ Retina
- ☐ Fovea centralis
- ☐ Optic disc
- ☐ Optic nerve

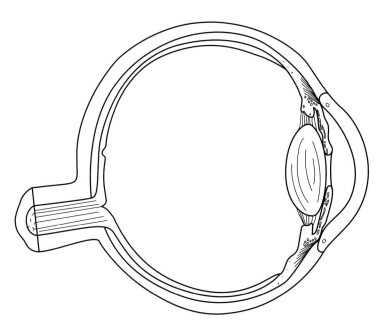

FIGURE **12.1** Sagittal section of the eyeball.

ACTIVITY 3 Nasal Anatomy

The nose contains millions of special sensory neurons that enable us to smell. Smell, or olfaction, relies on chemoreceptors to communicate information about the external environment to the brain. The roof of the nasal cavity houses olfactory chemoreceptors. Olfactory chemicals stimulate them, which results in conduction along the course of the olfactory nerve. Axons then travel through the cribriform plate and synapses with the olfactory bulb. Impulses are conducted along the axons of the olfactory nerve (CN I) to the olfactory cortex for interpretation.

Clinical Application

Anosmia

Anosmia is the loss of the ability to smell. This condition can be temporary (such as when a nose is full of mucus) or permanent (the destruction of nasal mucosa and olfactory neurons). Anosmia is sometimes viewed as trivial compared to losing the sense of vision or hearing. However, anosmia hinders a person's ability to smell fire, rotten food, and gas leaks. Furthermore, memory and smell are intimately linked and as such anosmia has led to depression in some patients because they can no longer have an "olfactory-memory" experience when returning to familiar smells such as their childhood home or smelling crayons (the smell most linked to childhood, according to research).

Weird and Wacky

Wake Up and Smell the Coffee!
Your nose possesses more than 5 million olfactory neurons that detect different smells. Rabbits, on the other hand, have more than 100 million, while dogs have more than 200 million olfactory neurons. However, even with millions of olfactory neurons, we cannot rely on our sense of smell to wake us up in the case of a fire. Research demonstrates that when we sleep our sense of smell diminishes to close to nothing, because our reticular activating system (part of the brain stem responsible for arousal from sleep) is neither connected to olfactory fibers nor to olfactory cortex. While sound, vision, vibration, and many other stimuli can wake us up, smell can't. An artificial nose (i.e., smoke alarm) needs to produce a sound and light to wake us up in the case of the fire. As in the saying, "wake up and smell the coffee," senses occur in just that order.

A hot cup of coffee.

1 Identify, label, and color the regions of the ear on Figure 12.2.

- ☐ Superior nasal concha
- ☐ Olfactory bulb
- ☐ Cribriform plate
- ☐ Olfactory nerve
- ☐ Olfactory tract

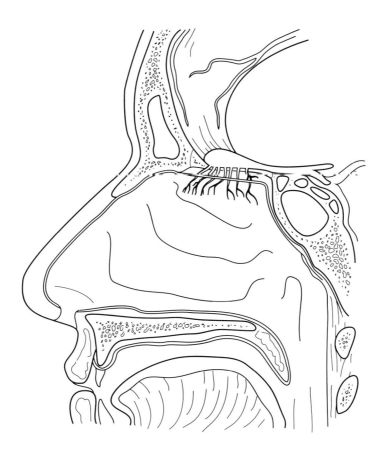

FIGURE **12.2** Sagittal section of the head.

UNIT 3: Coming to Your Senses

ACTIVITY 4 Ear Anatomy

The only place in our body where a specialized tissue is able to transduce sound waves into a nerve impulse is in our ears. The ear is the organ of hearing and equilibrium (balance), and consists of the following three regions:

1. **Outer ear.** Consists of the auricle and the external auditory canal. The auricle is a shell-shaped structure composed of elastic cartilage. It funnels sound into the external auditory canal, which extends from the outside of the ear to the tympanic membrane (more commonly referred to as the eardrum). The junction between the outer and middle ear is formed by the tympanic membrane.

2. **Middle ear.** Contains three ear ossicles and the opening to the auditory tube. The ear ossicles (malleus, incus, and stapes) transmit vibrations from the tympanic membrane to the inner ear through a structure called the oval window. The auditory (eustachian or pharyngotympanic) tube connects the nasopharynx with the middle ear.

3. **Inner ear.** Contains the spiral organ of Corti in the cochlea and vestibular apparatus in the semicircular canals. The spiral organ of Corti contains specialized hair cells that transmit sound impulses to the brain for hearing. Three semicircular canals are oriented at right angles to each other. They work together to sense rotational or centrifugal force along with acceleration and deceleration or force of inertia movements of the body, ultimately conveying dynamic equilibrium information to the brain. A vestibule has another kind of mechanoreceptor that responds to the pull of gravity onto crystals, called otoliths, that open mechanically gated channels when we bend our head. This is called a static equilibrium. These special sensory impulses from both receptor types are transmitted through the vestibulocochlear nerve (CN VIII) to the brain for interpretation.

Clinical Application

Music to My Ears

Musical ear syndrome is a condition where individuals with hearing loss develop auditory hallucinations and hear music when no music is playing. The cause of this condition is uncertain; however, one theory, called "the release phenomenon," is under consideration. The theory states that individuals with acquired deafness experience musical hallucinations because the brain imposes sound where sound no longer exists. In other words, the brain is accustomed to hearing sounds constantly, so when that stimulus is removed (in the form of acquired hearing loss), it is hard for the brain to interpret silence. With no input, the brain is theorized to create sounds because it has been so accustomed to hearing them.

Headphones used for listening to music.

 ## Weird and Wacky

The Sound of Your Own Voice

Have you ever winced when listening to a recording of your own voice? If so, you are not alone—the majority of people find the sound of their own voice strange. This is because of how our ears listen to the sounds coming from our own mouths. When you hear your own voice, two things occur. First, the sounds created by your own voice leave your vocal folds, exit the mouth, and become sound waves that hit your tympanic membrane, setting off the series of events to enable you to hear (external stimulus). The mechanical act of producing sounds involves the contraction of laryngeal muscles, which vibrates the vocal folds. The vibratory sensations are transmitted through the bones in the neck and head to the tympanic membrane and ear ossicles (internal stimulus). As such, when you speak, you deliver sound to your eardrum both externally (sound waves leaving your mouth) and internally (sound waves moving through your bones). You are the only person who hears your voice this way; everyone else only receives the external stimulus.

1 Identify, label, and color the regions of the ear on Figure 12.3.

- ☐ External ear
- ☐ Auricle
- ☐ External auditory canal
- ☐ Tympanic membrane
- ☐ Middle ear
- ☐ Malleus
- ☐ Incus
- ☐ Stapes
- ☐ Oval window
- ☐ Auditory tube
- ☐ Internal ear
- ☐ Vestibule
- ☐ Cochlea
- ☐ Semicircular canals
- ☐ Round window
- ☐ Vestibulocochlear nerve (CN VIII)
- ☐ Temporal bone

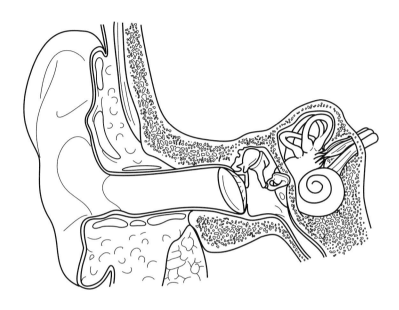

FIGURE **12.3** Coronal section of the ear.

ACTIVITY 5 Tongue Anatomy

The tongue contains special sensory neurons that enable us to taste. The sensation of taste relies on chemoreceptors on taste buds to interact with the chemicals in food to then send that information to the brain. These taste chemoreceptors are located on the taste buds housed on projections from the tongue papillae. Most papillae are scattered over the surface of the tongue, with the exception of the circumvallate papillae, which are located at the back of the tongue in a "V"-shape arrangement. Taste sensation is conveyed to the brain via branches from the facial nerve (CN VII) and glossopharyngeal nerve (CN IX).

Clinical Application

Sleep Apnea

Did you know that fat doesn't just accumulate in your belly? The tongue consists of muscle, connective tissue, and a high percentage of fat. Research has demonstrated that there is a correlation between obesity and tongue fat volume. Many obese patients with higher amounts of fat also have larger tongues, which puts them at a greater risk for obstructive sleep apnea. This is a sleep disorder that causes the individual to repeatedly stop and start breathing during sleep. The larger tongue seems to occlude the airway and may also prevent muscles that attach the tongue to bone from positioning the tongue away from the airway during sleep.

1 Identify, label, and color the connective tissue structures of the tongue on Figure 12.4.
- ☐ Fungiform papillae
- ☐ Circumvallate papillae
- ☐ Taste buds
- ☐ Epiglottis
- ☐ Palatine tonsils

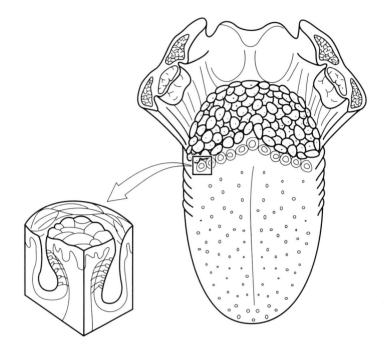

FIGURE **12.4** Tongue.

👁 OBSERVING

Complete the following hands-on laboratory activities to apply your knowledge of the special senses.

ACTIVITY 1 Anatomy of the Special Senses

1 Pair up with a classmate and use Figure 12.5 to identify the labeled structures on your lab partner's eye.

MATERIALS
Obtain the following items before beginning the laboratory activities:
- ❏ Snellen eye chart
- ❏ Tuning fork
- ❏ Lemon or lime juice
- ❏ Table sugar
- ❏ Tonic water

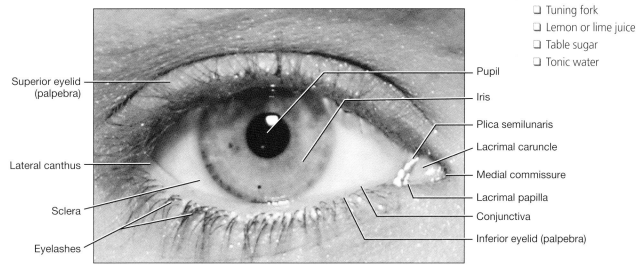

FIGURE **12.5** Anterior view of the eye.

2 Identify the structures of the eyeball listed in Figures 12.6 through 12.8.

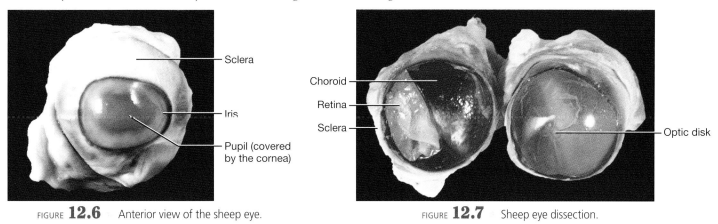

FIGURE **12.6** Anterior view of the sheep eye.

FIGURE **12.7** Sheep eye dissection.

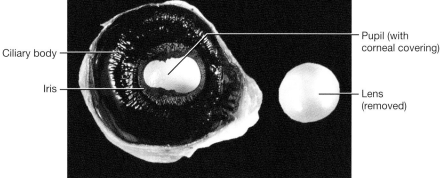

FIGURE **12.8** Sheep eye dissection with lens removed.

UNIT 3: Coming to Your Senses — CHAPTER 12: Special Senses

3 Pair up with a classmate and use Figure 12.9 to identify the following structures on your lab partner's ear.

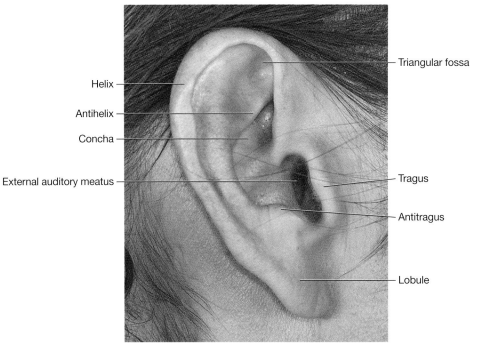

FIGURE **12.9** View of the external ear.

ACTIVITY 2 Tests for the Special Senses

With one or two of your classmates, complete the following activities to test your special senses.

Part A: Blind Spot Test
Complete the following and record your findings in the space provided:

1 Cover your left eye with your left hand.

2 Hold Figure 12.10 with your right hand about a foot and a half in front of your face.

3 Look at the dot with your right eye.

4 Bring the piece of paper slowly toward you while continually looking at the dot.

5 At a certain distance, the plus sign will disappear from sight.

6 In the space provided, hypothesize what happened to the plus sign.

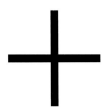

FIGURE **12.10** Blind spot test.

236 ■ *Discovering Anatomy: A Guided Examination of the Cadaver*

Part B: Visual Acuity Test

This test indicates the visual acuity of your lab partner relative to someone with perfect vision. For example, a person with 20/40 vision can see the chart from 20 feet away. However, someone with perfect vision could see it 40 feet away. Complete the following and record your findings in the space provided:

1 Put a Snellen eye chart on the wall and stand next to it.

2 Have your lab partner stand 20 feet away and read the chart. Verify that your lab partner is correctly reading the letters.

3 Record the number of the smallest line your lab partner can read without making a mistake.

Part C: Weber Test

The Weber test is a quick way to test hearing and detect unilateral (one-sided) conductive hearing loss (middle ear hearing loss) and unilateral sensorineural hearing loss (inner ear hearing loss). A normal hearing test results in the patient reporting that the sound quality is heard equally on both sides. Complete the following and record your findings in the space provided:

1 Have your lab partner sit in a chair facing forward with his/her eyes closed.

2 Strike a tuning fork lightly to start it ringing and place it on the top of his/her skull in the middle.

3 Instruct your lab partner to inform you in which ear the ring of the tuning fork is heard louder.

4 Which cranial nerve was being tested?

Part D: Rinne Test

This Rinne test is a hearing test primarily used for evaluating unilateral hearing loss. It compares the perception of sounds transmitted by air conduction to those transmitted by bone conduction through the mastoid process. Complete the following and record your findings in the space provided:

1 Have your lab partner sit in a chair facing forward with his/her eyes closed.

2 Strike a tuning fork lightly to start it ringing and place it on his/her mastoid process.

3 Instruct your lab partner to inform you when he/she can no longer hear the ringing of the tuning fork.

4 Once your lab partner can't hear it, move the still vibrating tuning fork 1 to 2 cm from the auditory canal.

5 Instruct your lab partner to inform you when he/she is unable to hear the tuning fork.

6 Air conduction is better than bone conduction. Therefore, a patient should be able to hear the tuning fork next to the auricle after they can no longer hear it when held against the mastoid process. What did you observe with your partner?

Part E: Romberg Test

The Romberg test is an exam used to assess neurological function. The exam is based on the premise that a person requires at least two of the following three senses to maintain balance while standing: proprioception (ability to know one's body in space), vestibular function (ability to know one's head position in space), and vision (which can be used to monitor and adjust for changes in body position). A patient who has a problem with one of these three senses (like proprioception) can still maintain balance by using vestibular function and vision. Complete the following and record your findings in the space provided:

1 Instruct your lab partner to stand erect with feet together.

2 Instruct your lab partner to close his/her eyes (stand alongside of them to ensure they do not fall over).

3 If your lab partner stands straight his/her vestibular and proprioceptive abilities are intact, indicating a negative sign. A positive sign is noted if he/she starts swaying. The essential feature is that the patient becomes more unsteady with eyes closed. What did your observe with your partner?

Part F: Tongue

This activity will test different parts of the tongue to see which are more sensitive to different characteristics of food: salty, bitter, sour, or sweet. Complete the following and record your findings in the space provided:

1 Place the following food items on the table:
- ☐ Table salt and water mixed together (salty taste).
- ☐ Table sugar and water mixed together (sweet taste).
- ☐ Lemon juice (sour taste).
- ☐ Tonic water (bitter taste).

2 Have your lab partner sit in a chair with his/her eyes closed.

3 Dip a toothpick into one of the solutions and place solution on different parts of the tongue. Record whether a region of the tongue is more sensitive to the different flavor.

4 Repeat this for each flavor of liquid (drink water between tests to clear flavors off from the tongue).

5 Using Figure 12.11, indicate areas of the tongue that are most sensitive to the different tastes. Compare the tongue drawings from other people.

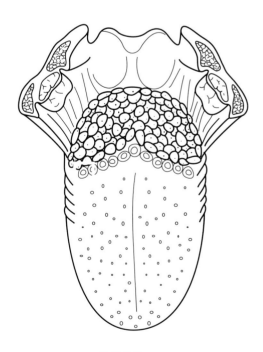

FIGURE **12.11** Tongue.

WRAPPING UP

Complete the following additional activities to help retain your knowledge of special senses.

Name _____

Date _____ Section _____

1. Match the structure in the right-hand column with its definition in the left-hand column.

   Definition

   _____ Contains chemoreceptors and is located along the superior nasal concha

   _____ Fills the space between the cornea and lens

   _____ Attaches to the tympanic membrane

   _____ Openings in the ethmoid bone for the olfactory nerves

   _____ Responsible for balance

   _____ Contains muscles that change the diameter of the pupil

   _____ The "whites" of the eyes

   _____ Responsible for hearing

   _____ Concentrated in the macula densa

   _____ Peripheral and dim light vision

   _____ Connects the pharynx to the middle ear

   _____ Located on the filiform and circumvallate papillae

   Structure

   a. Rods
   b. Cones
   c. Sclera
   d. Taste buds
   e. Olfactory mucosa
   f. Cribriform foramina
   g. Cochlea
   h. Semicircular canal
   i. Auditory tube
   j. Aqueous humor
   k. Iris
   l. Malleus

2. Macular degeneration is an eye condition where patients have a gradual vision loss for colors and reduction in high-acuity vision. Explain why.

3. Children who contract a throat infection are at risk of developing otitis media (middle-ear infection). Explain how a throat infection could spread to become a middle-ear infection.

UNIT 3: Coming to Your Senses

CHAPTER 12: Special Senses

4 Hypothesize the symptoms a patient with an inner-ear infection would most likely present with. Explain why?

5 Otosclerosis is a condition where a patient presents with progressive hearing loss due to abnormal bony remodeling on the stapes bone. Hypothesize the type of hearing loss a patient with otosclerosis would most likely present with: conductive or sensorineural? What results would you expect after conducting the Weber and Rinne tests on this patient?

6 Hypothesize a reason that people with a head cold and stuffy nose have a difficult time smelling.

7 A boy dries his tongue with a clean paper towel. If he places some table sugar on the top of his tongue, will he be able to taste the sugar? Explain.

4

Body Highways

CHAPTER 13	Endocrine System	**243**
CHAPTER 14	Cardiovascular System	**257**
CHAPTER 15	Lymphatic System	**285**
CHAPTER 16	Respiratory System	**297**
CHAPTER 17	Digestive System	**315**
CHAPTER 18	Urinary System	**339**

Endocrine System

CHAPTER 13

At the completion of this laboratory session, you should be able to do the following:

1. Compare and contrast how the endocrine and nervous systems regulate body tissues.
2. Describe the following endocrine glands and include their topography, hormones, target tissues, and effects:
   a. Pituitary gland.
   b. Hypothalamus.
   c. Thyroid gland.
   d. Parathyroid glands.
   e. Adrenal gland (cortex and medulla).
   f. Pancreas.
   g. Ovaries and testes.

The nervous and endocrine systems work closely together to regulate body processes and maintain homeostasis. The nervous system regulates these activities by way of electrochemical impulses transmitted by neurons. The response is usually immediate and brief in duration.

In contrast, the endocrine system administers its control of the body through the production and secretion of hormones. Hormones are molecules that are distributed via the circulatory system and affect distant cells by binding to specific receptors (much like a key fits into specific locks). Binding of the hormone to the receptor site initiates a regulatory effect on the cell and tissue. Traits of hormones include the following:

1. **Rate.** As a result of the hormone's distribution via the circulatory system, the speed of endocrine regulation is slower than that of nervous regulation.
2. **Distribution.** Hormones are distributed by the circulatory system and can be experienced anywhere in the body where a receptor site specific for the hormone resides.
3. **Duration.** Because the hormone can bind to a receptor site and not be degraded instantly, the duration of the effect on the cell can be longer lasting than that initiated by a neuron. Hormonal action requires seconds or days to elicit a response (in contrast to neurological responses measured in milliseconds).

There are two different classifications of glands in the human body: exocrine and endocrine. Exocrine glands produce substances and secrete them to an epithelial surface via ducts, like sweat glands in the skin. In contrast, endocrine glands are ductless and secrete their products into the circulatory system. As such, rich capillary beds are located within the tissue substance of each endocrine gland. The endocrine glands produce hormones and secrete them into the surrounding tissues where they are picked up by the capillary blood to be distributed to other tissues of the body via the circulatory vessels.

GETTING ACQUAINTED

Complete the following activities to become familiar with the endocrine system.

MATERIALS
Obtain the following items before beginning the laboratory activities:
- ❏ Textbook or access to Internet resources
- ❏ Colored pencils

ACTIVITY 1 Key Terms of the Endocrine System

The primary endocrine glands of the body are as follows:

1. **Pituitary gland.** The pituitary gland is the size of a pea and resides in the sella turcica of the sphenoid bone near the optic chiasma. It is often considered the most important endocrine gland because it produces hormones that control so many tissues. The pituitary gland consists of two lobes (anterior and posterior) based upon their embryonic origin (surface ectoderm and neuroectoderm respectively). Pituitary hormones from the two lobes are as follows:

    a. **Anterior lobe of the pituitary gland (adenohypophysis):**
        i. **Follicle stimulating hormone (FSH).** Gonadotropic hormone that stimulates ovarian follicle maturation in females and sperm production in males.
        ii. **Luteinizing hormone (LH).** Gonadotropic hormone that triggers ovulation and stimulates ovarian production of estrogen and progesterone in females. LH promotes testosterone production in males.
        iii. **Adrenocorticotrophic hormone (ACTH).** Promotes release of hormones produced by the adrenal cortex (e.g., cortisol).
        iv. **Thyroid-stimulating hormone (TSH).** Stimulates the thyroid gland to release T3 and T4 hormones.
        v. **Prolactin.** Targets glandular tissue in the breast to promote lactation.
        vi. **Growth hormone (GH).** Targets body cells, with particular effect on bone and muscle, stimulating growth.

    b. **Posterior lobe of the pituitary gland (neurohypophysis):**
        i. **Oxytocin.** Targets the uterus and mammary glands. In the uterus, oxytocin initiates labor and stimulates uterine contraction. In the mammary glands, oxytocin initiates the milk let-down reflex.
        ii. **Antidiuretic hormone (ADH).** Targets the nephrons in the kidney to reabsorb water to reduce urine volume output and increase fluid volume in the blood. ADH is also known as vasopressin.

Clinical Application

Oxytocin

During labor, the hormone oxytocin is responsible for uterine contractions to move the baby from the uterine cavity, into the birth canal, and ultimately to mother's arms. Following delivery, oxytocin continues to cause uterine contractions resulting in the compression of uterine vessels to reduce bleeding and to help the uterus return to its pre-pregnancy size. Oxytocin also causes the milk let-down reflex, enabling baby to nurse. Breastfeeding also benefits the mother, as it has been shown to reduce the risk of breast cancer and ovarian cancer. In addition to being released during childbirth and breastfeeding, oxytocin is also released during sexual intercourse. Some studies have shown it is released during the first months of dating and even while hugging. Oxytocin may have a role to play in romantic attachment and empathy.

2. **Hypothalamus.** Located below the thalamus, the hypothalamus serves an important role by linking the nervous system with the endocrine system (via both lobes of the pituitary gland). The hypothalamus produces oxytocin and antidiuretic hormone, which are transported and stored in the posterior lobe of the pituitary gland. The hypothalamus also produces releasing and inhibiting hormones that, to a large degree, control the release of anterior pituitary hormones via negative feedback mechanisms.

3. **Thyroid gland.** The thyroid gland is located on the trachea inferior to the thyroid cartilage. Thyroid follicular cells produce and secrete catabolic and calorigenic hormones called triiodothyronine (T3) and thyroxine (T4). T3 and T4 are synthesized from iodine and tyrosine and are regulated by TSH from the anterior pituitary gland. Additionally, thyroid parafollicular cells produce calcitonin, which plays a role in calcium homeostasis.

4. **Parathyroid glands.** These four glands are located along the posterior surface of the thyroid gland and play key roles in regulating calcium concentrations in blood and bone.

5. **Adrenal gland.** The adrenal gland is located on the superior pole of the kidney and consists of an outer cortex, which produces steroid hormones, and an inner medulla, which produces epinephrine and norepinephrine.

Clinical Application

Goiter

A goiter is a pathologic enlargement of the thyroid gland and presents as a swelling in the anterior part of the neck below the thyroid cartilage. A goiter is usually caused by iodine deficiency, which is an essential component of T3 and T4. When there is a deficiency of iodine, the thyroid produces plenty of T3 and T4 precursors but is unable to transport the precursors to the blood stream. As such, the body responds to the low levels of T3 and T4 in the blood by having the pituitary gland secrete more thyroid-stimulating hormone (TSH). This increase of TSH stimulates the thyroid gland to attempt to produce more T3 and T4. However, the thyroid only produces more of the precursors (which are not delivered into the blood stream) and this positive feedback loop causes the thyroid gland to swell.

Norwegian woman with a goiter.

   a. **Adrenal cortex.** The adrenal cortex is subdivided into three zones:
      i. **Zona glomerulosa.** Produces aldosterone, which regulates blood pressure and electrolytes.
      ii. **Zona fasciculata.** Produces cortisol, which regulates metabolism and immune suppression.
      iii. **Zona reticularis.** Produces androgens, which are sex hormones.
   b. **Adrenal medulla.** Produces catecholamines, such as adrenaline (epinephrine) and noradrenaline (norepinephrine), which function under the sympathetic response, also known as the fight-or-flight response.
6. **Pancreas.** The pancreas is located in the abdominal cavity deep to the stomach. It possesses both exocrine function (digestion) and endocrine function (blood sugar regulation). The pancreas possesses islets (of Langerhans), which secrete insulin and glucagon into the cardiovascular system. Insulin decreases blood glucose concentrations and glucagon increases blood glucose concentrations.
7. **Ovaries.** The ovaries secrete estrogen and progesterone and are located in the female pelvic cavity. Estrogen is responsible for the appearance of secondary sex characteristics during adolescence such as growth of body hair and breasts. Progesterone prepares the uterus for pregnancy and mammary glands for lactation in women. Estrogen and progesterone promote changes in the endometrium in the menstrual cycle.
8. **Testes.** The testes are located in the scrotum of the male. They produce the sex hormone testosterone, which is responsible for the secondary sex characteristics that appear during adolescence such as growing of body hair, deepening the voice, maintaining the sex drive, promoting the production of sperm, and maintaining muscle and bone mass.

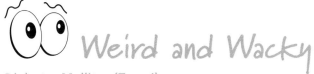

Diabetes Mellitus (Type I)

One of the important roles of insulin is to signal cells in the liver, muscle, and adipose tissue to import glucose from the blood to be used for energy. Diabetes mellitus (DM) is the most common endocrine disorder whereby the pancreatic islets no longer produce insulin. As a result, blood glucose levels become very high, as does the amount of glucose in the urine. This hyperglycemic (high blood glucose) condition results in increased hunger and thirst as well as frequent urination. Physicians today diagnose DM through the use of blood tests. However, DM has been written about for thousands of years. The early physicians would smell a patient's urine or breath to detect a pungent smell ("acetone breath"), or (brace yourself) taste a patient's urine to see if it tasted sweet. You see, diabetes mellitus is Latin for "excessive discharge of urine" (*diabetes*) and "honey" (*mel*). In other words, DM results in "excessive, honey-like urine." Type I diabetes is due to a lack of insulin. Type II diabetes occurs when body tissues become resistant to insulin.

1 Using your textbook or online resources, write definitions or descriptions of each of the terms in Table 13.1.

TABLE 13.1 Endocrine System Terms

Term	Definition
Endocrine gland	
Hormone	
Receptor molecule	
Negative feedback	
Positive feedback	
Pituitary gland	
FSH	
LH	
ACTH	
TSH	
Prolactin	
Growth hormone	
Oxytocin	
ADH	
Adrenal gland	
Adrenal cortex	
Adrenal medulla	
Zona glomerulosa	
Zona fasciculata	
Zona reticularis	

(continues)

TABLE 13.1 Endocrine System Terms *(cont.)*

Term	Definition
Aldosterone	
Cortisol	
Androgens	
Epinephrine/norepinephrine	
Pancreas	
Islets of Langerhans	
Insulin	
Glucagon	
Gonads	
Testosterone	
Estrogen	
Progesterone	
Thyroid and parathyroid glands	
T3 and T4	
Calcitonin	
Parathyroid hormone	

2 Identify, label, and color the following structures of the endocrine system on Figure 13.1.
- ☐ Hypothalamus
- ☐ Pituitary gland
- ☐ Thyroid gland
- ☐ Parathyroid glands
- ☐ Adrenal cortex
- ☐ Adrenal medulla
- ☐ Pancreas
- ☐ Islets of Langerhans
- ☐ Ovary
- ☐ Testis

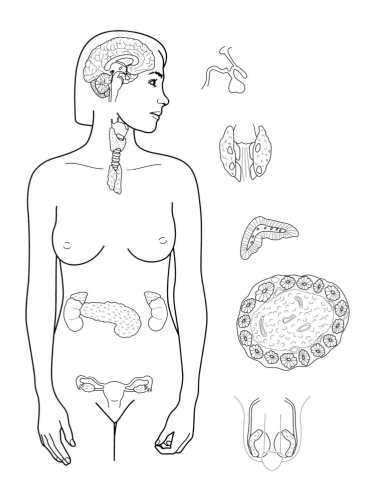

FIGURE **13.1** Structures of the endocrine system.

3 Identify and label the following structures on Figures 13.2 through 13.5.
- ☐ Pars distalis
- ☐ Pars intermedia
- ☐ Pars nervosa

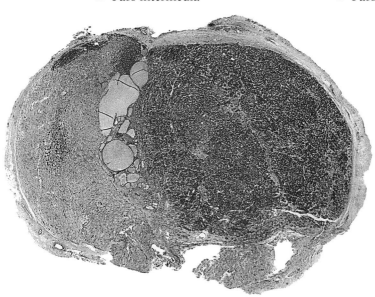

FIGURE **13.2** Pituitary gland.

- Capsule
- Zona glomerulosa
- Zona fasiculata
- Zona reticularis
- Adrenal medulla

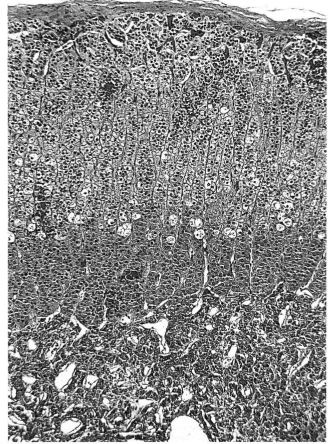

FIGURE 13.3 Adrenal gland.

- Follicular cells
- Parafollicular cells
- Colloid

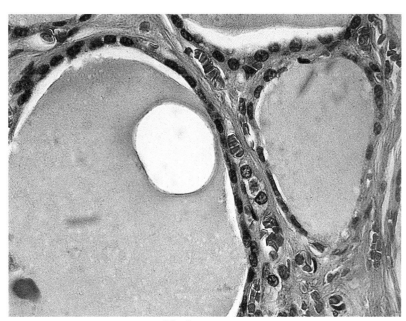

FIGURE 13.4 Thyroid gland.

- ☐ Exocrine tissue
- ☐ Islet of Langerhans
- ☐ Pancreatic duct

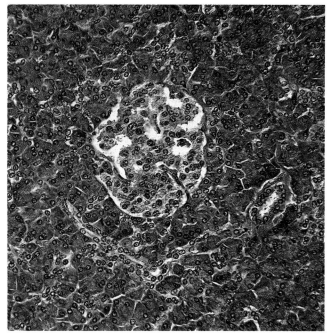

FIGURE **13.5** Pancreas.

👁 OBSERVING

Complete the following hands-on laboratory activities to apply your knowledge of the endocrine system.

ACTIVITY 1 Microscopic Structures of the Endocrine System

1. Obtain microscope slides of the thyroid, adrenal, pituitary, and pancreas from your instructor.

2. Place each slide under the microscope and start with the lowest power and advance to higher magnification to see more detail.

3. Once you have a view of the tissue, use your colored pencils to draw what you see in the field of view. Color in the different cells and layers and label the terms indicated.

   a. Thyroid gland
      Draw and label the following:
      - ☐ Thyroid follicle
      - ☐ Colloid
      - ☐ Follicular cells
      - ☐ Parafollicular cells (C cells)

MATERIALS

Obtain the following items before beginning the laboratory activities:

- ❏ Light microscope
- ❏ Microscope slides of the thyroid, parathyroid, adrenal, pituitary, and pancreas
- ❏ Colored pencils
- ❏ Cadaver
- ❏ Gloves
- ❏ Probe
- ❏ Protective gear (lab coat, scrubs, or apron)

b. Adrenal gland
   Draw and label the following:
   ☐ Zona glomerulosa
   ☐ Zona fasciculata
   ☐ Zona reticularis
   ☐ Adrenal medulla

c. Pituitary gland
   Draw and label the following:
   ☐ Pars distalis (adenophypophysis)
   ☐ Pars intermedia (Rathke's pouch)
   ☐ Pars nervosa (neurohypophysis)

d. Pancreas
   Draw and label the following:
   ☐ Exocrine portion (acinar cells)
   ☐ Islets of Langerhans

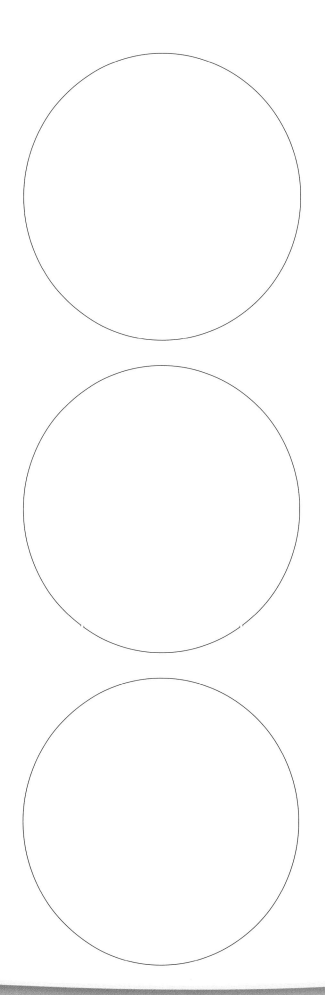

ACTIVITY 2 Gross Anatomy of the Endocrine System

1 Using Figures 13.6 through 13.8, identify the following structures on the cadaver.

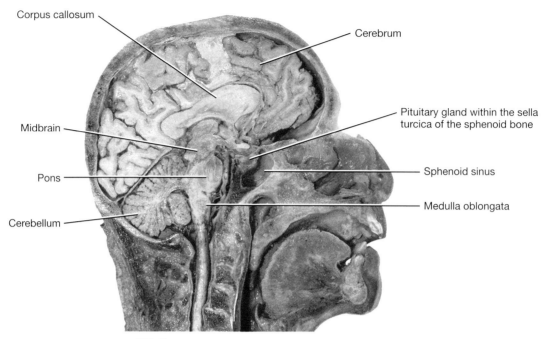

FIGURE **13.6** Medial view of the midsagittal section of the head.

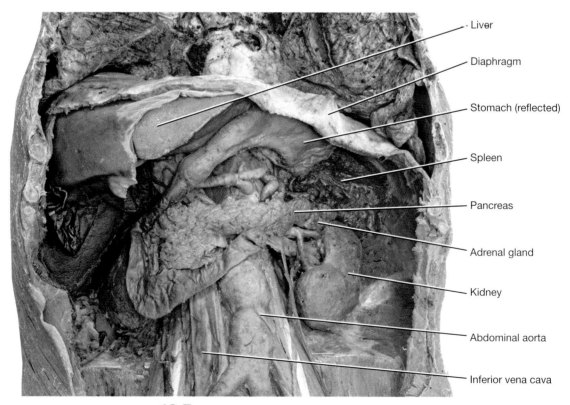

FIGURE **13.7** Abdominal cavity with the anterior wall removed.

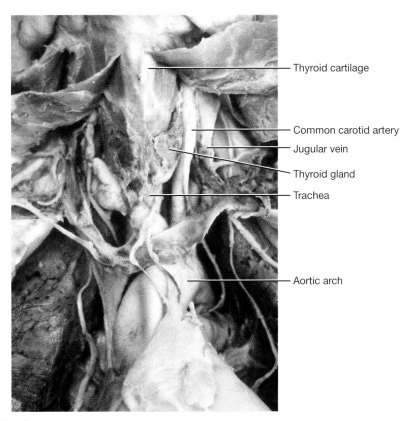

FIGURE 13.8 Neck dissection with the cervical muscles cut and reflected to show the thyroid gland.

WRAPPING UP
Complete the following additional activities to help retain your knowledge of the endocrine system.

Name _____

Date _____ Section _____

1. Endocrine glands produce and secrete hormones into the bloodstream, targeting cells with receptors specific for that hormone and resulting in some type of effect. Table 13.2 lists a number of these endocrine glands and their hormones; however, only part of the information is included for each line. Using the provided clues, fill in the remainder of the blank spaces.

TABLE 13.2 Table of Endocrine System

Endocrine Gland	Hormone Secreted	Target Tissue	Hormonal Effect
Pituitary gland	Follicle-stimulating hormone	Thyroid gland (follicular cells)	Thyroid gland secretes T3 and T4
	Insulin		
Pituitary gland		Muscles and bones	
Testis			
Pituitary gland		Adrenal cortex	
	Calcitonin		
			Urine volume decreases; blood volume increases
	Estrogen		
Parathyroid gland			
			Triggers ovulation
		Smooth muscle in uterus	Initiates labor
			Increases blood calcium concentration
	Norepinephrine		
	Cortisol		

2. Pitocin is a medicine given to pregnant women to initiate labor. Hypothesize the hormone this medicine mimics and explain how you came up with the answer.

3. A tumor of the parathyroid gland may present with a hypersecretion of its hormone and can eventually lead to osteoporosis. Provide a hypothesis explaining why.

4. Medicines similar in structure to the hormone calcitonin are prescribed to assist in the treatment of osteoporosis. Hypothesize why.

5. Type I diabetes mellitus is an autoimmune disorder where the host immune cells target and destroy beta cells within the islets of Langerhans. This results in a reduction in insulin secretion. Hypothesize the effect this has on blood glucose concentrations.

6. A man is taking an antiviral medication to treat an infection. An unintended side effect causes his kidney tubules to be unable to respond to ADH. Hypothesize what happens to the following: water consumption, urine output, blood osmolarity, and blood glucose concentration. Explain these findings.

CHAPTER 14

Cardiovascular System

At the completion of this laboratory session, you should be able to do the following:

1. Identify layers of the serous membranes surrounding the heart.
2. Describe the tissue layers that comprise the heart and vessels.
3. Identify and describe the roles of the chambers and valves of the heart.
4. Describe the coronary circulation and identify the primary coronary arteries and cardiac veins.
5. Describe the cardiac cycle and define the terms diastole and systole.
6. Describe the conduction system of the heart.
7. Identify the key systemic arteries and veins of the body.

The circulatory system consists of the blood, heart, lymph, and vessels, each of which is essential to the life of a complex multicellular organism. Blood, a specialized connective tissue, consists of formed elements that are suspended and carried in the plasma. These blood cells (including erythrocytes, leukocytes, and platelets) transport oxygen and carbon dioxide, provide immunity, and clot the blood.

GETTING ACQUAINTED

Complete the following activities to become familiar with the cardiovascular system.

MATERIALS

Obtain the following items before beginning the laboratory activities:
- Textbook or access to Internet resources
- Colored pencils

 Heart Layers

The heart is enclosed by a pericardial sac within the mediastinum. The sac is composed of the following layers:

1. **Parietal pericardium.** The parietal pericardium is the tough sac surrounding the heart and preventing it from overfilling. The external layer is made of dense fibrous connective tissue and its internal layer is composed of a serous membrane.
2. **Visceral pericardium.** The visceral pericardium is a serous layer and a continuation of the serous layer from the parietal pericardium that forms on the external surface of the heart. This layer combined with the underlying loose connective and adipose tissues is called the **epicardium**.
3. **Pericardial cavity.** The pericardial cavity is located between the parietal and visceral layers of the serous pericardium containing serous fluid, which reduces friction when the heart moves.

The wall of the heart is composed of the following histologic layers:

1. **Epicardium.** The epicardium is the thin external layer formed by the visceral layer of the serous pericardium, loose connective tissue, and adipose tissue. It is within the epicardium that the coronary arteries and cardiac veins course.
2. **Myocardium.** The thickest layer of the heart is the myocardium. It consists of cardiac muscle tissue, which contracts and pumps blood. There are two specializations to the myocardium in the ventricles:
   a. **Trabeculae carneae.** Irregular ridges of myocardium along the internal walls of the ventricles.
   b. **Papillary muscles.** Bundles of myocardium that project into the ventricles. These muscles serve as attachments for chordae tendineae to support atrioventricular valve function.

3. **Endocardium.** The endocardium internally lines the heart chambers and valves and consists of a thin layer of endothelium and loose connective tissue. The endocardium is continuous with the endothelium of the blood vessels entering and exiting the heart.

Clinical Application

Cardiac Tamponade

Pericardial fluid produced by the serous membrane lining the heart bathes and lubricates the organ as it pumps blood. Some diseases, such as cancer or pericarditis, may result in an excessive buildup of this fluid in the pericardial sac. This buildup of fluid (referred to as a pericardial effusion) results in cardiac tamponade. The fibrous layer of the pericardial sac is composed of dense irregular collagenous connective tissue and, as a result, does not easily stretch once fluid builds up the pericardium. Thus, pressure increases in the sac as fluid continues to build and, with each successive ventricular relaxation, less and less blood can enter the ventricles. In other words, if the pericardial sac is excessively filled with fluid (including blood or pus), the heart will have no room to beat. The result is a decrease in stroke volume, or the amount of blood exiting the heart during systole. If untreated, this condition may result in a cardiac arrest. A treatment for cardiac tamponade is a procedure called pericardiocentesis. A needle is inserted into the pericardial sac, usually through an infrasternal approach, and the excess fluid is removed.

1 Using your textbook or online resources, write a definition for each of the terms listed in Table 14.1.

TABLE **14.1** Layers of the Heart Wall

Term	Definition
Pericardial sac	
Pericardial cavity	
Myocardium	
Endocardium	

ACTIVITY 2 Chambers and Valves

One way to learn about the heart is to compare the function of the left side of the heart to the right:

1. **Left and right sides of the heart.** The heart is a four-chambered muscular pump. It can be functionally divided into two sides:
    a. **Right side (pulmonary circuit).** The right atrium receives deoxygenated blood from the systemic and coronary circulations and delivers this blood to the right ventricle. The right ventricle pumps this blood to the lungs for gas exchange via the pulmonary arteries.
    b. **Left side (systemic circuit).** The left atrium receives oxygenated blood from the lungs and delivers it to the left ventricle. The left ventricle pumps this blood through the aortic valve into the systemic and coronary circulations via the aorta.

Chambers

Another way to consider the heart is through its four chambers:

1. **Four chambers of the heart.** Each side of the heart has a receiving chamber (atrium) and a pumping chamber (ventricle). The four chambers of the heart include the following:

    a. **Right atrium.** Forms the right border of the heart and receives deoxygenated blood from the systemic circulation (superior vena cava and inferior vena cava) and coronary circulation (coronary sinus).

    b. **Right ventricle.** Forms most of the anterior surface of the heart and receives deoxygenated blood from the right atrium. Blood leaves the right ventricle through the pulmonary valve and into the pulmonary trunk and reaches the lungs for oxygenation.

    c. **Left atrium.** Forms the majority of the base and posterior surface of the heart and receives oxygenated blood from the four pulmonary veins.

    d. **Left ventricle.** Forms the apex of the heart, contains the thickest layer of myocardium, and receives oxygenated blood from the left atrium. Blood leaves the left ventricle through the aortic valve and enters the aorta. The left ventricle has a thicker wall than the right due to the generation of greater pressure from propelling blood to all portions of the systemic circulation.

Clinical Application

Patent Foramen Ovale

A patent foramen ovale (PFO) is a congenital defect in the interatrial septum. In essence, the PFO is a hole between the right and left atria. While a baby is completing fetal development, this hole is normal. During pregnancy, a mother's lungs exchange carbon dioxide for oxygen on behalf of the developing baby. Oxygen in the mother's bloodstream goes to the placenta and to the baby. In contrast, carbon dioxide from the baby's blood diffuses through the placenta into the mother's bloodstream where it eventually goes to the lungs and is exhaled. As such, the lungs of a fetus do not function in gas exchange. The foramen ovale enables an opening between the right atrium over to the left atrium, thus bypassing the lungs. However, following birth, the foramen ovale seals shut because of increased pressure in the left side of the heart. In approximately 20 percent of the population, the foramen ovale remains open and is thus called a patent foramen ovale.

Valves

The atria and ventricles on each side are connected by atrioventricular (AV) openings, and an associated valve prevents the backflow of blood into the atrium during ventricular contraction. The AV valves are attached to the papillary muscles via chordae tendineae. Papillary muscles contract when the valves close. Chordae tendineae tighten during systole, keeping the cusps from prolapsing and preventing regurgitation of blood into atria. The AV valves of the heart include the following:

1. **Tricuspid valve (right AV valve).** A three-cusped valve positioned between the right atrium and right ventricle. Closure of the tricuspid valve prevents blood from flowing from the right ventricle into the right atrium during systole and, along with the bicuspid valve, results in the S1 sound (AV valve closure) during the cardiac cycle.

2. **Bicuspid valve (mitral; left AV valve).** A two-cusped valve positioned between the left atrium and left ventricle. Closure of the bicuspid valve prevents blood from flowing from the left ventricle into the left atrium during systole and, when occurring together with closure of the tricuspid valve, results in the S1 sound (AV valve closure) during the cardiac cycle.

The semilunar valves are located between the ventricles and outgoing arteries and they prevent backflow of blood. The following are semilunar valves:

1. **Pulmonary semilunar valve.** The pulmonary semilunar valve consists of semilunar cusps between the right ventricle and the pulmonary trunk. Closure of the pulmonary valve prevents blood from flowing from the pulmonary trunk into the right ventricle during diastole and, when occurring together with closure of the aortic valve, results in the S2 sound (semilunar valve closure) of the cardiac cycle.

2. **Aortic semilunar valve.** The aortic semilunar valve consists of semilunar cusps between the left ventricle and the ascending aorta. Closure of the aortic valve prevents blood from flowing from the ascending aorta into the left ventricle during diastole and, when occurring together with closure of the pulmonary valve, results in the S2 sound (semilunar valve closure) during the cardiac cycle.

Clinical Application

Valvular Heart Disease

A primary function of heart valves is to ensure blood moves forward unimpeded during the cardiac cycle and not backward. Disease involving the heart valves usually causes them to fail in this function. Valvular disease is usually categorized as stenosis, insufficiency (through regurgitation or incompetence), or both. Stenosis is defined as the failure of a heart valve to open completely, which impedes forward flow by forcing blood through a smaller opening, making flow turbulent. In contrast, insufficiency results from failure of a valve to close completely, thereby allowing reversed blood flow.

1 Using your textbook or online resources, write a definition for each of the terms listed in Table 14.2.

TABLE **14.2** Heart Chambers, Valves, and Great Vessels

Term	Definition
Superior vena cava	
Inferior vena cava	
Right atrium	
Tricuspid valve	
Right ventricle	
Chordae tendineae	
Papillary muscles	
Pulmonary valve	
Pulmonary arteries	
Pulmonary veins	
Left atrium	
Bicuspid valve	
Left ventricle	
Aortic valve	
Aorta	

ACTIVITY 3 Cardiac Cycle and Coronary Circulation

Cardiac Cycle

The cardiac cycle refers to the events that occur from the beginning of one heartbeat to the beginning of the next. During the cardiac cycle, the function of the atria and ventricle are described by the following two processes:

1. **Systole.** During ventricular contraction (systole), blood is pumped from the left and right ventricles into the aorta and pulmonary trunk, respectively. Semilunar valves guard these two outflow tracks against backflow from the aorta and pulmonary arteries into the ventricles.
2. **Diastole.** During ventricular relaxation (diastole), blood passively fills the ventricles and then is pumped from the left and right atria. AV valves (bicuspid valve and tricuspid valve) guard against backflow from the ventricles into the atria.

And I Thought I Worked Hard
The human heart pumps on average 1.5 gallons of blood every minute, 40 million times a year, and will beat more than 2.5 billion times in an average lifetime.

Coronary Circulation

The heart has a high oxygen demand because it functions continuously to pump blood. The epicardium and myocardium both receive oxygenated blood from the following vessels and branches:

1. **Right coronary artery.** Courses through the coronary sulcus (circumscribed region of the heart between atria and ventricles) and supplies the right side of the heart via the right marginal, posterior descending, and sinoatrial (SA) nodal arteries.
2. **Left coronary artery.** Courses in through the coronary sulcus and supplies the left side of the heart via the left anterior descending and left circumflex arteries.

In a normal healthy heart, most of the arteries are true end arteries, which supply discrete regions of the myocardium with little overlap or few collaterals. Therefore, in the event of sudden-onset occlusion of a major branch of the coronary circulation, heart tissue becomes ischemic distal to the blockage causing angina pectoris. Therefore, blood circulates from the left ventricle to the myocardium via coronary arteries and back to the right atrium via cardiac veins, hence the coronary circulation. The three primary cardiac veins of the heart are as follows:

1. **Great cardiac vein.** Drains the left side of the heart.
2. **Middle cardiac vein.** Drains the posterior region of the heart.
3. **Small cardiac vein.** Drains the inferolateral region of the heart.

All three cardiac veins drain into the large coronary sinus located in the right atrium.

Clinical Application

Heart Attack
Coronary vessels are categorized as an "end circulation," which means coronary vessels do not form connections or collateral blood circuits with each other. As such, once a coronary artery is blocked (often by a blood clot), the heart muscle supplied by that vessel loses oxygen, becomes damaged, dies, and can no longer contract to pump blood throughout the body. This pathologic process is known by many names: myocardial infarction (or MI), coronary thrombosis, coronary occlusion, acute coronary syndrome, and, most commonly, a heart attack. Regardless of its name this process results in a medical emergency. Symptoms of a heart attack include chest and arm pain, tightness in the chest, fatigue, abnormal heart rate, and anxiety. Coronary bypass surgery, stents, medications, and drastic lifestyle changes to diet and exercise are the variety of treatment options for patients that survive a heart attack.

1 Using your textbook or online resources, write definitions or descriptions of each of the terms listed in Table 14.3.

TABLE **14.3** Coronary Circulation

Term	Definition or Description
Right coronary artery	
Left coronary artery	
Left anterior descending artery	
Coronary sinus	

ACTIVITY 4 Conduction System

The majority of cardiac muscle cells are contractile in nature (~99 percent). However, the remaining 1 percent of cardiac muscle cells are specialized. Known as pacemaker cells, these cardiac muscle cells have the unique ability of spontaneously depolarizing and generating action potentials. The action potentials they initiate trigger action potentials of the contractile cardiac muscle cells to contract.

The heart's pacemaker cells (SA node, AV node, Purkinje fibers, and Bundle of His) make up the conduction system and enable the atria and ventricles to contract spontaneously, which initiates and coordinates contraction. The cardiac conduction system functions in such a way that it allows only a unidirectional pathway of excitation and ensures coordinated contraction of atria and ventricles. The conducting branches are insulated from each other by connective tissue, thus decreasing inappropriate stimulation.

The parts of the conduction system of the heart are as follows:

1. **Sinoatrial (SA) node.** The SA node is considered the "pacemaker" because it initiates the heartbeat via a group of impulse-generating cardiac muscle cells located at the junction of the Superior vena cava and right atrium. The wave of depolarization sweeps down, causes the atrial walls to contract, and eventually influences the Atrioventricular node. External stimulation by autonomic nerves of the SA node influences heart rate (sympathetics increase and parasympathetics decrease heart rate).

2. **Atrioventricular (AV) node.** The AV node is a group of impulse-generating cardiac muscle cells located in the interatrial septum medial to the tricuspid valve. The AV node receives impulses from the SA node and conducts them to the bundle of His.

3. **Bundle of His.** The bundle of His descends from the AV node through the fibrous skeleton of the heart before branching into left and right bundles (of His), corresponding to the left and right ventricles. This divergent pathway ensures that ventricular contraction begins in the region of the apex.
   a. **Right bundle branch.** Courses along the right side of the interventricular septum and divides at the apex of the right ventricle. One part enters the septomarginal trabecula (moderator band) to reach the base of the anterior papillary muscle, and the other part continues in the Purkinje fibers to supply the ventricular and papillary myocardium.
      i. **Moderator band (septomarginal trabecula).** A band of cardiac muscle tissue in the right ventricle, which spans between the anterior papillary muscle and ventricular septum and contains a part of the right bundle branch.
      ii. **Purkinje fibers.** Modified ventricular muscle fibers that relay cardiac impulses responsible for contraction.
   b. **Left bundle branch.** Courses along the left side of the interventricular septum and descends to the apex of the left ventricle, where it continues as Purkinje fibers throughout the myocardium of the left ventricle.
      i. The AV bundle and its branches receive their arterial supply via the septal branches of the left coronary artery.

The autonomic nervous system (sympathetic and parasympathetic systems) is responsible for helping to regulate heart rate.

1. **Sympathetic innervation (cardiac splanchnic nerves).** Sympathetic innervation influences the SA and AV nodes, which increases heart rate. Sympathetic innervation also stimulates the myocardium to increase force of contraction and cardiac output and causes coronary vasodilation.
2. **Parasympathetic innervation (vagus nerves).** Parasympathetic innervation influences the SA and AV nodes and decreases heart rate, force of contraction, and cardiac output.

1 Using your textbook or online resources, write a definition for each of the terms listed in Table 14.4.

TABLE **14.4** Conduction System

Term	Definition
Sinoatrial node	
Atrioventricular node	
Atrioventricular bundle (of His)	
Purkinje fibers	

ACTIVITY 5 Structures of the Cardiovascular System

1 Identify, label, and color the following structures of the anatomy of the heart on Figure 14.1.

- ☐ Superior vena cava
- ☐ Inferior vena cava
- ☐ Pulmonary arteries
- ☐ Pulmonary veins
- ☐ Ligamentum arteriosum
- ☐ Aortic arch
- ☐ Left common carotid artery
- ☐ Left subclavian artery
- ☐ Left anterior descending artery
- ☐ Right coronary artery
- ☐ Great cardiac vein
- ☐ Right atrium
- ☐ Right ventricle
- ☐ Left ventricle
- ☐ Pulmonary trunk
- ☐ Ascending aorta
- ☐ Brachiocephalic trunk

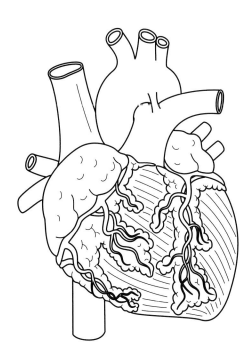

FIGURE **14.1** Anterior view of the heart.

UNIT **4:** Body Highways   CHAPTER **14:** Cardiovascular System ■ **263**

2 Identify, label, and color the following structures of the anatomy of the heart on Figure 14.2.
- ☐ Superior vena cava
- ☐ Inferior vena cava
- ☐ Pulmonary arteries
- ☐ Pulmonary veins
- ☐ Aortic arch
- ☐ Left circumflex coronary artery
- ☐ Right coronary artery
- ☐ Right posterior descending artery
- ☐ Coronary sinus
- ☐ Small cardiac vein
- ☐ Middle cardiac vein
- ☐ Right atrium
- ☐ Left atrium
- ☐ Right ventricle
- ☐ Left ventricle

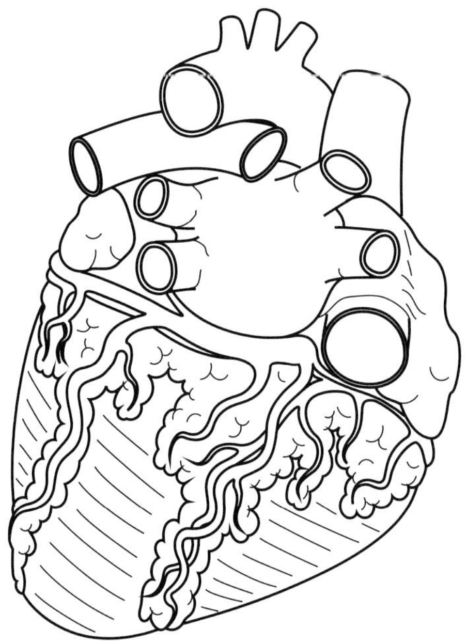

FIGURE **14.2** Posterior view of the heart.

3 Identify, label, and color the following structures of the anatomy of the heart on Figure 14.3.
- ☐ Superior vena cava
- ☐ Inferior vena cava
- ☐ Pulmonary arteries
- ☐ Pulmonary veins
- ☐ Aortic arch
- ☐ Right atrium
- ☐ Left atrium
- ☐ Right ventricle
- ☐ Left ventricle
- ☐ Tricuspid valve
- ☐ Bicuspid (mitral) valve
- ☐ Pulmonary (semilunar) valve

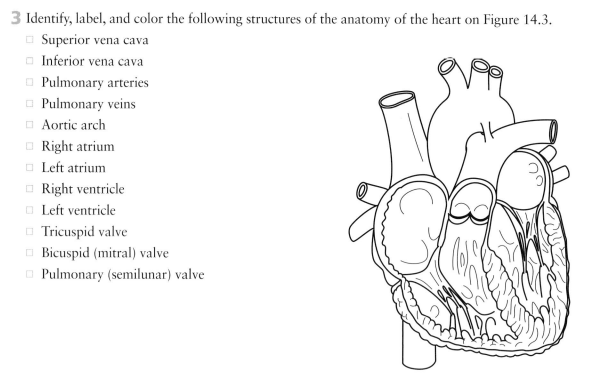

FIGURE **14.3** Anterior dissection of the heart.

Discovering Anatomy: A Guided Examination of the Cadaver

ACTIVITY 6 Pulmonary Circulation

Systemic circulation provides the blood supply, containing oxygen, nutrients, and hormones, to body tissues and carries away wastes such as carbon dioxide and metabolic byproducts. Freshly oxygenated blood is pumped out of the left ventricle into the aorta. Because all arteries arise directly or indirectly from this single great vessel, the blood can take various routes. The aorta arches upward from the heart and then curves and runs downward along the body midline to its terminus in the pelvis, where it bifurcates to supply the lower limbs. The branches of the aorta continue to branch, producing arterioles and, finally, the capillaries that course through the tissues. Veins inferior to the diaphragm return to the heart via the inferior vena cava, and organs above the diaphragm return to the heart via the superior vena cava.

Here are a few hints to help in memorizing the systemic arteries and veins of the body:

1. **Body region.** The name of the vessel often reflects the body region traversed (e.g., femoral artery), the organ supplied (e.g., renal artery), or the bone followed (e.g., radial artery).

2. **Neurovascular bundle.** Arteries and veins tend to course side by side and in many places, as do nerves.

3. **Bilateral symmetry and asymmetry.** Systemic vessels do not always match between right and left sides of the body. For example, vessels in the head and limbs are symmetrical but the vessels of the trunk are not.

1 Identify, label, and color the following structures of the anatomy of the heart on Figure 14.4 (use red for structures that transport oxygenated blood, purple for mixed blood, and blue for deoxygenated blood).

- ☐ Deoxygenated blood (blue)
  - Superior vena cava
  - Right atrium
  - Inferior vena cava
  - Right ventricle
  - Pulmonary trunk
  - Pulmonary arteries
- ☐ Mixed blood (purple)
  - Pulmonary capillaries
- ☐ Oxygenated blood (red)
  - Pulmonary veins
  - Left atrium
  - Left ventricle
  - Aortic arch

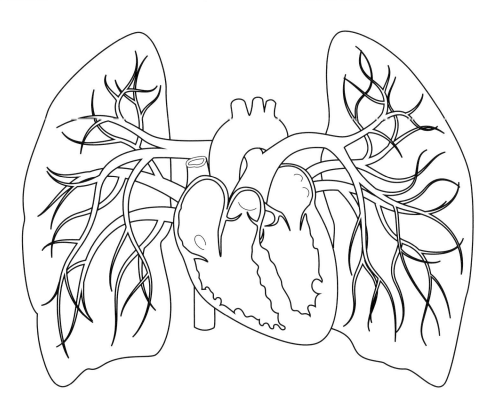

FIGURE 14.4 Pulmonary circulation.

ACTIVITY 7 Blood Vessels

The main types of blood vessels in the body are arteries, arterioles, capillaries, venules, and veins. Arteries transport blood away from the heart, while veins transport toward the heart. Most illustrations and models show vessels that transport oxygenated blood in red and deoxygenated blood in blue. Therefore, arteries are usually shown in red and veins in blue, with the exception of the pulmonary circulation. This activity will cover the various branches of the large systemic arterial and venous trees that supply and drain all the different regions of the body.

1. **Arteries.** Arteries are vessels that transport blood away from the heart. Compared to their venous counterparts, arteries have much thicker walls of smooth muscle. The systemic circuit begins with the largest artery in the body, the aorta, which directly or indirectly supplies all arteries of the body. Principal arteries to the following regions of the body include the following:

    a. **Brain.** The brain is a highly metabolic organ. Although it accounts for only about 2 percent of the body weight, it demands approximately 20 percent of the oxygen consumed by the body at rest. The brain requires a sufficient and constant circulation of blood. If blood flow slows, ischemia rapidly leads to loss of consciousness. The internal carotid arteries supply the anterior region of the brain while the vertebral arteries supply the posterior region of the brain. The paired internal carotid and vertebral arteries are connected by a cerebral arterial circle (of Willis). These extensive anastomotic connections allow for compensation of decreased blood flow by any one vessel with increased collateral blood flow from the others.

    b. **Upper limb.** The subclavian arteries supply blood to the upper limbs (right subclavian is a branch from the brachiocephalic artery and left subclavian arises directly from the aortic arch). The subclavian arteries become the axillary arteries at the lateral border of the first rib. The axillary arteries become the brachial arteries at the inferior border of the teres major muscle. In the cubital fossa, the brachial artery bifurcates into the radial and ulnar arteries, which supply the forearm and hand.

    c. **Abdomen.** After traversing the diaphragm, the abdominal aorta gives rise to the following main arteries to the GI tract:

        i. **Celiac trunk.** Supplies the stomach, liver, gallbladder, spleen, and part of the duodenum.

        ii. **Superior mesenteric arteries.** Courses within the mesentery and supplies the small intestines (part of duodenum, jejunum, and ileum) and part of the large intestines (cecum, appendix, ascending colon, and transverse colon).

        iii. **Renal arteries.** Supplies the kidneys and adrenal glands (adrenal glands receive arterial supply from three different arteries).

        iv. **Gonadal arteries.** Supplies the ovaries and veins.

        v. **Inferior mesenteric arteries.** Supplies the transverse colon, descending colon, sigmoid colon, rectum, and anus.

    d. **Pelvis and perineum.** The primary arterial supply to the pelvis and perineum is the internal iliac artery. In addition, it gives rise to branches that follow nerves into the gluteal region.

    e. **Lower limb.** The external iliac artery becomes the femoral artery after passing below the inguinal ligament. The femoral artery continues to the popliteal fossa (posterior knee), where it becomes the popliteal artery. Shortly thereafter, the popliteal artery divides into its two main branches: the posterior tibial artery and anterior tibial artery, which continues as the dorsalis pedis artery.

2. **Capillaries.** The systemic arteries arise from the aorta or branches of the aorta and transport to smaller vessels called arterioles. Arterioles transport blood into capillaries where diffusion and gas exchange occur. Capillaries give rise to venules, which exit the tissues and converge to form large vessels called veins. The veins return blood back to the heart.

3. **Veins.** Veins are vessels that transport blood toward the heart. Compared to their arterial counterparts, veins have much thinner walls of smooth muscle that range between 0.5 mm to 5 cm in diameter. Principal veins of the body include the following:
   a. **Head.** The head and neck are drained primarily by the internal and external jugular veins. The much smaller external jugular vein drains the face and the scalp, and the larger internal jugular vein, which travels in a sheath with the common carotid artery, drains the brain. The venous blood from the brain does not simply drain into one vein and exit the head. Instead, it drains into spaces between the two layers of the dura mater called the dural venous sinuses. Blood from the brain capillaries drains into the dural venous sinuses, which eventually drain into the internal jugular vein.
   b. **Upper limb.** Venous blood is drained from the upper limb through superficial and deep channels. Superficial veins (cephalic, basilic, and median cubital) drain cutaneous structures and are usually singular blood channels. Most deep veins (radial, ulnar, brachial, and axillary) course along their accompanying artery with the same name. Deep veins return blood to the heart via contraction of surrounding muscles.
   c. **Abdomen.** Veins draining blood from the organs of the abdomen are primarily named in parallel to the arteries that serve the same organs. For example, the renal veins drain the kidneys, the splenic vein drains the spleen, the gastric veins drain the stomach, the superior mesenteric vein drains the small intestines, and much of the large and the inferior mesenteric vein drains the rest of the large intestines. The veins draining the GI tract are unique compared to other system veins and are part of the hepatic portal system.
      i. **Hepatic portal system.** Even though the renal vein empties into the inferior vena cava, the blood from the GI tract does not drain into the inferior vena cava directly. Instead, veins from the GI tract and spleen drain into a common vein called the **hepatic portal vein**. Here, the nutrient rich blood from the GI tract filters through the hepatic sinusoids of the liver, where blood is processed, nutrients are removed, and toxins are broken down. In this way, everything we ingest, with the exception of lipids, filters through the liver prior to entering the systemic circulation. Once the blood has filtered through the hepatic portal system, it exits via hepatic veins and drains into the inferior vena cava, then into the right atrium of the heart.
   d. **Pelvis and perineum.** The primary venous drainage of the pelvis and perineum is the internal iliac vein.
   e. **Lower limb.** The deep structures of the lower limb are drained by the anterior and posterior tibial veins, which unite in the popliteal fossa to form the popliteal vein. In the distal thigh, the popliteal vein becomes the femoral vein, which becomes the external iliac vein after it passes under the inguinal ligament. The external iliac vein merges with the internal iliac vein, which drains pelvic structures, forming the common iliac vein. The two common iliac veins unite to form the inferior vena cava near the superior part of the pelvis. The largest superficial vein of the lower limb is the great saphenous vein, which drains the medial leg and thigh and empties into the femoral vein.

Far Out!

If all of the arteries, arterioles, capillaries, venules, and veins were laid end to end, they would extend close to 100,000 kilometers (enough to circle the earth more than two times).

1 Using your textbook or online resources, write a definition for each of the terms in Table 14.5.

TABLE **14.5** Arteries

Term	Definition
Aortic arch	
Brachiocephalic artery	
Subclavian artery	
Common carotid artery	
External carotid artery	
Internal carotid artery	
Vertebral artery	
Abdominal aorta	
Celiac trunk	
Renal artery	
Gonadal artery	
Superior mesenteric artery	
Inferior mesenteric artery	
Common iliac artery	
External iliac artery	
Internal iliac artery	
Femoral artery	
Popliteal artery	
Anterior tibial artery	

2 Using your textbook or online resources, write a definition for each of the terms in Table 14.6.

TABLE 14.6 Veins

Term	Definition
Superior vena cava	
Internal jugular vein	
Inferior vena cava	
Hepatic portal vein	
Splenic vein	
Superior mesenteric vein	
Inferior mesenteric vein	
Renal vein	
Left gonadal vein	
Right gonadal vein	
Common iliac vein	
External iliac vein	
Internal iliac vein	
Femoral vein	
Popliteal vein	
Great saphenous vein	

3 Identify, label, and color in the following vessels of the upper limb on Figures 14.5 and 14.6.

- ☐ Aortic arch
- ☐ Brachiocephalic artery
- ☐ Subclavian artery
- ☐ Axillary artery
- ☐ Brachial artery
- ☐ Radial artery
- ☐ Ulnar artery
- ☐ Palmar arches

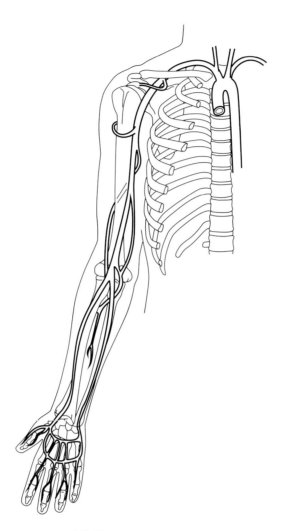

FIGURE **14.5** Arteries of the upper limb.

- Superior vena cava
- Brachiocephalic vein
- Subclavian vein
- Axillary vein
- Brachial vein
- Radial vein
- Ulnar vein
- Cephalic vein
- Basilic vein
- Median cubital vein
- Palmar venous arches

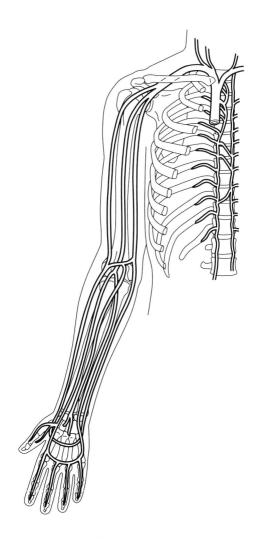

FIGURE 14.6 Veins of the upper limb.

4 Identify, label, and color in the following vessels of the abdomen on Figures 14.7 and 14.8.
- ☐ Abdominal aorta
- ☐ Celiac trunk
- ☐ Renal arteries
- ☐ Superior mesenteric artery
- ☐ Inferior mesenteric artery
- ☐ Common iliac arteries
- ☐ Internal iliac arteries
- ☐ External iliac arteries

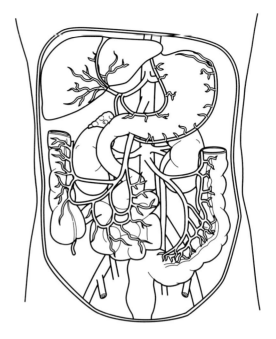

FIGURE **14.7** Arteries of the abdomen.

- ☐ Inferior vena cava
- ☐ Renal veins
- ☐ Suprarenal veins
- ☐ Common iliac veins
- ☐ External iliac veins
- ☐ Internal iliac veins
- ☐ Hepatic portal vein
- ☐ Gastric veins
- ☐ Superior mesenteric vein
- ☐ Inferior mesenteric vein
- ☐ Splenic vein

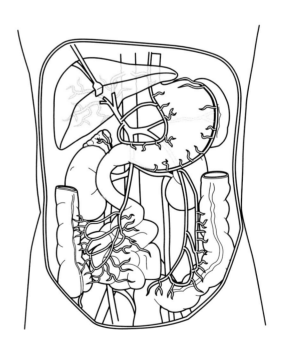

FIGURE **14.8** Veins of the abdomen.

5 Identify, label, and color in the following vessels of the lower limb on Figures 14.9 and 14.10.
- External iliac artery
- Femoral artery
- Popliteal artery
- Anterior tibial artery
- Posterior tibial artery
- Dorsalis pedis artery

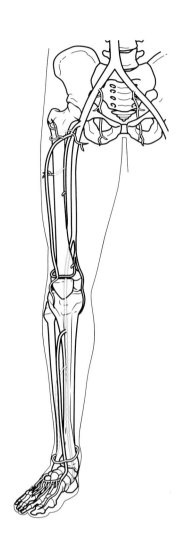

FIGURE **14.9** Arteries of the lower limb.

- ☐ External iliac vein
- ☐ Femoral vein
- ☐ Popliteal vein
- ☐ Anterior tibial vein
- ☐ Posterior tibial vein
- ☐ Great saphenous vein

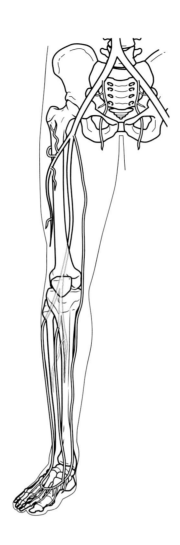

FIGURE **14.10** Veins of the lower limb.

OBSERVING

Complete the following hands-on laboratory activities to apply your knowledge of the cardiovascular system.

ACTIVITY 1 Cardiovascular System Anatomy

1 Using Figures 14.11 through 14.16, identify the following structures of the heart on the cadaver.

MATERIALS

Obtain the following items before beginning the laboratory activities:

- Cadaver
- Gloves
- Probe
- Protective gear (lab coat, scrubs, or apron)

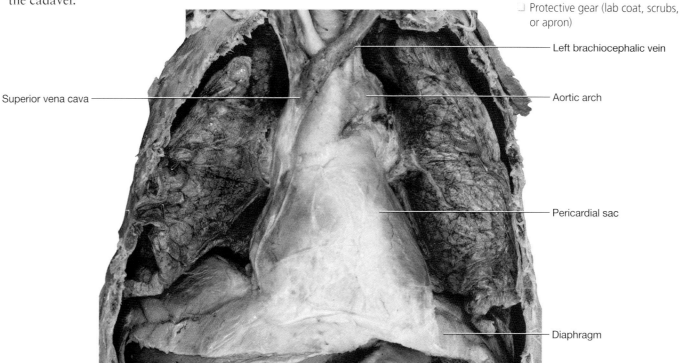

FIGURE **14.11** Pericardial sac and heart in situ with the sac closed.

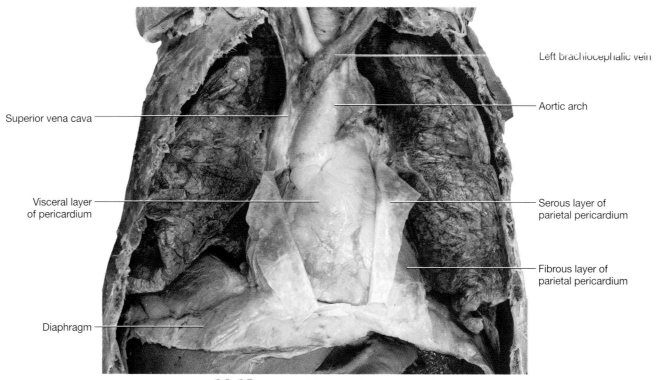

FIGURE **14.12** Pericardial sac and heart in situ with the sac open.

UNIT 4: Body Highways

CHAPTER 14: Cardiovascular System ■ **275**

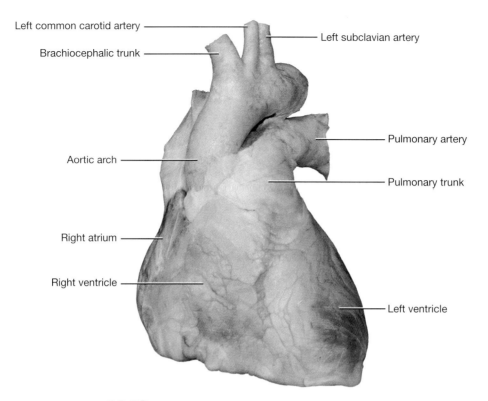

FIGURE **14.13** Anterior view of the superficial structures of the heart.

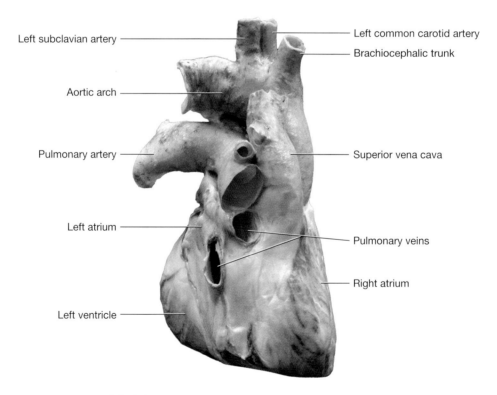

FIGURE **14.14** Posterior view of the superficial structures of the heart.

276 ■ *Discovering Anatomy: A Guided Examination of the Cadaver*

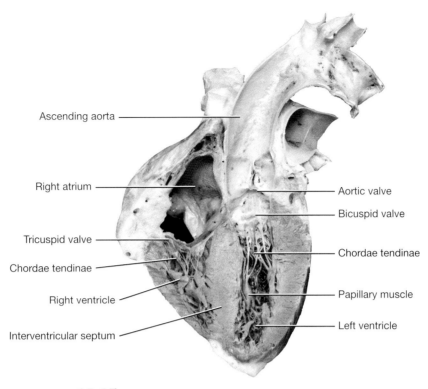

FIGURE **14.15** Coronal section of the internal chambers of the heart.

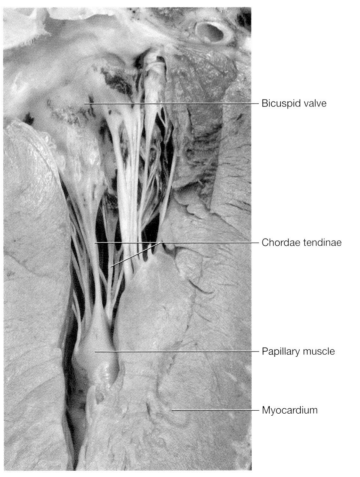

FIGURE **14.16** Close-up of the AV valve.

ACTIVITY 2 Blood Flow Tracing

1 Answer the following questions about blood flow through the heart and body.

Arteries and Veins:

a. Compare and contrast the wall of a vein with that of an artery (composition, thickness of wall, presence of valves).

b. Compare and contrast the direction of blood flow between arteries and veins.

c. Is arterial blood generally oxygenated or deoxygenated? Venous blood? Are there exceptions?

Heart Chambers:

a. Where does the right atrium pump blood once it contracts?

b. Where does the left atrium pump blood once it contracts?

c. Where does the right ventricle pump blood once it contracts?

d. Where does the left ventricle pump blood once it contracts?

e. The heart valves prevent the retrograde movement of blood flow during the cardiac cycle. Describe how each of the following heart valves prevents backflow of blood:
   i. Tricuspid valve

   ii. Bicuspid valve

   iii. Pulmonary valve

   iv. Aortic valve

In Question 2, you will trace the blood flow through various places in the body. As you trace, keep the following hints in mind:

- Arteries, capillaries, veins, and the heart make up the circulatory system. In other words, if you start in an artery, you must go through the arterial system, followed by the capillary, followed by the venous system, followed by the right side of the heart, followed by the pulmonary circulation, followed by the left side of the heart, and back again into arteries.
- Sometimes there is more than one correct route.
- The hepatic portal system drains blood from the GI tract.
- Venous blood from the brain drains into dural venous sinuses and then into the internal jugular vein.

Example:

Start: left radial vein. Stop: right radial artery

Route: left radial vein → left brachial vein → left axillary vein → left subclavian vein → left brachiocephalic vein → superior vena cava → right atrium → tricuspid valve → right ventricle → pulmonary arteries → lungs → pulmonary veins → left atrium → bicuspid valve → left ventricle → aortic valve → aorta → brachiocephalic artery → right subclavian artery → right axillary artery → right brachial artery → right radial artery.

2 Trace blood flow through the following circuits:

a. Start: left great saphenous vein. Stop: right common carotid artery.
   Route:

b. Start: left side of the heart. Stop: right dorsalis pedis pulse.
   Route:

c. Start: jejunum (small intestine). Stop: frontal lobe of cerebrum.
   Route:

d. Start: descending colon (large intestine). Stop: appendix.
   Route:

e. Start: pituitary gland. Stop: thyroid gland.
   Route:

WRAPPING UP
Complete the following additional activities to help retain your knowledge of the cardiovascular system.

Name _____

Date _____ Section _____

1. Identify the following vessels in Figure 14.17.

    Abdominal aorta

    Left renal artery

    Right renal artery

    Hepatic artery proper

    Splenic artery

    Left common iliac artery

    Right common iliac artery

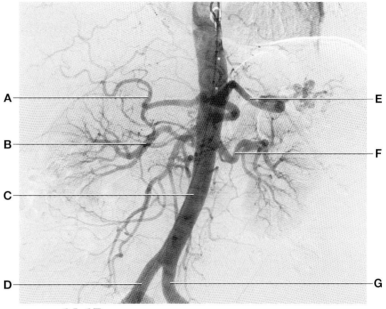

FIGURE **14.17** Angiogram of the abdominal aorta and associated branches, anterior view.

2. Figure 14.18 shows an angiogram of the aortic arch. Using your knowledge of the primary branches arising from the aortic arch, record what anomaly you observe in this patient.

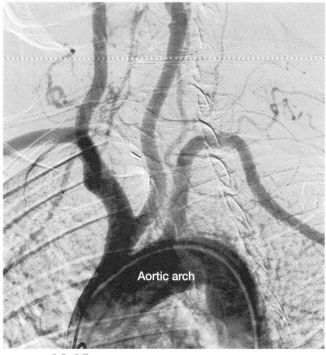

FIGURE **14.18** Angiogram of the aortic arch, anterior view.

3. A 56-year-old man with an insufficient aortic valve is scheduled for surgery. Identify the chamber of the heart that the regurgitating blood will most likely enter. Will the blood regurgitate primarily during systole or diastole? Explain why.

4. Cardiac tamponade is a critical clinical condition where the pericardial sac fills with blood. Hypothesize why this condition leads to decrease in stroke volume (stroke volume is the amount of blood pumped from the left ventricle with each heartbeat). Why would this be so serious?

5. An angiography of a 62-year-old man was ordered and the subsequent image demonstrated stenosis of his left anterior descending artery (LAD). Trace the route the catheter would most likely take, starting from his femoral artery and finishing with the injection of a contrast into his LAD.

6. The interatrial septum is the tissue that separates the right and left atria. A congenital defect in this septum, known as an atrial septal defect (ASD), enables blood to shunt from the left atrium to the right atrium. When a patient presents with a large ASD, there is a significant left-to-right atrial shunt of blood. Hypothesize how a large ASD could result in pulmonary arterial hypertension (high blood pressure in the pulmonary arteries).

7. A congenital heart defect causes one of the tricuspid valve leaflets to form lower than the other leaflets. This misalignment of valve leaflets decreases the efficiency of the heart. Hypothesize how this would affect blood flow during diastole and systole.

WRAPPING UP
(Continued)

Name _____

Date _____ Section _____

8 A serious heart murmur results when the aortic valve becomes calcified and is not able to open the way it should. Would this be considered a systolic murmur or diastolic murmur? Explain why.

9 A deep vein thrombosis (DVT) can form in the veins of the lower limb if no movement occurs over a long period of time (such as during a long plane ride). If left atrial fibrillation occurs, the heart does not pump blood out of the left atrium as it should and a clot can form. Explain why a DVT would most likely cause a pulmonary embolism and a blood clot in the left atrium would most likely cause a stroke.

10 Imaging reveals a large blood clot lodged in the superior mesenteric artery. Identify the organs most likely affected as a result of this clot.

11 The first-pass effect is a phenomenon of drug metabolism whereby the concentration of a drug is greatly reduced before it reaches the systemic circulation. It is the fraction of lost drug during the process of absorption, which is generally related to the function of the gut wall and liver. Contrast how drugs digested orally and drugs administered sublingually differ in their concentration when they enter the bloodstream.

CHAPTER 15
Lymphatic System

At the completion of this laboratory session, you should be able to do the following:
1. Describe the functions of the lymphatic system.
2. Identify key lymphatic organs, lymph nodes, lymphatic vessels, and the location of drainage.

When listing the systems of the body, it seems that the lymphatic system is rarely at the top of the list. However, the lymphatic system provides essential and supportive roles to the cardiovascular, digestive, and immune systems of the body. In the cardiovascular system, lymphatics maintain the homeostasis of extracellular fluid. In the digestive system, they absorb fats from the small intestines and transport them to the circulatory system. For the immune system, the different parts of the lymphatic system (lymph nodes, thymus, spleen, tonsils, and mucosa-associated lymphoid tissue, or MALT, help to combat pathogens.

GETTING ACQUAINTED

Complete the following activities to become familiar with the lymphatic system.

The lymphatic system consists of a diverse group of organs with the following functions:

1. **Collects excess interstitial fluid.** Blood circulates throughout the body from the heart to the capillaries and back to the heart. Gases, nutrients, and wastes are exchanged between the blood and interstitial fluid at the capillaries. This exchange occurs because the hydrostatic and osmotic pressures force fluid out of the blood at the proximal (arterial) ends of the capillary beds and cause most of the fluid to be reabsorbed at the distal (venous) ends. Some fluid that exited the capillaries remains behind in the tissue spaces (~3 L/day), essentially becoming interstitial fluid. As a result, excess fluid and escaped plasma proteins continually accumulate in the interstitial space. The lymphatic system solves this expanding tissue problem, known as edema, by collecting the excess interstitial fluid and returning it to the bloodstream. This transport system begins in blind-ended lymphatic capillaries, which are located in the interstitial space between cells and blood capillaries. Lymphatic capillaries have specialized one-way swinging doors allowing excess interstitial fluid to enter the capillary when tissue pressures increase but close to ensure fluid does not flow back into the interstitial space. Once interstitial fluid enters the lymphatics, it is called lymph. The lymphatic vessels form a one-way system in which lymph flows only toward the heart. Lymph travels along a series of lymphatic vessels, into trunks and eventually drains into two large lymphatic ducts:
    a. **Right lymphatic duct.** Drains lymph from the right arm, and right side of the head, neck, and thorax into the right subclavian vein at the junction of the right internal jugular vein.
    b. **Thoracic duct.** Drains lymph from the rest of the body into the left subclavian vein at the junction of the left internal jugular vein.
2. **Absorbs fats from intestines.** The fats ingested through our diet do not enter directly into the bloodstream. They are first absorbed into lymphatic capillaries known as central lacteals. The emulsified fats and free fatty acids absorbed into the lymphatic system mix with the clear lymph and become a milky white fluid called chyle. The chyle continues on

MATERIALS
Obtain the following items before beginning the laboratory activities:
- Textbook or access to Internet resources
- Colored pencils

into small lymphatic vessels, which merge with larger lymphatic vessels into the cisterna chyli. Lymphatic vessels ultimately drain into the thoracic duct, which empties the lymph that is rich in fat into the junction of the left internal jugular vein and subclavian vein.

3. **Activates the immune system.** A primary function of the lymphatic system is to activate the immune system. This occurs through lymphatic organs such as lymph nodes, thymus, spleen, tonsils, and the specialized tissue called MALT.

    a. **Lymph nodes.** When systemic tissues are inflamed, the openings in lymphatic capillaries may permit the uptake of large proteins, cell debris, pathogens (e.g., bacteria and viruses), and cancer cells. As a result, pathogens and cancer cells could use the lymphatics to travel throughout the body to infect other areas. However, this problem is partly resolved by the fact that the lymph within lymphatic vessels must percolate through lymph nodes. Lymph nodes are composed of lymphatic tissue surrounded by a dense connective tissue encasing and beaded along lymphatic vessels like pearls on a necklace. As lymph in the lymphatic vessels percolates through the lymph nodes, it removes toxins, cancer cells, pathogens, and cells infected with pathogens.

    b. **Thymus.** Located anterior to the heart and posterior to the sternum in the thorax. The thymus helps to form elements of the immune system in the early years of life. The thymus functions strictly in the maturation of T-lymphocytes, which play a central role in cell-mediated immunity. As such, the thymus is the only lymphatic tissue that does not directly fight pathogens with antigens. In fact, there is a "blood-thymus barrier" that stops antigens in the blood from diffusing into the thymus in order to prevent premature activation of im-mature lymphocytes. The thymus is most active during childhood and begins to atrophy during adolescence. By adulthood the thymus is no longer required and is replaced by connective tissue.

    c. **Thymic (Hassals) corpuscles.** Small granular cells within the thymic medulla with concentric layers of modified epithelial cells forming layers around their centers.

Clinical Application

Edema

"Edema" is the term used to describe an excessive accumulation of watery fluid in the tissues of the body and usually manifests as swelling. Edema may occur for many different reasons. One reason edema occurs is a blockage of the lymphatic system resulting in excessive interstitial fluid accumulating in the tissues. Obstruction may occur from cancer, enlarged lymph nodes, destruction of lymph vessels, or infiltration of lymphatics by an infection. One example of lymphatic obstruction is elephantiasis. Elephantiasis is characterized by gross enlargement of an area of the body (e.g., lower limb or scrotum). The condition is spread through mosquito bites. When a mosquito that is carrying the worm parasite *Wuchereria bancrofti* bites a human, the larvae may be injected into the skin. The tiny larvae locate in the interstitial fluid and flow into the lymphatic capillaries where they become trapped in lymph nodes and block the flow of lymph. As such, the excess interstitial fluid collects in the tissues and causes the tissues to swell.

Clinical Application

Swollen Lymph Nodes

Excess interstitial fluid flows into lymph capillaries and on into larger lymphatic vessels. The lymph flows through lymph nodes, which trap bacteria, viruses, cancer, and other pathogens. Lymph nodes may be the size of the head of a pin or an olive. When lymph nodes trap an excessively large amount of pathogens (due to injury, infection, or tumor), the nodes may swell. Some people refer to these as "swollen glands," though they are not glands (they do not produce and secrete a product). Lymph arises from tissues and then flows downstream. As such, when a doctor detects a swollen lymph node during a physical exam, he/she will often look in other areas of the body for the problem. For example, swollen lymph nodes in the neck may indicate an infection in the oral cavity or on the back of the scalp.

d. **Spleen**. An organ located in the upper left quadrant of the abdomen where lymphocyte proliferation, as well as immune surveillance and response, occurs. However, another primary role of the spleen is that it filters blood. As blood flows through the splenic sinuses, aged and defective blood cells, platelets, viruses, bacteria, and toxins are removed from the blood. The spleen contains both white pulp and red pulp. **White pulp** describes the regions within the spleen that are composed primarily of lymphocytes suspended in reticular fibers and serve the immune functions of the spleen. **Red pulp** describes the regions composed of venous sinuses, splenic cords, reticular tissue, and abundant red blood cells and is concerned with the disposing of old blood cells and blood-borne pathogens.

e. **Tonsils**. Groups of unencapsulated lymphatic tissue that form a ring found at the entrance to the pharynx. The tonsils are named according to their location.

   i. **Pharyngeal (adenoid) tonsils**. Located in the posterior region of the nasopharynx.
   ii. **Palatine tonsils**. Located in the posterior region of the oropharynx.
   iii. **Lingual tonsils**. Located at the base of the tongue.

   Tonsils contain regions where the overlying epithelium invaginates into the interior and forms blind-ended structures called crypts. The crypts trap bacteria, and as the bacteria move through the mucosal epithelium into the lymphoid tissue, the immune cells destroy them.

f. **Mucosa-associated lymphatic tissue (MALT)**. MALT are clusters of unencapsulated lymphatic tissue scattered throughout the body within the walls of organs. The MALT tissues are prominent in the wall of the ileum (Peyer's patches), appendix, and respiratory tract.

Clinical Application
Tonsillectomy

Tonsils act as filters that trap germs that enter the mouth and nose. A tonsillectomy is the surgical removal of the tonsils often due to excessive swelling or growth caused by recurrent infections. Although this jargon is not absolute, the use of the term "tonsillectomy" often refers to the palatine tonsils, whereas an adenoidectomy refers to the removal of the pharyngeal tonsils. A tonsillectomy will be generally suggested if the patient has had five to seven episodes of tonsil enlargement in the preceding twelve months. Tonsillectomies have been practiced for thousands of years. Records indicate that tonsils were at one time removed using fingernails and hooks. This advanced to scissors, scalpels, forceps, and wire loops. Contemporary surgeons now adopt the use of electrical or thermal energy as the means of separating the tonsils from surrounding tissue to reduce bleeding, minimize surrounding tissue damage, and improve recovery time.

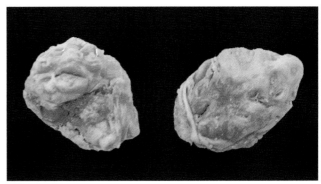

Palatine tonsils that have been removed in a tonsillectomy.

ACTIVITY 1 Lymphatic Structures and Organs

1 Using your textbook or online resources, write a definition for each of the terms in Table 15.1.

TABLE **15.1** Lymphatic Structures and Organs

Term	Definition
Lymphatic structures	
Interstitial space	
Systemic capillary	
Lymphatic capillary	
Plasma	
Interstitial fluid	
Lymph	
Lymphatic vessel	
Right lymphatic duct	
Thoracic duct	
Cisterna chyli	
Lymphatic organs	
Lymph node	
Thymus	
Spleen (red and white pulp)	
Tonsils	
MALT	

2 Identify, label, and color the following structures on Figure 15.1.
- Cervical lymph nodes
- Axillary lymph nodes
- Lymph nodes of the GI tract
- Inguinal lymph nodes
- Cubital lymph nodes
- Pharyngeal (adenoid) tonsils
- Palatine tonsil
- Lingual tonsil
- Spleen
- Thoracic duct
- Right lymphatic duct
- Cisterna chyli

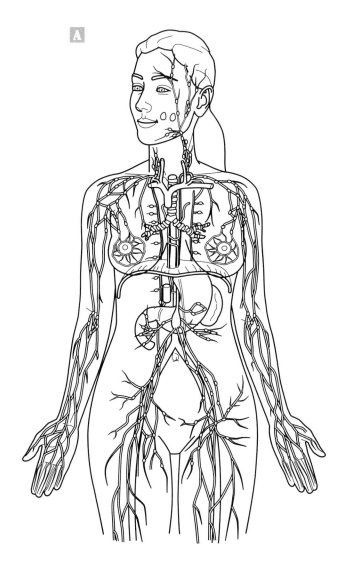

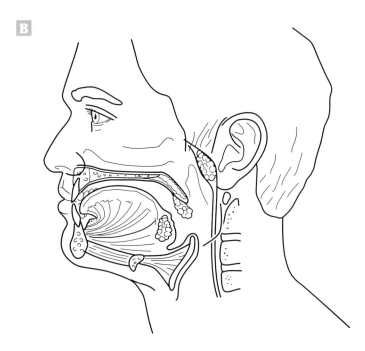

FIGURE **15.1** Lymphoid organs: (**A**) overview of the lymphatic organs and vessels; (**B**) tonsils.

3 Label the following structures in Figures 15.2 through 15.5.
- ☐ White pulp
- ☐ Red pulp
- ☐ Central artery
- ☐ Germinal center
- ☐ Splenic nodule

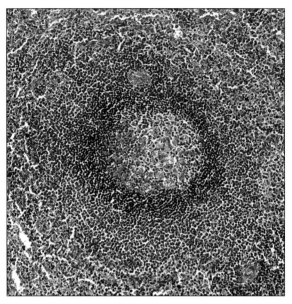

FIGURE **15.2** Spleen.

- ☐ Oral mucosa
- ☐ Lymphatic nodule
- ☐ Germinal center

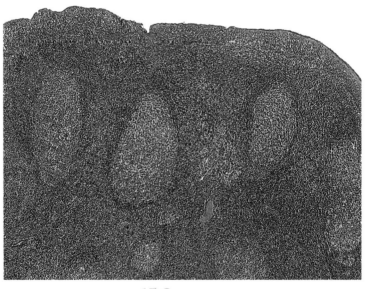

FIGURE **15.3** Palatine tonsil.

- ☐ Capsule
- ☐ Cortex
- ☐ Medulla
- ☐ Lymphatic nodule

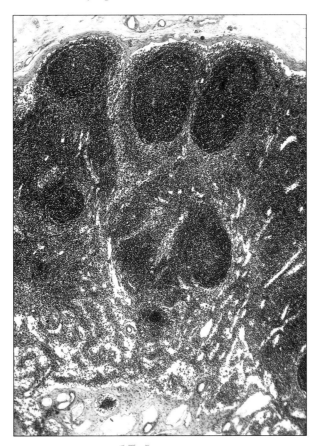

FIGURE 15.4 Lymph node.

- ☐ Mucosa
- ☐ Submucosa
- ☐ Peyer's patches
- ☐ Muscularis externa

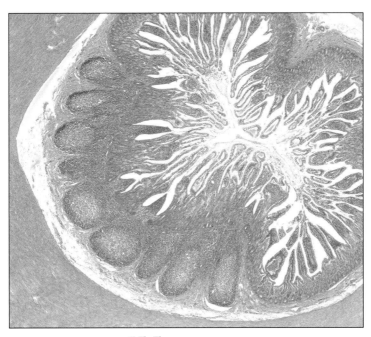

FIGURE 15.5 Ileum and Peyer's patches.

OBSERVING

Complete the following hands-on laboratory activities to apply your knowledge of the lymphatic system.

ACTIVITY 1 — Microscopic Structures of the Lymphatic System

During this laboratory session, you will use microscope slides, photos, cadavers, and your classmates to study the lymphatic system.

1. Obtain microscope slides of a lymph node, ileum, spleen, and tonsil from your instructor.

2. Place each slide under the microscope and start with the lowest power and advance to higher magnification to see more detail.

3. Once you have a view of the tissue, use your colored pencils and draw what you see in the field of view, color in the different cells and layers, and label the terms indicated.

MATERIALS

Obtain the following items before beginning the laboratory activities:

- Colored pencils
- Light microscope
- Prepared slides of the lymph node, ileum, spleen, and tonsil
- Cadaver
- Gloves
- Probe
- Protective gear (lab coat, scrubs, or apron)

a. Lymph node

Draw and label the following:
- Capsule
- Cortex of a lymph node
- Medulla of a lymph node
- Lymphatic nodule

b. Ileum

Draw and label the following:
- Mucosa (simple columnar epithelium)
- Submucosa
- Peyer's patch
- Muscularis externa

c. Spleen

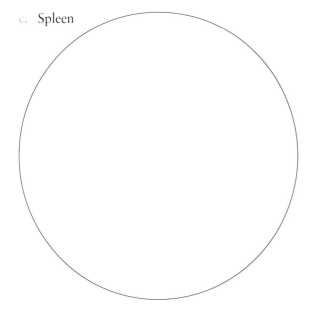

Draw and label the following:
- Central artery
- Splenic nodule
- Germinal center
- Red pulp
- White pulp

d. Tonsil

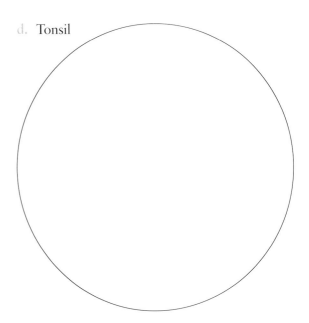

Draw and label the following:
- Oral mucosa
- Lymphatic nodule
- Germinal center
- Tonsillar crypt

ACTIVITY 2 Gross Anatomy of the Lymphatic System

1 Identify the following structures in Figure 15.6.

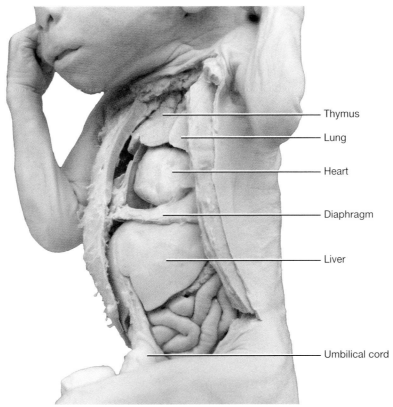

FIGURE **15.6**   Thymus within a fetus during the third trimester.

UNIT **4: Body Highways** CHAPTER **15:** Lymphatic System ■ **293**

2 Identify the following structures in Figure 15.7.

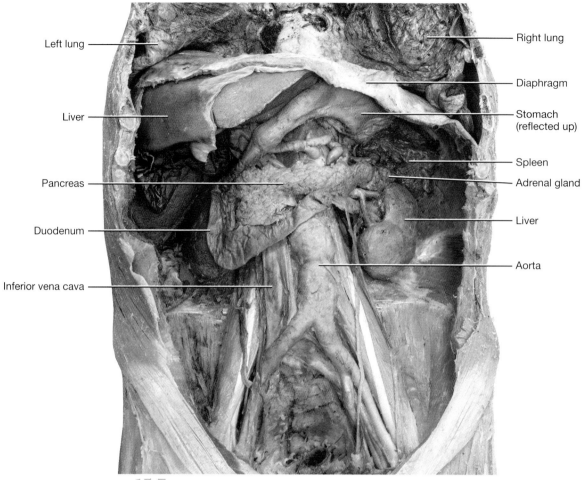

FIGURE **15.7** Spleen of the abdominal cavity with the overlying viscera removed.

3 Identify the following structures in Figure 15.8.

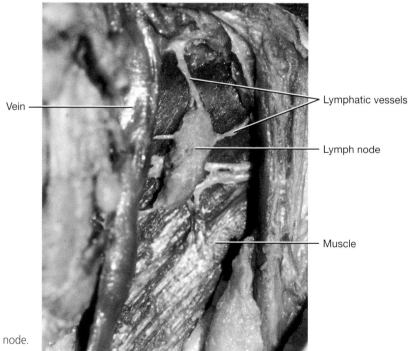

FIGURE **15.8** Lymph node.

WRAPPING UP

Complete the following additional activities to help retain your knowledge of the lymphatic system.

Name _____

Date _____ Section _____

In this activity you will trace the flow of lymph through various places in the body. As you trace, keep the following hints in mind:

- Remember, the blood exiting capillaries is transported to the heart by way of veins. Arteries transport blood away from the heart and to the capillaries. Thus, the cardiovascular system is a circulatory system.
- The lymphatic system is not a circulatory system. Lymphatic vessels arise in the interstitial space and transport lymph to the heart.
- For example, trace lymph from the right hand:
  - Right hand lymphatic capillaries and lymphatic vessels → right axillary lymph nodes → right lymphatic duct → junction of the right internal jugular and subclavian veins.

1. Trace lymph from the following body locations to the heart:

   a. Left hand
      Route:

   b. Left ear
      Route:

   c. Right ear
      Route:

   d. Left lung
      Route:

   e. Right lung
      Route:

   f. Left foot
      Route:

UNIT 4: Body Highways   CHAPTER 15: Lymphatic System ■ 295

g. Right foot
Route:

h. Small intestines
Route:

2. Compare and contrast tonsils, lymph nodes, and MALT.

3. A possible symptom of a foot infection is swollen inguinal lymph nodes. Why might an infection in the foot cause lymph nodes in the inguinal region to swell?

4. A 48-year-old woman undergoes a radical mastectomy in which her left breast is removed as well as her surrounding lymphatic vessels and lymph nodes (including her axillary lymph nodes). Explain why lymphatic vessels and nodes were removed. Explain why she will likely have edema in her left upper limb.

5. A patient suffering from lymphoma presents with a pleural effusion. Examination demonstrates an accumulation of fluid in the pleural cavity is milky white and contains a high concentration of fat. Hypothesize what structure was most likely injured, resulting in this leak. Explain your answer.

CHAPTER

Respiratory System

16

At the completion of this laboratory session, you should be able to do the following:

1. Compare and contrast the conduction and respiratory regions of the respiratory system.
2. Identify and describe key respiratory organs (airways, alveoli, pleural membranes) and describe their functions.
3. Explain the mechanics of breathing.

The respiratory system is responsible for taking oxygen from the air and supplying it to our cells, as well as taking carbon dioxide from our bloodstream and transporting it to the outside of the body. The process is accomplished by a series of airways (tubes) that are classified as either conduction airways or respiratory airways.

GETTING ACQUAINTED

Complete the following activities to become familiar with the respiratory system.

MATERIALS
Obtain the following items before beginning the laboratory activities:

- Textbook or access to Internet resources
- Colored pencils

Activity 1 Conduction Airways

Conduction airways consist of the nasal and oral cavities, nasopharynx, larynx, trachea, bronchi, and terminal bronchioles. These structures do not directly participate in gas exchange; instead they "conduct" air in and out of the body (hence the name). During the process of conduction, the air is warmed, moistened, and cleaned in order to protect the delicate nature of the alveoli where gas exchange occurs. Our focus will be from the trachea down as follows:

1. **Trachea.** The trachea begins below the larynx and courses down to the T4–T5 vertebral level where it bifurcates at the **carina** into the left and right primary bronchi. It consists of the same mucosal layers as the GI tract (mucosa, submucsoa, muscularis externa, and adventitia), including 18 to 20 horizontally oriented "C"-shaped cartilaginous rings that open posteriorly. The cartilaginous rings prevent the trachea from collapsing while breathing. The trachea is anterior to the esophagus.

2. **Bronchial tree.** The trachea branches into the left and right primary (principal) bronchi, which supply the left and right lung, respectively. Each lung is organized around its system of airways called the bronchial tree where the primary bronchus is the trunk and the bronchi and bronchioles are the branches. In contrast to the "C"-shaped cartilage rings in the trachea, the bronchi have plates of cartilage.

   a. **Primary (principal) bronchi.** The right primary bronchus is shorter, wider, and more vertical than the left primary bronchus. It branches into superior, middle, and inferior secondary (lobar) bronchi, corresponding to superior, middle, and inferior lobes of the right lung, respectively. The left primary bronchus branches into superior and inferior secondary bronchi, corresponding to superior and inferior lobes of the left lung, respectively.

   b. **Secondary (lobar) bronchi.** From the primary bronchi branch the secondary bronchi, which supply each lobe of the lungs. There are three secondary bronchi corresponding to the three lobes of the right lung and two secondary bronchi corresponding to the two lobes of the left lung.

297

c. **Tertiary (segmental) bronchi.** Secondary bronchi divide into tertiary bronchi, which supply bronchopulmonary segments. A bronchopulmonary segment consists of a lobar (segmental or tertiary) bronchus, a corresponding branch of the pulmonary artery, and the supplied segment of the lung tissue, all surrounded by a connective tissue septum. A bronchopulmonary segment refers to the portion of the lung supplied by each segmental bronchus and segmental artery. The pulmonary veins lie between bronchopulmonary segments.

3. **Terminal bronchioles.** After about ten generations of branching, the conduction airways lose cartilage and glands and are referred to as bronchioles. The terminal bronchioles are the smallest and last portion of the conduction airways.

Clinical Application

COPD

Chronic bronchitis and emphysema are the most common conditions that cause chronic obstructive pulmonary disease (COPD). Chronic bronchitis describes a condition with inflammation of the airways and excessive secretion of mucus into their lumen. The swelling and excessive mucus obstruct the airways (hence COPD) and the body responds to this obstruction by coughing, sometimes violently, to clear the airways. Chronic bronchitis is defined as a cough that produces mucus and occurs for numerous months in a row. Emphysema is a chronic, debilitating disease characterized by destruction of the wall of alveolar ducts and sacs, as well as individual alveoli. The destruction of alveoli results in a permanent dilation of air spaces, with a reduction in surface area for gas exchange. Smoking is the leading cause of chronic bronchitis and emphysema.

1 Using your textbook or online resources, write a definition for each of the structures listed in Table 16.1.

TABLE 16.1 Structures of the Respiratory System

Term	Definition
Larynx	
Trachea	
Primary bronchi	
Secondary bronchi	
Tertiary bronchi	
Terminal bronchioles	
Lungs	
Lobes	

2 Identify, label, and color the following structures of the respiratory system in Figure 16.1.

- Nasal cavity
- Hard palate
- Oral cavity
- Choana
- Nasopharynx
- Oropharynx
- Laryngopharynx
- Larynx
- Trachea
- Carina
- Left primary bronchus
- Right primary bronchus
- Secondary bronchi
- Tertiary bronchi
- Diaphragm
- Parietal pleura
- Visceral pleura
- Pleural cavity

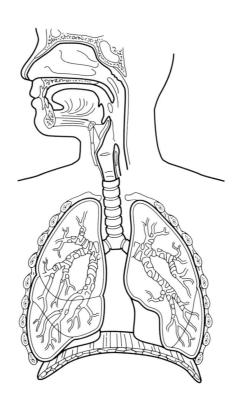

FIGURE 16.1 Gross anatomy of the respiratory system.

ACTIVITY 2 Respiratory Airways

The primary function of the respiratory system is gas exchange. The following structures assist in this gas exchange process:

1. **Respiratory bronchioles.** Each terminal bronchiole branches and gives rise to two to five respiratory bronchioles, which are the initial location for respiration (gas exchange) to occur. It is the presence of alveoli along the walls that determines the title "respiratory bronchiole."

2. **Alveolar ducts.** Respiratory bronchioles branch and create airways, called alveolar ducts, completely lined with alveoli. Alveolar ducts branch and terminate in clusters of alveoli called alveolar sacs.

3. **Alveoli.** Alveoli are saclike structures where gases are exchanged with the pulmonary capillaries. As such, they are the structural and functional units of the respiratory system. There are approximately 300 million alveoli packed into each lung, producing a total surface area of approximately 140 m^2 for gas exchange to occur. The alveoli consist of three different cells:

    a. **Type I pneumocytes.** The largest but least numerous cells. They line the alveoli and provide the thin area for gas exchange to occur. These cells share a basement membrane with the endothelial cells from the pulmonary capillaries.

    b. **Type II pneumocytes.** The smallest but most numerous cells lining the alveoli. They produce and secrete surfactant, which reduces the surface tension of the alveoli.

    c. **Alveolar (dust) cells.** The macrophage providing immune surveillance of the alveoli.

Clinical Application

Respiratory Distress Syndrome

One of the last tissues to mature during fetal development is type II pneumocytes, the cells responsible for the production of surfactant. When babies are born prematurely (before their type II pneumocytes have fully developed), they often have difficulty breathing because of the lack of surfactant. Without surfactant, the baby's tiny airways collapse with each exhalation. It is laborious and difficult for the airways to open back up. This condition is known as respiratory distress syndrome (RDS) and presents shortly after birth. Symptoms include rapid heart rate, nasal flaring, cyanosis (blue discoloration of skin), and labored breathing. Fortunately, symptoms for many RSD patients can be alleviated by the administration of glucocorticoids, which induce the synthesis of surfactant.

Controver-cell!

Club cells are dome-shaped cells found in bronchioles that have a variety of functions. They protect bronchiolar epithelium, reduce the viscosity of surface mucus, and detoxify inhaled air. Club cells were previously known as "Clara cells," after the German anatomist Max Clara (1899–1966). Controversy has followed Clara throughout history because of his close ties to the Nazi Party. The work that led to Clara's discovery of Clara cells was based on tissue samples obtained from prisoners executed by the Nazi Party. Eponyms (a person after whom a scientific discovery is made) are meant to honor the work conducted by individual scientists. Therefore, numerous scientific societies determined that using the eponym "Clara cells" was the equivalent of honoring him and the name was formally changed to "club cells" in 2010.

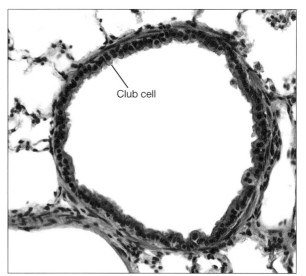

Photomicrograph of bronchiole from the lung showing club cells.

1. Using your textbook or online resources, write a definition for each of the terms listed in Table 16.2.

TABLE 16.2 Structures of the Respiratory System

Term	Definition
Respiratory bronchioles	
Alveolar ducts	
Alveoli	
Type I pneumocytes	
Type II pneumocytes	
Dust cells	

2. Identify, label, and color the following microscopic structures in Figure 16.2.
- Bronchiole
- Smooth muscle
- Elastic fibers
- Terminal bronchiole
- Respiratory bronchiole
- Alveolar duct
- Alveolar sac
- Alveoli
- Visceral pleura
- Lymphatic vessel
- Pulmonary venule
- Pulmonary arteriole
- Pulmonary capillary

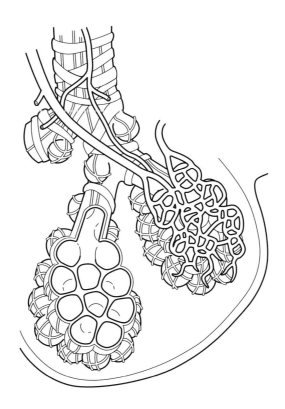

FIGURE 16.2 Microanatomy of the respiratory system.

3 Label Figures 16.3 through 16.6 using the terms provided.

- ☐ Respiratory epithelium
- ☐ Goblet cells
- ☐ Basement membrane
- ☐ Seromucous gland and duct
- ☐ Hyaline cartilage
- ☐ Perichondrium

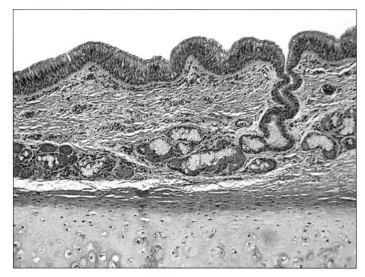

FIGURE **16.3** Tracheal wall.

- ☐ Pulmonary arteriole
- ☐ Smooth muscle
- ☐ Bronchiole
- ☐ Alveoli

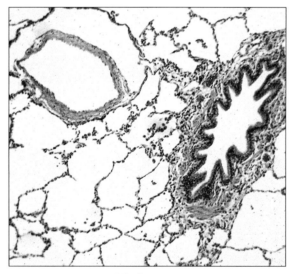

FIGURE **16.4** Bronchiole.

- Terminal bronchiole
- Respiratory bronchiole
- Alveolar duct
- Alveoli

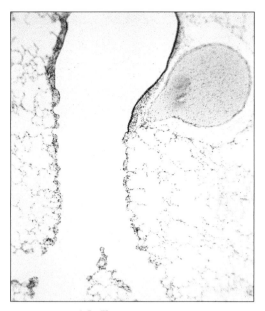

FIGURE 16.5 Terminal bronchiole.

- Alveolus
- Pulmonary capillary
- Dust cell
- Type I pneumocyte
- Type II pneumocyte

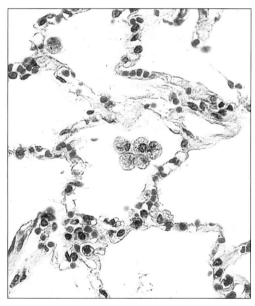

FIGURE 16.6 Pulmonary alveoli.

ACTIVITY 3 Serous Membranes

Pleural Sac

Each lung is enveloped in a serous membrane lining the lung and internal surface of the thoracic cavity pleural sac. The pleural sacs occupy a large portion of the thoracic cavity and are positioned on either side of the heart. Each pleural sac contains an outer parietal layer and an inner visceral layer, which join together at the hilum as the visceral layer reflects from the lung onto the thoracic wall as the parietal layer.

1. **Parietal pleura.** The parietal pleura is the external layer of the pleural sac that lines the internal surface of the thoracic wall, the superior surface of the diaphragm, and the lateral walls of the mediastinum. They are loosely bound to the endothoracic fascia by a thin layer of connective tissue.
2. **Visceral pleura.** The internal layer of the pleural sac that lines the lobes and fissures of the lungs is called the visceral pleura. They are contiguous with the parietal pleura at the site where bronchi, vessels, nerves, and lymphatics pass from the mediastinum into the root of the lung.
3. **Pleural space.** Each pleural cavity has a pleural space between the parietal and visceral pleura, which is filled with pleural fluid. There is no communication between the left and right pleural spaces. The serous fluid lubricates the lungs to decrease friction while breathing and form surface tension between the lung and thoracic wall. As such, the lungs adhere tightly to the internal surface of the rib cage.

Clinical Application

Pneumothorax

Imagine placing a drop of water between two plates of glass and then placing the plates of glass together. The water enables the glass plates to move across each other but also keeps them adhered to one another due to the hydrostatic pressure. The two glass plates are like parietal and visceral pleura. Small amounts of fluid enable the pleura to move along each other but also causes hydrostatic pressure to keep the lungs adhered to the internal surface of the rib cage. If air is introduced into the pleural space as a result of chest trauma, such as a knife wound, the coupling between the parietal pleura and visceral pleura may be broken, causing the lung to collapse. This pathology is called a **pneumothorax**. When blood fills the pleural space, the pathology is called a hemothorax.

Mechanics of Breathing

Pulmonary ventilation is the process of air entering and exiting the lungs due to contraction and relaxation of respiratory muscles. According to Boyle's law, volume and pressure are inversely related. As such, when muscles cause the volume of the lungs to increase, the pressure of the gases in the lungs decreases, allowing air to enter the lung. Conversely, when muscles cause the volume of the lungs to decrease, the pressure of the gases in the lungs increases, forcing air out of the lungs. Boyle's law is demonstrated when defining inhalation and exhalation:

1. **Inhalation (inspiration).** As the muscles of inhalation contract, the volume of the lungs increases in all dimensions. The increased volume causes the gas pressure to decrease. When atmospheric pressure is greater than the pressure within the lungs, air rushes in.
    a. **Diaphragm.** The diaphragm contracts and flattens, thus increasing the size of the thoracic cavity and, therefore, the size of the lungs as well. This results in air being sucked from the atmosphere into the lungs. During quiet breathing, inhalation is accomplished by the diaphragm.
    b. **External intercostal muscles.** Contraction results in expansion of the rib cage and therefore assists in inhalation.
2. **Exhalation (expiration).** During quiet or passive breathing, exhalation occurs as the diaphragm relaxes. As the muscles of exhalation contract, the volume of the lungs decreases. The decreased volume causes the gas pressure to increase. When atmospheric pressure is lower than the pressure in the lungs, air rushes out.
    a. **Diaphragm.** The diaphragm relaxes and becomes dome-shaped, thus decreasing the size of the thoracic cavity and therefore the lungs as well. This results in air being pushed out of the lungs into the atmosphere.
    b. **Internal intercostal muscles.** Contraction results in constriction of the rib cage and therefore assists in exhalation.

1. Using your textbook or online resources, write a definition for each of the following anatomical structures in Table 16.3.

TABLE 16.3 Serous Membranes

Term	Definition
Parietal pleura	
Visceral pleura	
Pleural space	
Pleural fluid	
Diaphragm	
Phrenic nerve	
Boyle's law	
Inspiration	
Expiration	

OBSERVING

Complete the following hands-on laboratory activities to apply your knowledge of the respiratory system.

ACTIVITY 1 Microscopic Structures of the Respiratory System

1. Obtain microscope slides of the trachea, bronchiole, terminal bronchiole, and alveolar sac from your instructor.
2. Place each slide under the microscope and start with the lowest power and advance to higher magnification to see more detail.
3. Once you have a view, use your colored pencils and draw what you see in the field of view. Color in the different cells and layers and label the terms indicated.

MATERIALS

Obtain the following items before beginning the laboratory activities:

- Colored pencils
- Light microscope
- Prepared microscope slides of the trachea, bronchiole, terminal bronchiole, and alveolar sac
- Cadaver
- Gloves
- Probe
- Protective gear (lab coat, scrubs, or apron)

a. Trachea

b. Bronchiole

Draw and label the following:
- Respiratory epithelium
- Goblet cells
- Basement membrane
- Seromucous gland and duct
- Hyaline cartilage
- Perichondrium

Draw and label the following:
- Pulmonary arteriole
- Smooth muscle
- Bronchiole
- Alveoli

c. Terminal bronchiole

d. Alveolar sac

Draw and label the following:
- Terminal bronchiole
- Respiratory bronchiole
- Alveolar duct
- Alveoli

Draw and label the following:
- Alveolus
- Pulmonary capillary
- Type I pneumocyte
- Type II pneumocyte
- Dust cell

ACTIVITY 2 Gross Anatomy of the Respiratory System

1 Identify the following structures in Figure 16.7.

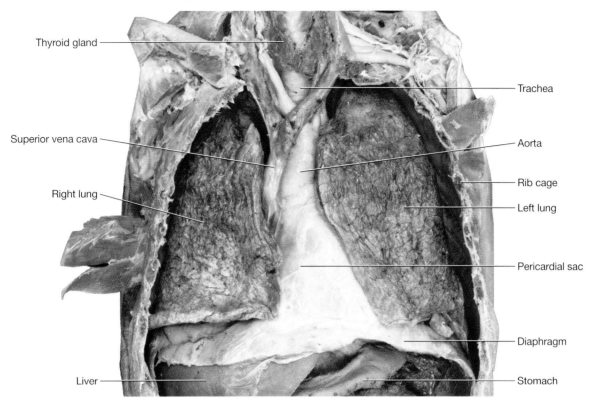

FIGURE 16.7 Lungs in situ with the anterior thoracic wall removed.

2 Identify the following structures in Figure 16.8.

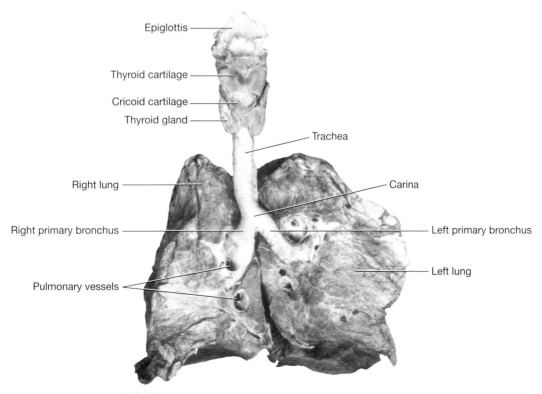

FIGURE 16.8 Anterior view of the larynx, trachea, bronchial tree, and lungs.

3 Identify the following structures in Figure 16.9.

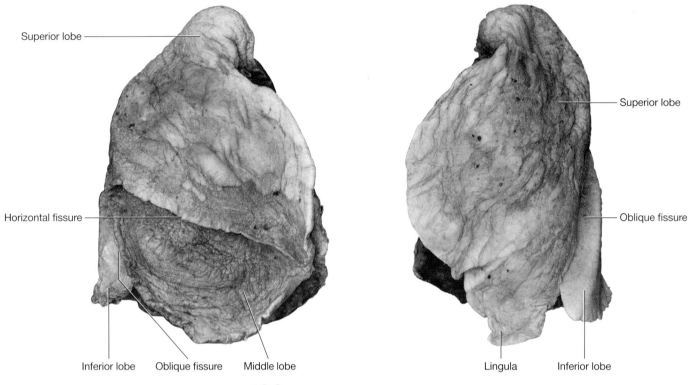

FIGURE **16.9** Anterior view of the right and left lungs.

4 Identify the following structures in Figure 16.10.

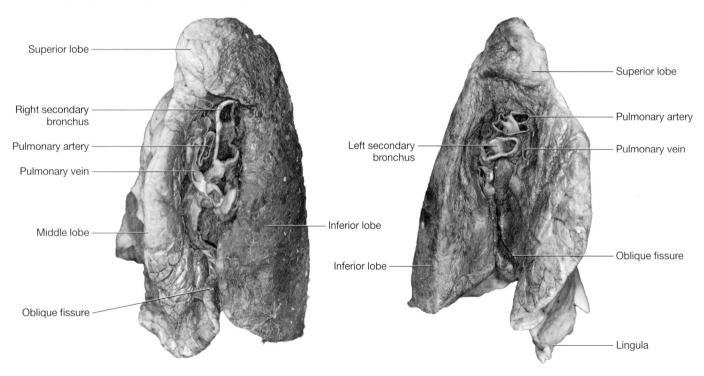

FIGURE **16.10** Medial view of the right and left lungs.

5 Identify the following structures in Figure 16.11.

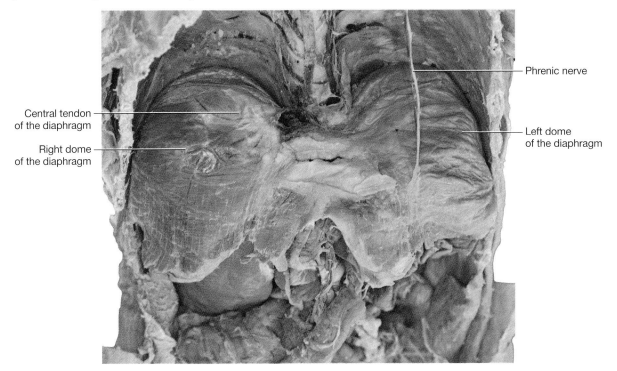

FIGURE 16.11 Anterior view of the diaphragm with the anterior trunk wall and lungs removed.

WRAPPING UP

Complete the following additional activities to help retain your knowledge of the respiratory system.

Name _____

Date _____ Section _____

1. Trace an inhaled molecule of oxygen from the larynx to the pulmonary capillaries.

2. Trace an inhaled molecule of oxygen from the pulmonary capillaries to the heart and then to the left rectus femoris muscle.

3. Muscle cells in the rectus femoris make ATP from the molecule of oxygen, and carbon dioxide is given off as a waste product. Trace a molecule of carbon dioxide from the rectus femoris muscle to the heart and lungs.

4. Trace the molecule of carbon dioxide from the lungs to the outside of the body.

5. The trachea is composed of horizontally oriented "C" shaped cartilaginous rings. What role does the hyaline cartilage skeleton play in the trachea? Hypothesize why the rings are "C"-shaped and not a complete circle.

6. Provide an explanation for why cartilage lines the bronchial tree. Why does it disappear for the bronchioles? What replaces its function in the alveolar sacs?

7 Pleurisy is a condition in which the pleural membranes (parietal pleura and visceral pleura) are inflamed. This may result in a reduction of pleural fluid production. Hypothesize what problems this may result in.

8 A 25-year-old man is injured in a motorcycle accident. A metal rod punctured through his rib cage into his thoracic cavity. Hypothesize why he presented with a pneumothorax.

9 A 64-year-old smoker presents with emphysema, a disease caused by the destruction of the walls between alveoli. Hypothesize the effect emphysema has on gas exchange.

10 Provide an explanation for why epithelium in the trachea is pseudostratified ciliated columnar epithelium. What purpose do goblet cells play? Why are there no goblet cells in the bronchioles?

11 Cigarette smoke destroys the cilia in respiratory epithelium. Hypothesize what effect this would have on the patient.

12 Provide an explanation for why epithelium in the alveoli is simple squamous epithelium.

WRAPPING UP
(Continued)

Name _____

Date _____ Section _____

13 Identify the gases most likely represented by the arrows going to and from "a" and "b" in Figure 16.12.

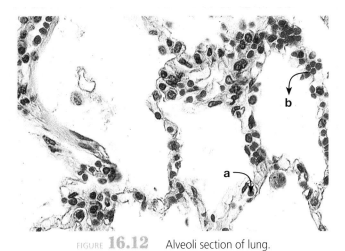

FIGURE **16.12** Alveoli section of lung.

a. _____

b. _____

Digestive System

CHAPTER 17

At the completion of this laboratory session, you should be able to do the following:

1. Describe the overall function of the digestive system.
2. Differentiate between organs of the gastrointestinal (GI) tract and accessory digestive organs.
3. List and describe the major processes occurring in each organ of the digestive system.
4. Describe the location and function of major peritoneal organs.
5. Outline the tissue layers of the GI tract.

The food we ingest contains carbohydrates, proteins, and lipids, which are the nutrients our body cells use for energy, building, and repair. Therefore, a mouthful of turkey sandwich contains these ingredients but they are too large for our microscopic cells to use. As such, we must break down the food into smaller pieces through the process called mechanical digestion. Mechanical digestion makes big pieces of food (turkey sandwich) into smaller pieces of food (very small pieces of turkey sandwich). Even when the food is mechanically broken down into smaller pieces, they are still too large for the body to absorb and utilize. As such, the very small pieces of that same turkey sandwich must be broken down into the building blocks of the macromolecules in a process called chemical digestion. This process of breaking down foods into smaller substances (carbohydrates into mono- and disaccharides, proteins into amino acids, and lipids into glycerol and free fatty acids) is what enables the digestive system to absorb the nutrients. In review, the digestive system mechanically and chemically digests food in order for the nutrients to be absorbed into the bloodstream (to be dispersed to all body cells) and to excrete indigestible substances.

GETTING ACQUAINTED

Complete the following activities to become familiar with the digestive system.

MATERIALS

Obtain the following items before beginning the laboratory activities:

- Textbook or access to Internet resources
- Colored pencils

ACTIVITY 1 Gastrointestinal Tract

The digestive system is composed of two classifications of organs: (1) organs of the gastrointestinal (GI) tract through which food travels (also known as the alimentary canal), and (2) accessory digestive organs, which are associated with the GI tract and take care of mechanical or chemical digestion.

The GI tract consists of the following organs:

1. **Oral cavity.** The oral cavity is the digestive system's entryway for food where it is mechanically broken down into smaller pieces by the teeth.
    a. **Teeth.** Used for biting or chewing. Humans have a variety of teeth that provide distinct functions: incisors (cutting), canines (holding food in order to tear it apart), and molars (grinding and crushing).
    b. **Palate.** The "ceiling" of the mouth. The hard palate separates the oral and nasal cavities. The soft palate rises during swallowing in order to close off the oral from the nasal cavity. Food is manipulated against these hard and soft surfaces and mechanically ground into smaller pieces.

c. **Tongue.** A muscular organ that manipulates food, senses its temperature and texture, and tastes flavors. Its external surface possesses **filiform papillae** that provide a coarse surface that helps to grip and manipulate food. The **circumvallate, foliate,** and **fungiform papillae** contain taste buds.

2. **Pharynx.** Once the bolus (chewed-up piece of food packed together by saliva) is swallowed, it enters the pharynx where pharyngeal constrictor muscles contract and transport the bolus into the esophagus. During the swallowing process, the pharynx, larynx, and epiglottis move to ensure food enters the esophagus and not the trachea.

3. **Esophagus.** The esophagus is a tubular organ that arises from the pharynx, vertically descends posterior to the trachea and heart, and traverses the diaphragm to join the stomach. The gastroesophageal sphincter (cardiac sphincter) restricts regurgitation of gastric contents back into the esophagus.

   a. **Peristalsis.** The esophagus contains both skeletal and smooth muscle fibers that contract rhythmically through a process called peristalsis, which propels the bolus along its length.

4. **Stomach.** The stomach is divided into the cardia (surrounds the gastroesophageal sphincter), fundus (superior dome-shaped region), body (middle portion), lesser curvature (smaller curve on top), greater curvature (larger curve on bottom), and pylorus (contains the pyloric sphincter, which prevents regurgitation of duodenal contents back into the stomach). The stomach mucosa possesses gastric glands with many secretions. Hydrochloric acid has the following properties:

   a. **Hydrochloric (HCl) acid.** Produced and secreted by parietal cells lining the stomach mucosa. HCl assists in the protection of the internal environment from external pathogens and in activating pepsinogen.

      i. **Mucus.** Secreted by numerous goblet cells to line the stomach, protecting it from the HCl.

      ii. **Protein digestion.** Chief cells in the gastric glands that produce and secrete the pre-enzyme pepsinogen. Pepsinogen becomes pepsin when it is activated by HCl. Pepsin initiates the chemical digestion of proteins into smaller peptides and amino acids.

      iii. **Gastrin.** Hormone secreted by G-cells of the stomach and duodenun. The vagus nerve (CN X) assists gastrin in this function, which increases pepsinogen and HCl secretions.

Clinical Application

GERD

A function of the stomach is the production of hydrochloric acid. If the cardiac sphincter fails to contain the acidic chyme produced by the stomach, the acid moves into the esophagus, irritating its mucosal lining and causing gastroesophageal reflux disease (GERD). The irritation presents as an uncomfortable, burning sensation in the region of the esophagus, deep to the heart. As a result, this condition is also referred to as "heartburn." The most common causes of GERD are smoking, obesity, medicines that relax smooth muscle, hiatal hernias, and pregnancy.

5. **Small intestine.** The small intestine (SI) is the section of the GI tract where chemical digestion and absorption primarily occurs. The mucosae of the SI possess plicae circulares (circular folds) to increase surface area for absorption. These circular folds possess villi and enterocytes possess microvilli to further increase surface area. Intestinal glands are formed by the epithelia lining the small intestine. The glands contain goblet cells (produce mucus), Paneth cells (innate immune system), and stem cells (replicating cells that give rise to goblet and Paneth cells). The SI has the following three distinct parts:

   a. **Duodenum.** The initial portion of the SI (first 10 inches) that makes a "C"-shaped curve around the head of the pancreas. The duodenum contains the hepatopancreatic ampulla (of Vater) where digestive secretions from the liver, gallbladder, and pancreas enter the GI tract.

   b. **Jejunum.** The middle portion of the SI (middle 8 feet) where the mucosal lining is specialized for absorption by enterocytes. Primarily located in the upper left quadrant of the abdominal cavity.

   c. **Ileum.** The terminal portion of the SI (last 12 feet) located in the lower right quadrant of the abdomen. There is no demarcation between the jejunum and ileum. However, the ileum generally has more mesenteric fat, possesses fewer circular folds, and contains Peyer's patches (MALT).

That's a Lot of Space!

The small intestine is approximately 7 meters long. Based upon this length, the surface area of the small intestine lumen should be about 0.6 square meters. However, when macroscopic and microscopic structural features are considered, the surface area is actually about 250 square meters, or the size of a tennis court. This occurs because circular folds increase the absorptive surface area of the luminal wall by 3-fold, villi increase it 8- to 10-fold, and microvilli by 20-fold. All of these structural features assist in one main function: increase the surface area for food absorption.

6. **Large intestine.** The large intestine (LI) is named for its large luminal diameter and not its length (it is only about 5 feet long). Much of the LI features bands of longitudinal smooth muscle called **taeniae coli** that have a drawstring effect on the LI, pulling it into pouches called **haustra**. **Epiploic appendages** are small pouches of peritoneum located along the LI that are filled with fat. The LI consists of the following five regions:

    a. **Cecum** (pouch). Located in the lower right quadrant of the abdomen and receives contents from the ileum. The cecum is separated from the ileum by the ileocecal valve and transports food into the ascending colon.

    b. **Vermiform appendix.** The narrow, blind-ended extension of the cecum. The vermiform appendix possesses the same histologic structure as the large intestine but with fewer glands and a high concentration of lymphoid follicles in the mucosa and submucosa.

    c. **Colon.** The colon consists of four sections: ascending colon, transverse colon, descending colon, and sigmoid colon. The colon reabsorbs water and electrolytes from the solid wastes before they are defecated.

    d. **Rectum.** The portion of the large intestine that courses vertically in front of the sacrum and serves as a temporary storage for feces. The walls of the rectum expand as it is filled with feces, causing stretch receptors to stimulate the desire to defecate.

    e. **Anus.** The terminal portion of the large intestine possessing the smooth muscle **internal anal sphincter** and skeletal muscle **external anal sphincter**. The internal anal sphincter is responsible for involuntary contraction or relaxation, while the external anal sphincter is responsible for voluntary contraction or relaxation.

Clinical Application

Celiac Disease

Celiac disease, also known as celiac sprue, is a chronic immune-mediated intolerance to gluten (a protein found in wheat, barley, and rye) that induces intestinal mucosal lesions in genetically predisposed individuals. This immune response causes damage to the microvilli of mucosal epithelium. These microscopic fingerlike projections promote nutrient absorption and, when damaged, result in malabsorption, weight loss, bloating, flatulence, diarrhea, fatigue, vitamin deficiencies, and anemia. Currently, there is no cure for celiac disease and those with this condition are instructed to follow a diet of food absent in gluten.

Wheat.

Clinical Application

Who Cut the Cheese?

The term for intestinal gas is *flatus*, which comes from the Latin "to blow." Flatus comes from two primary sources: gases swallowed during eating and breathing, and gases produced by the fermentation by bacteria in the GI tract. Foods such as beans, lentils, dairy, and cabbage are rich in substances, such as polysaccharides, that are not broken down by digestive enzymes from the pancreas. As such, these undigested food substances are pushed through the small intestine into the large intestine. At this point, bacteria in the colon get to work, releasing a variety of gases in the process, including carbon dioxide, hydrogen, methane, and hydrogen sulfide (which gives flatulence its rotten-egg smell). The majority of intestinal gas (~99%) is non-smelly. The other 1% gives flatus its reputation. The volume of flatus ranges from 450 mL to 1500 mL in a 24-hour period with a range of 8 to 20 flatus episodes per day.

Clinical Application

Appendicitis

When the appendix becomes inflamed, it results in a condition called appendicitis, which often causes pain in the lower right abdomen. However, in many people this pain begins as dull, achy pain around the umbilicus. How can the appendix, which is positioned in the lower right corner of the abdomen, cause the pain to be felt around the navel? There is a hypothesized reason for why this occurs. Visceral sensory neurons from the visceral peritoneum of the appendix travel to the T10 spinal cord level. Additionally, intercostal nerves (somatic sensory neurons) conduct sensation from skin around the umbilicus also into the T10 spinal cord level. Therefore, visceral sensory neurons from the inflamed appendix and somatic sensory neurons from the skin around the umbilicus enter the T10 spinal cord level and synapse in the same region. The brain has difficulty interpreting where the inflamed tissue is arising and thus often perceives the pain originating in the region around the navel. This phenomenon of referred pain occurs in a similar manner for heart attacks.

1 Using your textbook or online resources, write a definition for each of the structures in Table 17.1.

TABLE **17.1** Digestive System Organs

Term	Definition
Teeth	
Palate	
Tongue	
Taste buds	
Pharynx	
Esophagus	
Gastroesophageal sphincter	
Stomach	
Pyloric sphincter	
Peristalsis	
Duodenum	
Jejunum	
Ileum	
Cecum	
Vermiform appendix	
Colon	
Taeniae coli	
Haustra	
Rectum	
External anal sphincter	

2 Identify, label, and color the following structures of the digestive system in Figure 17.1.

- ☐ Oral cavity
- ☐ Hard palate
- ☐ Teeth
- ☐ Tongue
- ☐ Pharynx
- ☐ Parotid gland
- ☐ Submandibular gland
- ☐ Sublingual gland
- ☐ Esophagus
- ☐ Stomach
- ☐ Pancreas
- ☐ Liver
- ☐ Gallbladder
- ☐ Duodenum
- ☐ Jejunum
- ☐ Ileum
- ☐ Cecum
- ☐ Appendix
- ☐ Ascending colon
- ☐ Transverse colon
- ☐ Descending colon
- ☐ Sigmoid colon
- ☐ Rectum
- ☐ Taeniae coli
- ☐ Haustra

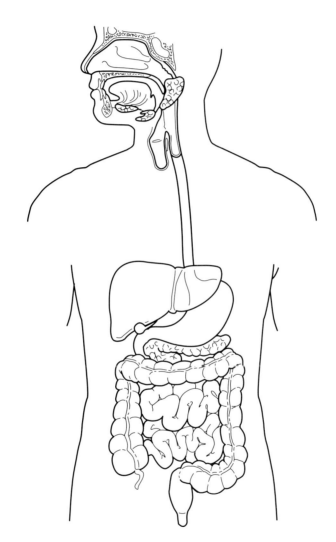

FIGURE 17.1 Gross anatomy of the digestive system.

3 Using your textbook or online resources, write a definition for each of the terms in Table 17.2.

TABLE **17.2** Concepts of the Digestive System

Term	Definition
Mechanical digestion	
Chemical digestion	

ACTIVITY 2 Accessory Digestive Organs

The accessory digestive organs consist of the following:

1. **Salivary glands.** Salivary glands produce saliva and transport the secretions into the oral cavity via ducts. Saliva contains water to moisten food, mucus to bind chewed-up (masticated) food into a slippery bolus, the enzyme salivary amylase to chemically digest carbohydrates, and lysozyme to prevent overgrowth of oral microbial populations. The oral cavity is constantly washed with saliva, which carries away food particles and keeps the mouth relatively clean. The three paired salivary glands include the following:
   a. **Submandibular salivary gland.** Located below the mandible and produces the majority of total salivary volume (60 to 70 percent).
   b. **Parotid gland.** Located superficial to the masseter and beside the ear. It is the largest salivary gland in size.
   c. **Sublingual salivary gland.** The smallest of the glands located under the tongue. It produces the least amount of total salivary volume (3 to 5 percent).
2. **Liver and gallbladder.** The liver is the largest gland in the body and consists of four lobes: right, left, caudate, and quadrate. Located in the upper right quadrant of the abdomen, it is encased in Glisson's capsule, a thin connective tissue membrane, which is further covered by visceral peritoneum. The liver is composed of plates of liver cells, called **hepatocytes,** which are interspersed between capillaries called **hepatic sinusoids.** Recall that blood draining the spleen and abdominal digestive organs percolates through the hepatic sinusoids where hepatocytes filter and process the blood before it enters the systemic circulation. The liver also produces bile, which exits the liver by way of the common hepatic duct and is stored and concentrated in the gallbladder. Bile emulsifies fat and assists in lipid digestion.
   a. **Hepatocytes.** Organized in a lobular fashion. At the center of each lobule is a **central vein,** which drains into hepatic veins and ultimately into the inferior vena cava. Portal triads contain the following three structures:
      i. **Bile duct.** Transports bile from hepatocytes to the hepatic ducts.
      ii. **Portal vein/venule.** A tributary off the hepatic portal vein that delivers nutrient-rich deoxygenated blood from the GI tract to the liver for processing and detoxification.
      iii. **Hepatic artery/arteriole.** Delivers oxygen-rich blood to the hepatocytes.

Clinical Application

Gallstones

As an individual ages, the probability for bile salts normally excreted in bile to precipitate and form gallstones increases. The formation of gallstones is twice as likely to occur in women than men. Gallstones become a problem when they are small enough to exit the gallbladder, but too large to fit through the bile ducts. When this occurs the flow of bile is obstructed, resulting in inflammation and enlargement of the gallbladder and often a great deal of pain and discomfort. The most common method of treatment is surgical removal of the gallbladder through a cholecystectomy procedure. Following this procedure, bile is delivered directly from the hepatocytes to the duodenum.

3. **Pancreas.** The pancreas is located deep to the stomach and functions as an exocrine gland (digestion) and endocrine gland (metabolism). The exocrine portion produces pancreatic secretions that are transported via the pancreatic duct into the duodenum via the hepatopancreatic ampulla. Pancreatic secretions contain bicarbonate to neutralize the acid produced by the stomach as well as the following digestive enzymes:
   a. **Pancreatic amylase.** Chemically digests carbohydrates into simple sugars to be absorbed by the small intestine into the hepatic portal system.
   b. **Pancreatic trypsin and chymotrypsin.** Chemically digests proteins into polypeptides to be absorbed by the SI into the hepatic portal system.
   c. **Pancreatic lipase.** Chemically digests fat into monoglycerides and free fatty acids to be absorbed by the small intestine into the lymphatic system.

Clinical Application

Cystic Fibrosis

Cystic fibrosis results in, among other things, thick and dehydrated pancreatic secretion, which blocks the pancreatic ducts. Consequently, secretion of the pancreatic digestive enzymes is impaired, which impedes intestinal absorption of nutrients. If left untreated, this will eventually result in stunted development and a shortened lifespan. One method of treatment is oral pancreatic enzyme supplements taken with each meal. However, another problem associated with cystic fibrosis and the pancreas is that the pancreas continues to make digestive enzymes even if the pancreatic ducts are blocked. Over time, these abundant enzymes damage the pancreas, cause fibrosis, and render the pancreas unable to function properly.

1 Using your textbook or online resources, write a definition for each of the terms in Table 17.3.

TABLE **17.3** Accessory Digestive Organs

Term	Definition
Salivary glands	
Liver	
Gallbladder	
Bile	
Emulsification	
Bile duct	
Portal vein	
Hepatic artery	
Pancreas	
Pancreatic amylase	
Pancreatic trypsin	
Pancreatic lipase	

2 Identify, label, and color the following structures in Figure 17.2.

- ☐ Liver
- ☐ Falciform ligament
- ☐ Left lobe
- ☐ Right lobe
- ☐ Gallbladder
- ☐ Common hepatic duct
- ☐ Cystic duct
- ☐ Common bile duct
- ☐ Pancreas
- ☐ Main pancreatic duct
- ☐ Duodenum
- ☐ Hepatopancreatic ampulla
- ☐ Sphincter of Oddi

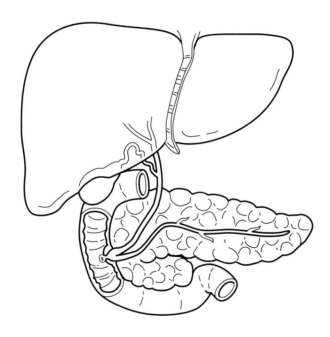

FIGURE **17.2** Liver, gallbladder, pancreas, and duodenum.

ACTIVITY 3 Peritoneum

Much of the abdominopelvic portion of the GI tract and accessory digestive organs reside in the peritoneal cavity. Like the pleural and pericardial cavities, the peritoneal cavity is the space between a double-layered serous membrane. Serous fluid is secreted into this space, enabling organs to move over each other without friction. The two peritoneal layers include the following:

1. **Parietal peritoneum.** The parietal peritoneum lines the internal surface of the abdominopelvic cavity.
2. **Visceral peritoneum.** Lining the surface of many digestive organs is the visceral peritoneum. The visceral peritoneum around the intestines folds over on itself to form a double-layered membrane known as the mesentery.
   a. **Mesentery.** Houses blood vessels, nerves, and lymphatic vessels and anchors these structures and the intestines in place.

1 Using your textbook or online resources, write a definition for each of the terms in Table 17.4.

TABLE 17.4 Peritoneal Layers and Structures

Term	Definition
Parietal peritoneum	
Visceral peritoneum	
Mesentery	

2 Identify, label, and color the following structures in Figure 17.3.
- Liver
- Pancreas
- Duodenum
- Transverse colon
- Small intestines
- Rectum
- Bladder
- Lesser omentum
- Greater omentum
- Transverse mesocolon
- Visceral peritoneum
- Mesentery
- Parietal peritoneum
- Peritoneal cavity

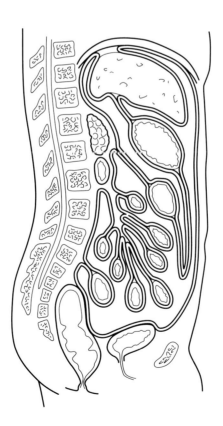

FIGURE 17.3 Peritoneal membranes and peritoneal cavity.

ACTIVITY 4 Tissue Layers of the GI Tract

The organs of the GI tract are the same for other hollow organs and include the following:

1. **Mucosa.** The mucosa lines the internal surface of the tubular GI tract and consists of epithelium overlying a lamina propria and a thin layer of smooth muscle called the muscularis mucosa. The epithelium is either stratified squamous epithelium (for the esophagus) or simple columnar epithelium (for the rest of the GI tract). The mucosa is generally covered with a layer of mucus that helps protect the underlying epithelium from the effects of the acid and digestive enzymes.
2. **Submucosa.** The submucosa is a layer of dense irregular collagenous connective tissue deep to the mucosa containing blood vessels, nerves, lymphatics, and elastic fibers.
3. **Muscularis externa.** The muscularis externa contains two layers of smooth muscle: an inner circular layer and an outer longitudinal layer. These two layers work in concert to produce the rhythmic contractions of peristalsis.
4. **Serosa.** The serosa is the outer layer composed of the visceral peritoneum and connective tissue. In the thoracic cavity and pelvis it is called the adventitia.

1 Using your textbook or online resources, write a definition for each of the terms in Table 17.5.

TABLE **17.5** Histology of the GI Tract

Term	Definition
Mucosa	
Lamina propria	
Submucosa	
Muscularis externa	
Serosa	

2 Identify, label, and color the following structures in Figure 17.4.

- ☐ Muscularis externa
- ☐ Submucosa
- ☐ Mucosa
- ☐ Lumen
- ☐ Plicae circulares
- ☐ Villi
- ☐ Blood vessels
- ☐ Lacteal
- ☐ Enterocytes
- ☐ Microvilli

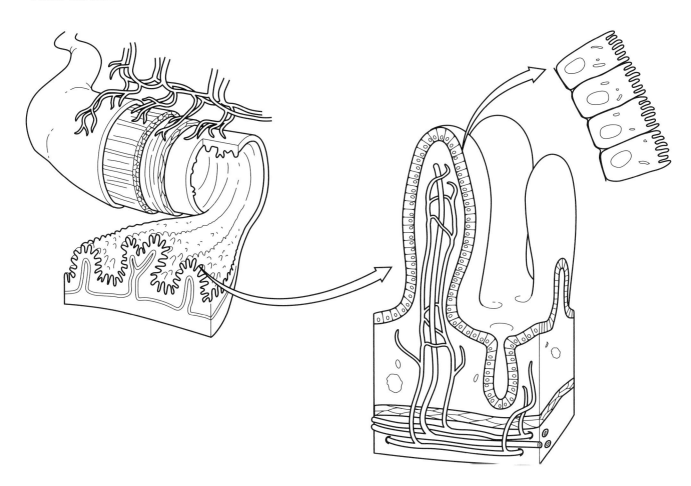

FIGURE **17.4** Tissue layers of the digestive system.

3 Label Figures 17.5 through 17.8 using the terms provided.
- ☐ Stratified squamous epithelium
- ☐ Submucosa
- ☐ Muscularis externa
- ☐ Esophageal gland

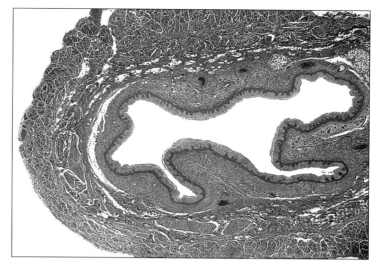

FIGURE **17.5** Esophagus.

- ☐ Mucosa
- ☐ Lamina propria
- ☐ Submucosa
- ☐ Muscularis externa
- ☐ Serosa

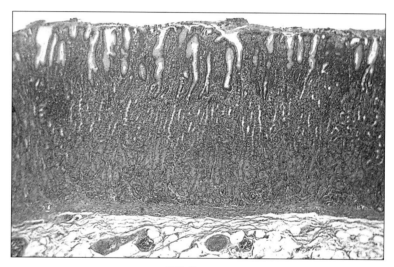

FIGURE **17.6** Stomach.

- ☐ Villi
- ☐ Plicae circulares
- ☐ Submucosa
- ☐ Muscularis externa

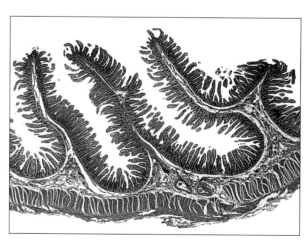

FIGURE **17.7** Jejunum.

- ☐ Liver lobule
- ☐ Central vein
- ☐ Portal triad
  - ▪ Bile duct
  - ▪ Portal vein
  - ▪ Hepatic artery

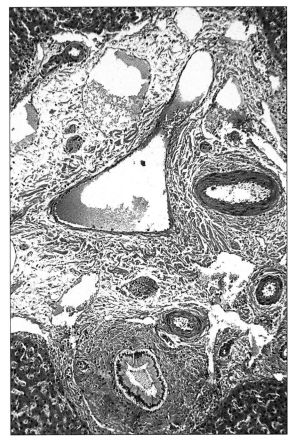

FIGURE **17.8** Liver.

ACTIVITY 5 Retroperitoneal and Intraperitoneal Organs

The peritoneal cavity is a fluid-filled space between the parietal and visceral peritonea (serous membranes lining the internal surface of the abdominal cavity and external surface of organs). The peritonea are serous membranes given different names (parietal and visceral) based upon their location (lining the wall or lining organs). Similar to the pericardial and pleural cavities, the peritoneal cavity is derived from the coelomic cavity during embryonic development. This serous membrane secretes about 50 mL of fluid per day, which possesses anti-inflammatory properties and serves as a lubricant.

Structures positioned deep to the peritoneum are referred to as retroperitoneal. Retroperitoneal organs are as follows:

- Adrenal glands
- Kidneys
- Ureter
- Aorta
- Inferior vena cava
- Rectum
- Pancreas (head, neck, and body; the tail is considered intraperitoneal)
- Duodenum
- Ascending and descending colon

All other digestive organs are considered intraperitoneal because the organs are enclosed by the peritoneum (i.e., stomach, jejunum, ileum, and transverse colon).

1. Using your textbook or online resources, write a definition for each of the following terms in Table 17.6.

TABLE 17.6 Peritoneum

Term	Definition
Peritoneal cavity	
Parietal peritoneum	
Visceral peritoneum	
Mesentery	
Peritoneal fluid	
Intraperitoneal organ	
Retroperitoneal organ	

OBSERVING

Complete the following hands-on laboratory activities to apply your knowledge of the digestive system.

ACTIVITY 1 Digestive Tissue

1. Obtain microscope slides of the esophagus, stomach, small intestine, and liver from your instructor.

2. Place each slide under the microscope and start with the lowest power and advance to higher magnification to see more detail.

3. Once you have a view of the tissue, use your colored pencils to draw what you see in the field of view. Color in the different cells or layers and label the terms indicated.

   a. Esophagus
   Draw and label the following:
   - Stratified squamous epithelium
   - Submucosa
   - Muscularis externa
   - Esophageal gland

MATERIALS

Obtain the following items before beginning the laboratory activities:

- Colored pencils
- Light microscope
- Microscope slides of the esophagus, stomach, small intestine, and liver
- Cadaver
- Gloves
- Probe
- Protective gear (lab coat, scrubs, or apron)

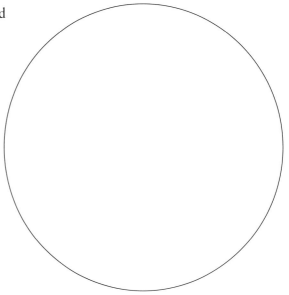

328 ■ *Discovering Anatomy: A Guided Examination of the Cadaver*

b. Stomach

Draw and label the following:
- ☐ Mucosa
- ☐ Lamina propria
- ☐ Submucosa
- ☐ Muscularis externa
- ☐ Serosa

c. Jejunum

Draw and label the following:
- ☐ Villi
- ☐ Plicae circulares
- ☐ Submucosa
- ☐ Muscularis externa

d. Liver

Draw and label the following:
- ☐ Liver lobule
- ☐ Central vein
- ☐ Portal triad
  - Bile duct
  - Portal vein
  - Hepatic artery

ACTIVITY 2 Gross Anatomy of the Digestive System

1 Identify the structures labeled in Figure 17.9 on the cadaver.

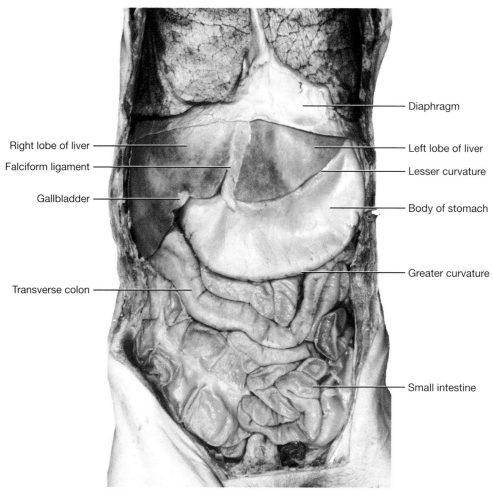

FIGURE 17.9   Abdominal cavity with the anterior abdominal wall removed.

2 Identify the structures labeled in Figure 17.10 on the cadaver.

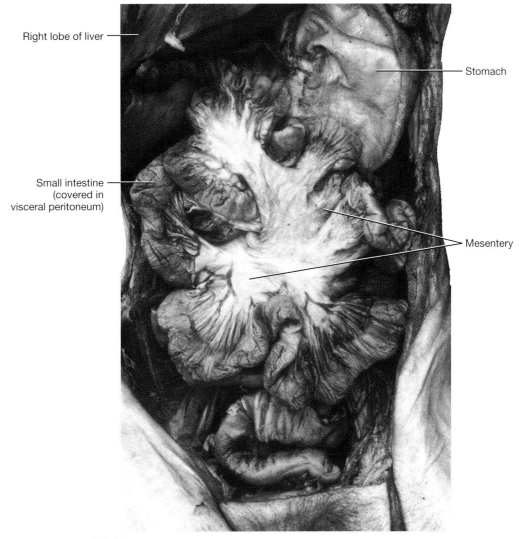

FIGURE **17.10** Small intestines and mesentery. The small intestines are splayed out to reveal the mesenteries, which anchor them to the posterior abdominal wall.

3 Identify the structures labeled in Figure 17.11 on the cadaver.

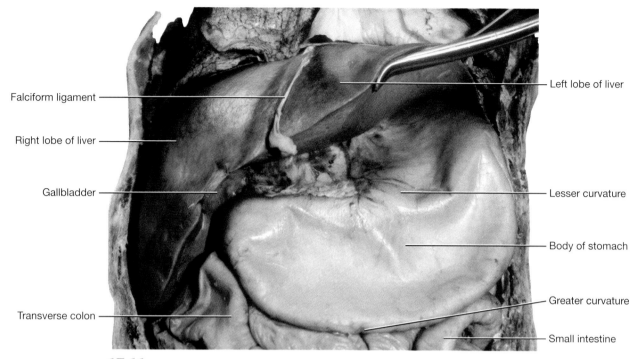

FIGURE **17.11** Stomach in situ. The liver is reflected to show the lesser curvature and lesser omentum.

4 Identify the structures labeled in Figure 17.12 on the cadaver.

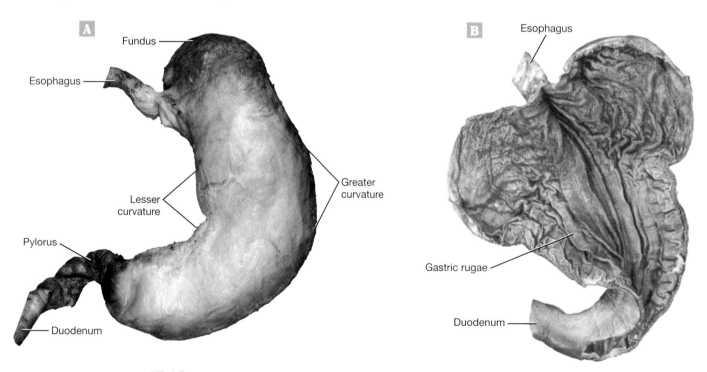

FIGURE **17.12** Stomach removed from abdominal cavity: (**A**) external features; (**B**) internal features.

5 Identify the structures labeled in Figure 17.13 on the cadaver.

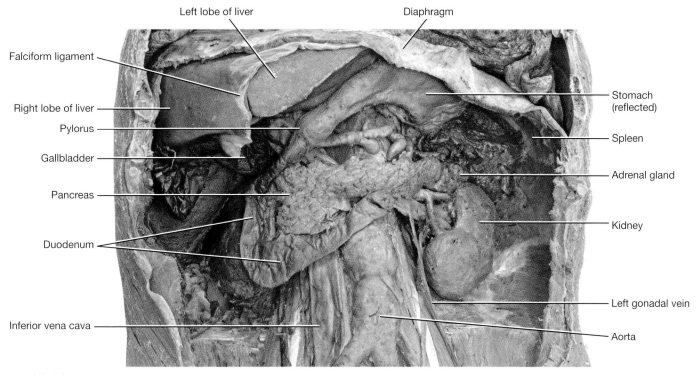

FIGURE 17.13   Duodenum and pancreas. Stomach is reflected to show that the underlying pancreas and small and large intestines have been removed.

6 Identify the structures labeled in Figure 17.14 on the cadaver.

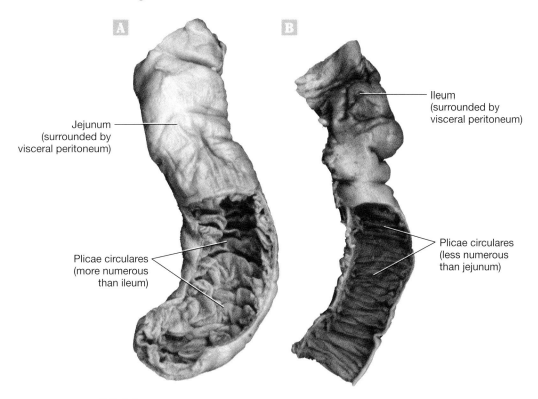

FIGURE 17.14   (A) Jejunum; (B) ileum. Observe the difference in the plicae circulares.

7 Identify the structures labeled in Figure 17.15 on the cadaver.

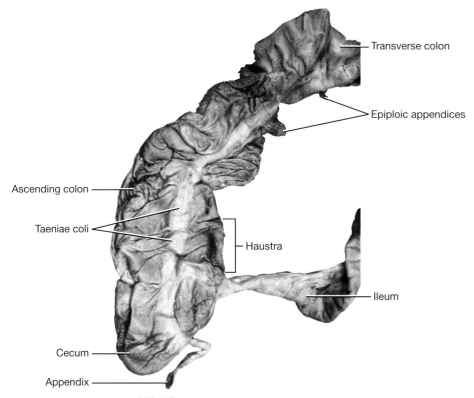

FIGURE **17.15** Proximal portion of the large intestines.

8 Identify the structures labeled in Figure 17.16 on the cadaver.

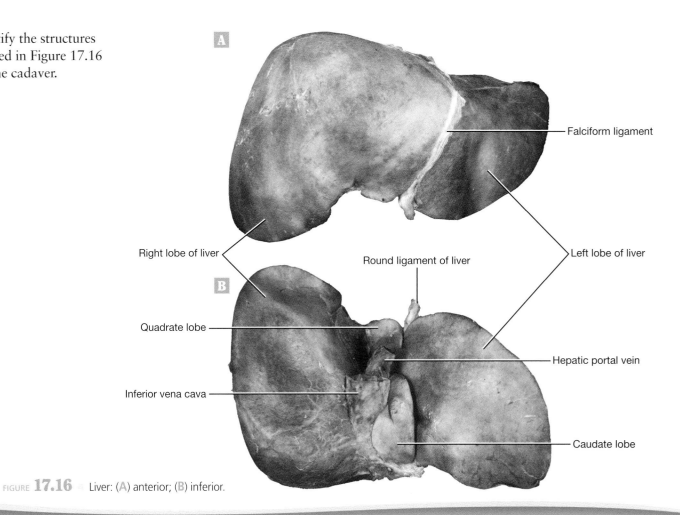

FIGURE **17.16** Liver: (**A**) anterior; (**B**) inferior.

WRAPPING UP

Complete the following additional activities to help retain your knowledge of the digestive system.

Name _____

Date _____ Section _____

1. Draw the GI tract on Figure 17.17 and include what (if any) digestive process occurs (peristalsis, mechanical digestion, chemical digestion, absorption, secretion) when you ingest a turkey sandwich (be sure to include when and where chemical digestion and absorption occurs for carbohydrates, proteins, and fats). Record your description of each stage in the space provided.

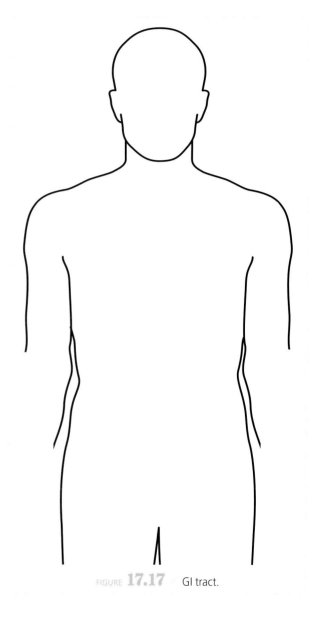

FIGURE 17.17 GI tract.

a. Oral cavity

b. Pharynx

c. Esophagus

d. Stomach

e. Duodenum

f. Liver

g. Gallbladder

h. Pancreas

i. Jejunum/ileum

j. Large intestines

k. Rectum/anus

WRAPPING UP
(Continued)

Name _____

Date _____ Section _____

2. Identify the following structures on Figure 17.18.

   a.

   b.

   c.

   d.

   e.

   f.

   g.

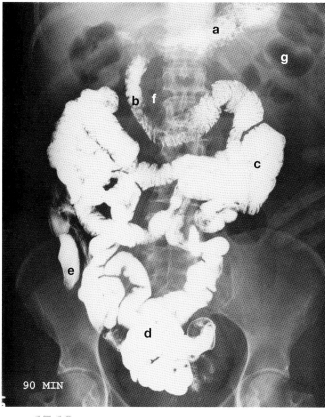

FIGURE 17.18 Contrast-enhanced abdominal X-ray (barium meal).

3. Provide an explanation for why "b" and "c" in Figure 17.18 have a feathery appearance, whereas "d" does not.

4. Hypothesize why starchy foods like pasta and potatoes taste so good to most of us as we chew them. In other words, how might their taste change the longer they are in our mouths?

5. During sleep the flow of saliva is reduced considerably. Provide an explanation for how this helps explain why we have halitosis (bad breath) in the morning.

6. Xerostomia describes a condition where a patient has a chronic reduction in the production and secretion of saliva. Provide an explanation why patients with xerostomia eventually have an increase in cavities (dental caries).

7. What type of epithelium lines the mucosa of the esophagus? The small intestines? Provide an explanation why these two organs of the GI tract have different epithelium.

8. Gastroesophageal reflux disease (GERD) is a chronic disease of the GI tract where stomach acid regurgitates into the esophagus and irritates the mucosal lining. Hypothesize why the acid irritates the mucosa of the esophagus but not the stomach.

9. The law of Courvoisier states that a painless enlargement of the gallbladder is most likely due to carcinoma of the head of the pancreas and not from a gallstone (the gallbladder is usually scarred and shrunken in the case of stones). Why would the gallbladder be distended in the case of pancreatic cancer? (*Hint:* Draw a picture if that would help explain your reasoning.)

10. Describe three structures that the small intestine possesses that increase its surface area for absorption.

Urinary System

CHAPTER 18

At the completion of this laboratory session, you should be able to do the following:

1. List the major functions of the urinary system.
2. Describe the gross anatomy of the kidney including vascular supply.
3. Identify and describe the function of each part of a nephron.
4. Identify the ureter, bladder, and urethra and describe their functions.

The organs of the urinary system include two kidneys, two ureters, one urinary bladder, and one urethra. The urinary system forms an extensive association with the cardiovascular system, much like the respiratory and digestive systems. The kidneys filter blood plasma and then return most of the water and solutes back into the bloodstream. The remaining water and solutes becomes urine, which is transported by the ureters into the urinary bladder to be stored. Urine is eventually excreted from the body via the urethra.

The urinary system is responsible for the following essential homeostatic functions:

- Regulates blood volume, total body water content, blood pH (acid/base balance), and electrolyte content of the blood plasma.
- Excretes waste products from the blood.
- Produces the enzyme renin, which helps regulate blood pressure and kidney function.
- Produces the hormone erythropoietin, which stimulates red blood cell formation in bone marrow.
- Metabolizes vitamin D into its active form.
- Detoxifies certain compounds (as does the liver).
- Makes glucose during times of starvation.

GETTING ACQUAINTED
Complete the following activities to become familiar with the urinary system.

Approximately one-fourth of total cardiac output (1200 mL of blood) is delivered to the kidneys each minute. As such, the kidneys filter several liters of fluid from the bloodstream every day into filtrate (up to 180 L a day). This enables toxins, metabolic wastes, and excess ions to be excreted in the form of urine all while providing the ability to reabsorb needed substances back into the bloodstream. As the kidneys perform these excretory functions, they simultaneously regulate the volume and chemical makeup of the blood, maintaining the proper balance between water and salts and between acids and bases. The kidneys are the functional unit of the urinary system and the other organs are the ureters, urinary bladder, and urethra.

MATERIALS
Obtain the following items before beginning the laboratory activities:
- Textbook or access to Internet resources
- Colored pencils

ACTIVITY 1 Kidney Anatomy

Each kidney weighs about 150 grams, is approximately the size of a bar of soap (12 × 6 × 3 cm), and is located in the retroperitoneal space (deep to peritoneal membranes) between the T12–L3 vertebral levels. The right kidney is slightly lower than the left because of the liver.

Supportive Kidney Tissues

The following three connective tissue layers encase each kidney:

1. **Renal fascia.** The renal fascia is an external layer of dense irregular connective tissue that surrounds and anchors each kidney and adrenal gland.
2. **Adipose tissue.** Adipose tissue composes the middle layer of tissue that helps position the kidney against the posterior abdominal wall and provides some cushioning protection.
3. **Renal capsule.** The renal capsule is an internal, transparent layer of connective tissue that encases each kidney like plastic wrap and prevents infections in surrounding regions from spreading to the kidneys.

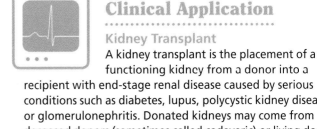

Clinical Application

Kidney Transplant

A kidney transplant is the placement of a functioning kidney from a donor into a recipient with end-stage renal disease caused by serious conditions such as diabetes, lupus, polycystic kidney disease, or glomerulonephritis. Donated kidneys may come from deceased donors (sometimes called cadaveric) or living donors who are genetically related or unrelated. The donated kidney is usually placed in the right or left lower abdominal cavities in the recipient. The renal artery and vein of the donated kidney are connected to the external iliac artery and vein of the recipient, though the common iliac vessels may also be used. A transabdominal Doppler ultrasound is performed following surgery to look for vascular complications such as renal vein thrombosis or arterial stenosis, as well as structural complications such as ureteral obstruction and hematoma.

Kidney Regions

Each kidney has the following three distinct regions when viewed in a coronal section:

1. **Renal cortex.** The renal cortex is the kidney's most superficial territory and consists of numerous blood vessels, causing its dark brown appearance. These blood vessels serve the nephrons, which are millions of tiny blood-filtering structures within each kidney. Renal columns project inward from the cortex and separate the renal pyramids from each other. Like the rest of the renal cortex, the renal columns contain many blood vessels and appear dark brown.
2. **Renal medulla.** The renal medulla consists of 12 to 18 conical structures, called renal pyramids, organized in the middle of the kidney. The medulla consists of collecting ducts, the loop of Henle, and capillary beds. Extensions of the medulla, called medullary rays, are organized around the collecting tubules and project into the cortex. These tubes course from base to apex of the renal pyramid and thus give it a striped or striated appearance. The renal medulla is mostly comprised of collecting ducts and the loops of Henle (glomeruli are located in the cortex, not the medulla). As such, the renal cortex and renal medulla are quite easy to distinguish from one another under the microscope.
3. **Calyces.** The collecting ducts exit at the tips of the renal pyramids into spaces called minor calyces, into which urine flows. The minor calyces merge together to form the major calyces, which in turn merge in the renal pelvis. The renal pelvis narrows to become the ureter.

Kidney Vasculature

The vascular supply of the kidney is a portal system in which blood flows through two consecutive capillary beds. The following arteries, arterioles, capillaries, venules, and veins enable blood to flow into, through, and out of the kidneys in order to maintain blood homeostasis.

1. **Renal arteries.** The renal arteries arise as 90-degree branches from the abdominal aorta and deliver about 25 percent of the total systemic cardiac output to the kidneys. Upon entering the hilum of the kidney, the renal arteries divide into progressively smaller arteries as they pass through the medulla to the cortex. The branches are as follows:
    a. **Segmental arteries.** Located in the renal pelvis.
    b. **Interlobar arteries.** Branches along renal columns between the medullary pyramids.
    c. **Arcuate arteries.** Branches that curve around the top of the renal pyramids.
    d. **Interlobular (cortical radiate) arteries.** Terminal branches in the renal cortex that divide into afferent glomerular arterioles.
    e. **Afferent glomerular arterioles.** Supply the glomerulus.
2. **Glomerulus.** The glomerulus is a capillary positioned at the start of the nephron. The glomerulus possesses fenestrated epithelium thus enabling plasma and small solutes to flow from the bloodstream into Bowman's capsule becoming **filtrate**. The Bowman's capsule transports the filtrate into the renal tubules where the physiological processes of reabsorption and secretion occur. Blood that remains in the glomerulus flows into the efferent glomerular arteriole.

a. **Efferent glomerular arteriole.** Transports blood from the glomerulus into a second capillary bed known as peritubular capillaries (cortical nephrons) or vasa recta (juxtaglomerular nephrons).
3. **Peritubular capillaries.** These capillaries surround the proximal and distal convoluted tubules of the nephron. They provide renal tubules with oxygen and nutrients and also take substances reabsorbed by the tubules (like glucose and water) back into the blood.
4. **Vasa recta.** In some nephrons, another set of capillaries is found surrounding the loop of Henle. This is called the vasa recta.
5. **Renal veins.** The peritubular capillaries and vasa recta deliver blood into venules and then into interlobular veins, followed by arcuate veins, followed by interlobar veins, and ultimately into the renal vein.

To summarize, the glomerular capillaries and their specialized fenestrations allow a great deal of solutes such as sodium, potassium, and water to be filtered out of the plasma and into the glomerular capsular space. Once the fluid enters the capsular space, it is no longer referred to as plasma, but is now referred to as filtrate. It is this filtrate that will begin to flow into the looped tubular structure that we call the nephron. The majority of the filtrate is reabsorbed back into the bloodstream by way of the peritubular capillaries and vasa recta. Some compounds are actively excreted from the blood and into the filtrate for elimination (e.g., hydrogen ions, organic acids, urea). Whatever filtrate is left at the end of the nephron is excreted as urine.

1 Using your textbook or online resources, write a definition for each of the structures listed in Table 18.1.

TABLE **18.1** Anatomy of the Urinary System

Term	Definition
Renal cortex	
Renal column	
Renal medulla	
Medullary pyramid	
Minor and major calyces	
Renal pelvis	
Ureter	
Urinary bladder	
Internal urethral sphincter	
External urethral sphincter	
Urethra	

2 Identify, label, and color the following structures on Figure 18.1.

- Urinary system organs:
  - Kidney
  - Ureter
  - Urinary bladder
  - Urethra
- Other organs:
  - Diaphragm
  - Aorta
  - Renal arteries
  - Common iliac arteries
  - Inferior vena cava
  - Renal veins
  - Common iliac veins
  - Adrenal glands
  - T10–T12 vertebrae
  - L1–L2 vertebrae
  - Floating ribs

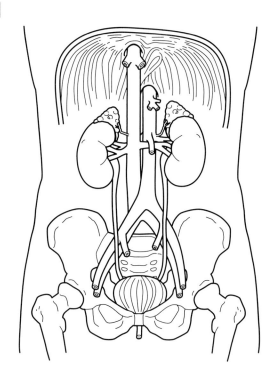

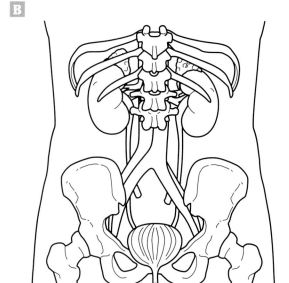

FIGURE 18.1 Organs of the urinary system: (A) anterior; (B) posterior.

3 Identify, label, and color the following structures on Figure 18.2.

- Renal cortex
- Renal column
- Renal medulla
- Renal pyramid
- Renal papilla
- Minor calyx
- Major calyx
- Renal pelvis
- Ureter
- Renal capsule
- Renal artery
- Renal vein

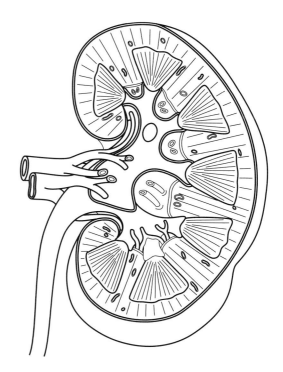

FIGURE 18.2   Gross anatomy of the kidney.

ACTIVITY 2 Structures of the Nephron

Each kidney consists of about a million nephrons, which serve as the functional unit of the urinary system. Blood is delivered under high pressure into each nephron's glomerulus via the afferent glomerular arteriole. Here, the blood plasma (water and small solutes) is forced out of the glomerulus into Bowman's space within Bowman's capsule.

Bowman's capsule is an enlarged, cup-shaped extension of the renal tubule and completely surrounds the glomerulus (like a baseball glove surrounding a ball). Together the glomerulus and its surrounding glomerular capsule are referred to as the **renal corpuscle**. The capsule has the following two layers:

1. **Parietal (external) layer.** The parietal layer is composed of simple squamous epithelium.
2. **Visceral (inner) layer.** The visceral layer is composed of cells (podocytes) that surround the glomerular capillaries. The podocytes form narrow filtration slits that prevent large products in the blood, such as blood cells and large proteins, from exiting into Bowman's capsule.

Once blood plasma exits the glomerular capillary and enters the capsular space, it becomes known as filtrate. Filtrate then flows through the series of nephron tubules where most of the filtrate is reabsorbed. The majority of the renal tubules are confined to the renal cortex; only the nephron loops of certain nephrons dip down into the renal medulla. These tubes are collectively called the **renal tubule** and consist of the following parts:

1. **Proximal convoluted tubule (PCT).** The proximal convoluted tubule consists of simple cuboidal epithelial cells with microvilli that reabsorb substances from the filtrate into the peritubular capillaries. The PCT reabsorbs the following from the filtrate under normal conditions:
    a. 66 percent of the salt, water, and potassium.
    b. All of the glucose and amino acids.
    c. 50 percent of the urea.
    d. 80 percent of the phosphates.
    e. 70–90 percent of the citrate.

2. **Loop of Henle.** The "U"-shaped loop of Henle has descending and ascending limbs. The initial segment of the descending limb is simply a continuation of the PCT and, as such, its epithelial construct is the same. The loop of Henle has the following two segments:
   a. **Thin segment.** The remainder of the descending limb, the loop, and the initial segment of the ascending limb. This part of the loop of Henle is composed of simple squamous epithelium and is freely permeable to water.
   b. **Thick segment.** The epithelium of the ascending limb becomes cuboidal and is not permeable to water.
   c. **Macula densa.** At the meeting of the ascending limb of the loop of Henle and the distal convoluted tubule, there is a group of tall, closely packed cells known as the macula densa.
      i. The macula densa is in contact with a portion of the afferent glomerular arteriole that contains specialized cells called juxtaglomerular (JG) cells.
      ii. The JG cells and macula densa together form the juxtaglomerular apparatus (JGA), which helps control the flow of filtrate through the nephron and the blood pressure within the glomerulus.
      iii. Granular cells produce the most effective blood pressure raising hormone—renin.
3. **Distal convoluted tubule (DCT).** Of the original blood filtered at the glomerulus approximately 25% of water and 10% of sodium chloride remain by the time the filtrate reaches the DCT. From this point on the recovery of water and ions is continuously regulated by hormones (i.e., aldosterone for sodium, ADH for water and PTH for calcium) to meet the body's requirements. If necessary, most of the water and sodium in the DCT and collecting ducts can be reclaimed. The DCTs possess a larger lumen than the PCTs because the simple cuboidal cells are shorter and DCTs lack a brush border.
4. **Collecting ducts.** Each collecting duct receives urine from numerous DCTs. The collecting ducts course from the base to the apex of the medullary pyramids, which gives them their striped appearance. As the collecting ducts approach the apex of the medullary pyramid, they fuse to form larger papillary ducts. Once the filtrate leaves the papillary ducts and flows into the minor calyx, it is referred to as urine. The collecting ducts are relatively impermeable to water. However, the hormone ADH makes the collecting ducts more permeable to water by inserting aquaporins into the luminal membranes and reabsorption of water occurs.

Other structures of the nephron include the following:
1. **Calyces.** The calyces collect urine, which drains continuously from the papillae, and transport the urine into the renal pelvis. Urine flows from minor calyces into major calyces, which in turn terminate in the renal pelvis.
2. **Renal pelvis.** The renal pelvis is a funnel-shaped tube continuous with the ureter, ultimately collecting all of the urine produced from each of the million nephrons within that kidney.

Clinical Application

Kidney Stones

Kidney stones (urolithiasis) are small, hard, mineral deposits of calcium, oxalate, and uric acid that form inside the kidneys. Kidney stones have many causes that can affect any part of the urinary system, from the kidneys to the bladder. Often, stones form when the urine becomes concentrated, allowing minerals to crystallize and stick together. Patients with kidney stones may be asymptomatic if the stones form and pass without any obstruction. However, in cases where the stone obstructs the renal pelvis or ureter, the pain is usually excruciating and radiates from the flank to the groin. It has been described as one of the strongest pain sensations known, even greater than childbirth! The pain caused by the lodged stone typically lasts from 20 minutes to an hour and is caused by peristaltic contractions of the ureter as it attempts to expel the stone. A treatment method for larger kidney stones is extracorporeal shock wave lithotripsy (ESWL), which uses shock waves to break the stone into smaller pieces to make it easier to pass. On a side note, it is reported that the actor William Shatner (who portrayed Captain James T. Kirk in the original *Star Trek* TV series) sold one of his kidney stones for $25,000 for a charity event.

Kidney stones from two different patients.

Weird and Wacky

What's in a Name?

It has been said that the sweetest word to anyone is his or her own name. If that is the case, the German anatomist Friedrich Gustav Jakob Henle (1809–1885) is smiling sweetly. Due to his contributions to the study of the nephron loop, this structure was titled the "loop of Henle." However, use of his name didn't stop there; the crypts of Henle (located in the conjunctiva), Henle's fissures (fibrous tissue between cardiac muscle fibers), Henle's ampulla (ampulla of the oviduct), Henle's ligament (tendon of the transverse abdominis muscle), Henle's sheath (the perineurium surrounding a nerve fascicle), and Henle's spine (spine on the mastoid process) are all named after his contribution to their discovery. He is also credited for research that helped to discover the germ theory (microorganisms cause infectious diseases) but, as far as we know, his name was not used in this area of science.

1 Using your textbook or online resources, write a definition for each of the structures listed in Table 18.2.

TABLE 18.2 Structures of the Nephron

Term	Definition
Nephron	
Afferent glomerular arteriole	
Glomerulus	
Efferent glomerular arteriole	
Peritubular capillaries	
Vasa recta	
Proximal convoluted tubule	
Loop of Henle	
Distal convoluted tubule	
Collecting duct	

2 Identify, label, and color the following structures on Figure 18.3.

- ☐ Afferent glomerular arteriole
- ☐ Glomerulus
- ☐ Efferent glomerular arteriole
- ☐ Peritubular capillaries
- ☐ Vasa recta
- ☐ Bowman's capsule
- ☐ Bowman's space
- ☐ Proximal convoluted tubule
- ☐ Loop of Henle
- ☐ Distal convoluted tubule
- ☐ Collecting duct
- ☐ Arcuate artery
- ☐ Interlobular artery
- ☐ Interlobular vein
- ☐ Arcuate vein
- ☐ Renal cortex
- ☐ Renal medulla

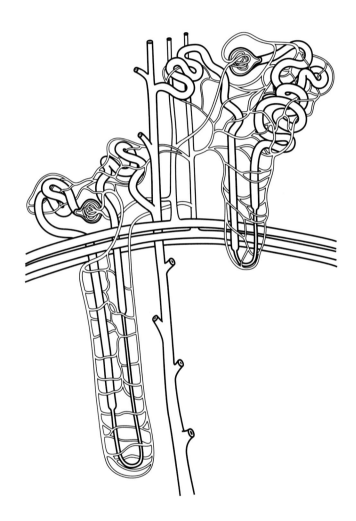

FIGURE **18.3** Schematic of the nephron.

3 Identify, label, and color the following structures on Figure 18.4.

- Afferent glomerular arteriole
- Glomerulus
- Efferent glomerular arteriole
- Lumen of glomerulus
- Bowman's capsule (parietal)
- Bowman's capsule (visceral)
- Podocyte
- Bowman's space
- Proximal convoluted tubule (PCT)
- Distal convoluted tubule (DCT)
- Macula densa cells
- Granular cells
- Red blood cell

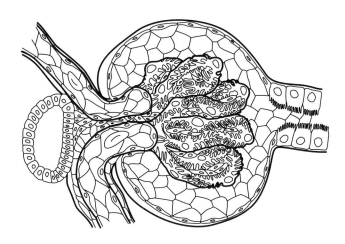

FIGURE 18.4   Schematic of juxtaglomerular apparatus.

ACTIVITY 3 Ureters, Urinary Bladder, and Urethra

The anatomy and function of the ureters, urinary bladder, and urethra are as follows:

1. **Ureters.** Ureters have thick smooth muscular walls and transport urine from the renal pelvis to the urinary bladder by way of peristalsis.
2. **Bladder.** The urinary bladder is an expandable sac that temporarily stores urine. It is a retroperitoneal organ located anterior to the rectum in males and anterior to the vagina and uterus in females. The smooth muscle lining the wall of the urinary bladder is called the detrusor muscle.
    a. **Trigone.** The smooth triangular area on the inferior portion of the urinary bladder wall, which contains the two openings for the ureters and the urethral orifice.
    b. **Empty bladder.** When the urinary bladder contains little to no urine, it collapses into a pyramidal shape and its walls become thick and full of rugae (folds).
    c. **Full bladder.** As urine slowly fills the urinary bladder, it expands and stretches into the abdominal cavity. A full bladder holds approximately 0.5 L of urine but can accommodate more if necessary.
3. **Urethra.** Urine is transported from the urinary bladder to outside the body by the urethra. In males the urethra is quite long and shares function with the reproductive system (transports semen). In females, the urethra is quite short. The urethra contains two sphincter muscles:
    a. **Internal urethral sphincter.** Composed of smooth muscle and is innervated to involuntarily relax (let urine flow) or contract (keep urine in the bladder) based on autonomic innervation.
    b. **External urethral sphincter.** Composed of skeletal muscle and is innervated to voluntarily relax (let urine flow) or contract (keep urine in the bladder) based on innervation from the pudendal nerve.

Like other hollow organs, the ureters and urinary bladder are composed of the following layers:

1. **Mucosa.** The mucosa is the innermost layer composed of transitional epithelium and a thin layer of loose connective tissue. Transitional epithelium is stratified epithelium with dome-shaped cells at the apical surface and cuboidal cells at the basal surface. The apical cells can change shape from dome-shaped to squamous-shaped based upon the amount of stretching the ureter and bladder do to accommodate the volume of urine.
2. **Submucosa.** The submucosa is the middle layer composed of collagenous connective tissue and glands that secrete a watery mucus, which prevents urine from damaging and ulcerating the epithelium.
3. **Muscularis externa.** The muscularis externa is composed of smooth muscle that stretches to accommodate the increased volume of urine or contracts to propel the urine forward through the urinary system.
4. **Adventitia.** The adventitia is the outermost layer of dense irregular collagenous connective tissue.

Asparagus Anyone?

Have you ever eaten asparagus only to realize your urine has a funny smell afterwards? If that has happened you are in good company because two out of every four people report the same thing. Asparagus contains a compound called asparagusic acid that is broken down during digestion into sulfur-containing compounds (i.e., dimethyl sulfide). This usually happens within 20 minutes after eating. The sulfur compounds are volatile and thus vaporize from the urine into a gaseous state once they exit the urethra and enter the atmosphere. The odor floats up, dissolves in the nasal mucosa, and binds to olfactory neurons where the message is then sent to the brain. The process of creating sulfur compounds from ingested asparagus seems to occur in everyone. However, only half of us can smell it. Why? Research discovered that there is a special gene that is required to be able to smell this sulfur-containing compound created by asparagus. In other words, the question is not whether the asparagus odor is present in the urine or not, it is whether you are *able* to smell it.

1 Identify, label, and color the following structures on Figure 18.5.
- Ureter
- Ureteric orifices
- Peritoneum
- Urinary bladder
- Rugae
- Detrusor muscle
- Trigone
- Urethra
- Internal urethral sphincter
- External urethral sphincter
- Urogenital diaphragm

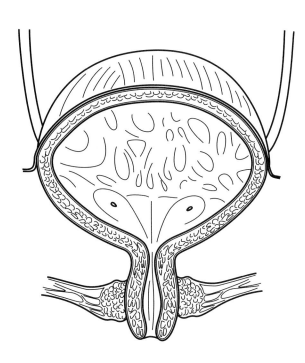

FIGURE 18.5 Female ureter, urinary bladder, and urethra.

2 Label Figures 18.6 through 18.10 using the list of structures provided for each figure.
- ☐ Glomerulus
- ☐ Renal tubules
- ☐ Capsular space

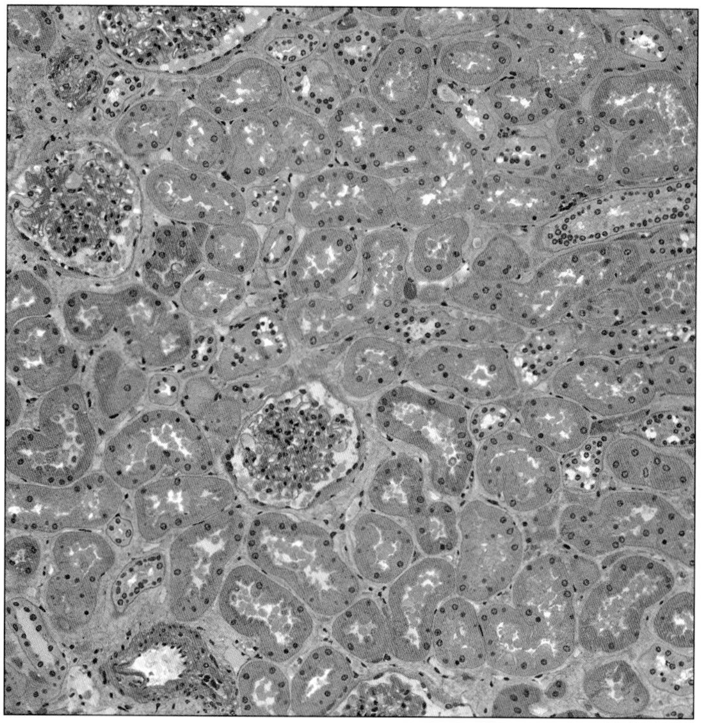

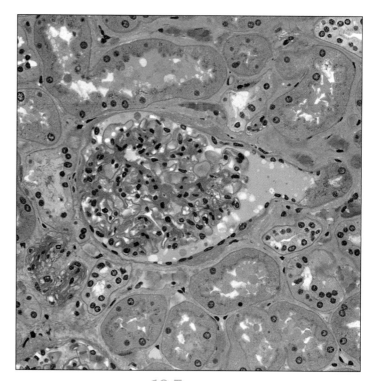

FIGURE **18.6** ■ Renal cortex.

- ☐ Bowman's capsule (parietal layer)
- ☐ Bowman's capsule (visceral layer)
- ☐ Bowman's space
- ☐ Macula densa
- ☐ Glomerular capillaries

FIGURE **18.7** ■ Glomerulus.

- ☐ Collecting ducts
- ☐ Loop of Henle
- ☐ Vasa recta

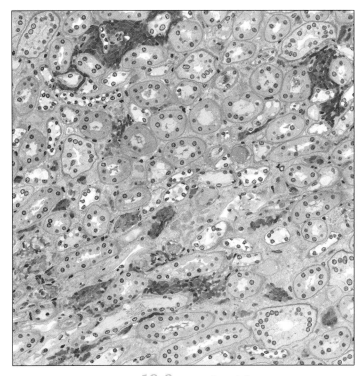

FIGURE 18.8 Renal medulla.

- ☐ Transitional epithelium
- ☐ Submucosa
- ☐ Muscularis externa

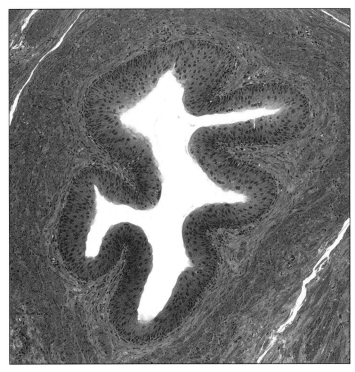

FIGURE 18.9 Ureter.

- Transitional epithelium
- Submucosa
- Smooth muscle (detrusor muscle)

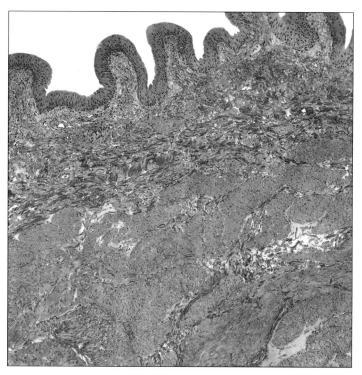

FIGURE 18.10 Urinary bladder.

OBSERVING

Complete the following hands-on laboratory activities to apply your knowledge of the urinary system.

MATERIALS

Obtain the following items before beginning the laboratory activities:

- Prepared microscope slides of the renal cortex, renal medulla, ureter, and bladder
- Light microscope
- Colored pencils
- Cadaver
- Gloves
- Probe
- Protective gear (lab coat, scrubs, or apron)

ACTIVITY 1 Microscopic Structures of the Urinary System

1. Obtain microscope slides of the renal cortex, renal medulla, ureter, and bladder from your instructor.

2. Place each slide under the microscope and start with the lowest power and advance to higher magnification to see more detail.

3. Once you have a view of the tissue, use your colored pencils and draw what you see in the field of view. Color in the different cells or layers and label the terms indicated.

a. Renal cortex
Draw and label the following:
- Glomerulus
- Bowman's capsule
- Bowman's space
- Urinary tubules

352 ■ *Discovering Anatomy: A Guided Examination of the Cadaver*

b. Renal medulla

Draw and label the following:
- ☐ Loops of Henle
- ☐ Vasa recta
- ☐ Collecting ducts

c. Ureter

Draw and label the following:
- ☐ Lumen
- ☐ Transitional epithelium
- ☐ Submucosa
- ☐ Muscularis externa

d. Urinary bladder

Draw and label the following:
- ☐ Lumen
- ☐ Transitional epithelium
- ☐ Submucosa
- ☐ Muscularis externa (detrusor muscle)

ACTIVITY 2 Gross Anatomy of the Urinary System

1 Identify the structures labeled in Figure 18.11 on the cadaver.

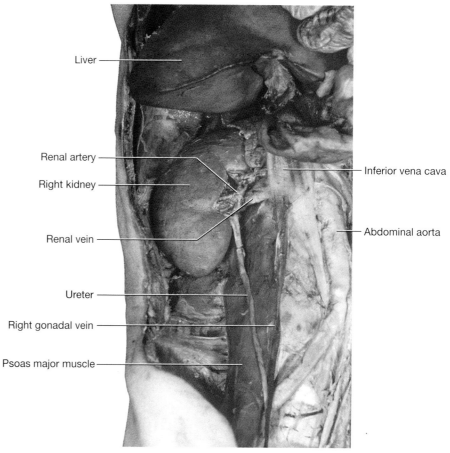

FIGURE 18.11 Abdominal cavity with the anterior abdominal wall removed.

2 Identify the structures labeled in Figure 18.12 on the cadaver.

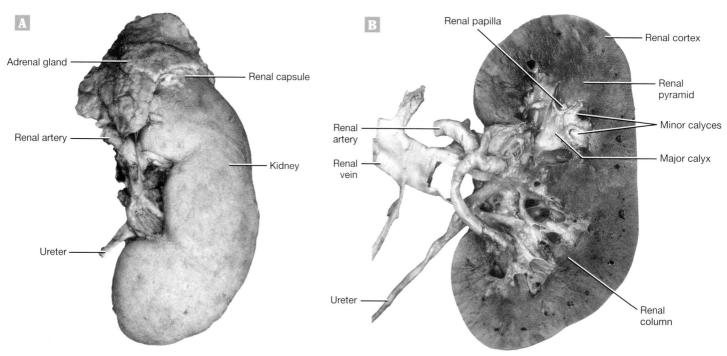

FIGURE 18.12 Kidney: (A) anterior view; (B) coronal section.

WRAPPING UP

Complete the following additional activities to help retain your knowledge of the urinary system.

Name _____

Date _____ Section _____

1. Identify the following structures on Figure 18.13.

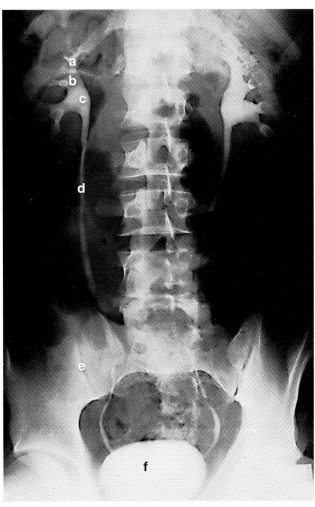

FIGURE 18.13 Pyelogram of the urinary system.

a.

b.

c.

d.

e.

f.

UNIT 4: Body Highways · CHAPTER 18: Urinary System · 355

2. The adipose tissue surrounding each kidney is vital in securing it to the posterior abdominal wall. If a patient has rapid weight loss and loses this adipose tissue, the kidneys may experience a shift to a lower position. This shifting of the kidney may result in a kinked ureter, which prevents urine from draining from the kidney into the urinary bladder. Provide an explanation of how a kinked ureter results in a condition called hydronephrosis. Hypothesize why this condition is severe and damaging to the patient.

3. Provide an explanation for why each nephron needs two sets of capillary beds in sequence. In other words, why is a glomerulus needed as well as the peritubular capillaries?

4. Hypothesize why the histological organization of the PCT requires simple epithelium whereas the ureter and bladder require stratified epithelium.

5. Fanconi syndrome is a disease inhibiting the function of the PCT. Compare and contrast urine concentration for patients with normal kidney function with that of patients with Fanconi syndrome for the following: glucose, amino acids, uric acid, phosphate, and bicarbonate.

5
The Next Generation

CHAPTER 19 Male Reproductive System **359**

CHAPTER 20 Female Reproductive System **371**

Male Reproductive System

CHAPTER 19

At the completion of this laboratory session, you should be able to do the following:

1. List the major functions of the male reproductive system.
2. Describe the structure and function of the testes.
3. Outline the location, structure, and function of male accessory organs.
4. Describe the structure and function of the penis.
5. List the sources and functions of the components of semen.

The male reproductive system consists of the primary sex organs (paired testes, which produce sperm) and secondary sex organs. Secondary sex organs include the epididymis, ductus deferens, ejaculatory duct, seminal vesicles, prostate gland, bulbourethral gland, urethra, and penis.

GETTING ACQUAINTED

Complete the following activities to become familiar with male reproductive system.

MATERIALS
Obtain the following items before beginning the laboratory activities:
- ❏ Textbook or access to Internet resources
- ❏ Colored pencils

Testes and Associated Structures

Each testis (male gonad) is approximately 4 × 3 cm in size and is considered the primary sex organ of the male reproductive system because it produces sperm. In other words, the testes are the gamete (sex cell)-producing organs of the male reproductive system. The following structures are associated with the testes:

1. **Tunica vaginalis and albuginea.** A two-layered mesothelium sheath called the tunica vaginalis surrounds each testis. Deep to this is a white fibrous capsule called the tunica albuginea. This sheath divides the interior of the testes into approximately 250 to 300 wedge-shaped compartments called **lobules**.

2. **Seminiferous tubules.** Each testicular lobule contains one to three tightly coiled seminiferous tubules where sperm develop. The seminiferous tubules converge on the superior region of the testicles to form the rete testis where fluid is absorbed from the lumen. Sperm leave the rete testis through the efferent ductules and proceed into the epididymis.

3. **Interstitial (Leydig) cells.** Surrounding the seminiferous tubules is loose connective tissue. Within the loose connective tissue resides a group of cells appropriately called interstitial or Leydig cells. Interstitial cells produce and secrete testosterone into the surrounding tissues and fluid. As such, different cell populations carry out the sperm-producing and hormone-producing functions of the testes.

4. **Scrotum.** The testes reside outside the body in the scrotum, which is a sac composed of skin, connective tissue, and muscle. The proposed reason that the testes are located outside the body is because sperm production requires a temperature of 34°C (about 94°F), which is 3°C degrees lower than body temperature (37°C). Spermatogenesis is less efficient at lower and higher temperatures. If the temperature drops, the cremaster muscle lining the spermatic cord and scrotum contracts and pulls the testis closer to the body, which provides slightly more warmth to maintain optimal testicular temperature. When cooling is required, the cremaster muscle relaxes and the testicle is lowered away from the warm body.

5. **Spermatic cord.** The spermatic cord is a continuation of the scrotum into the body wall. It contains the ductus deferens, testicular artery, testicular veins (pampiniform plexus of veins), lymphatics, and nerves.

1 Using your textbook or online resources, write a definition for each of the terms listed in Table 19.1.

TABLE **19.1** Anatomy of the Scrotal Sac

Term	Definition
Tunica vaginalis	
Testis	
Seminiferous tubules	
Interstitial (Leydig) cells	
Scrotum	
Spermatic cord	

2 Identify, label, and color the following structures on Figure 19.1.
- ☐ Scrotum
- ☐ Spermatic cord
- ☐ Pampiniform plexus of veins
- ☐ Testicular artery
- ☐ Testis
- ☐ Ductus deferens
- ☐ Epididymis
- ☐ Tunica vaginalis
- ☐ Tunica albuginea
- ☐ Seminiferous tubules
- ☐ Rete testis

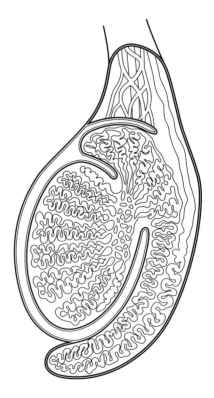

FIGURE **19.1** Testis, midsagittal section.

360 ■ *Discovering Anatomy: A Guided Examination of the Cadaver*

ACTIVITY 2 Additional Organs of the Male Reproductive System

The Male Duct System

Sperm travel from the testes to the exterior of the body through a system of tubes. In order from proximal to distal, the tubes are as follows:

1. **Epididymis.** The epididymis, located on the superior and posterolateral surface of the testis, is approximately 3.5 cm long and receives sperm from the efferent ductules. The epididymis provides nutrients to the sperm temporarily stored within the lumen, absorbs excess testicular fluid, and provides a location and environment for immature sperm to fully develop. During male ejaculation, sperm are ejected from the epididymis into the ductus deferens, not from the testes as many believe. Sperm are stored in the epididymis for several months and are eventually phagocytized if not ejaculated.

2. **Ductus deferens.** The ductus deferens is approximately a half meter in length and ascends within the spermatic cord, traverses the inguinal canal, courses over the urinary bladder, and finally merges with the seminal vesicle to form the ejaculatory duct. The ductus deferens is easily palpated as it passes anterior to the pubic bone due to its thick smooth muscle layer wall. The ductus deferens propels live sperm into the urethra from their storage sites in the epididymis and distal part of the ductus deferens. At the moment of ejaculation, the thick layers of smooth muscle in its walls create strong peristaltic waves that rapidly squeeze the sperm forward.

3. **Ejaculatory ducts.** Each ejaculatory duct passes through the prostate gland, where it joins with the prostatic urethra.

4. **Urethra.** The urethra is the terminal portion of the male duct system and serves both the urinary and reproductive systems (conveys both urine and semen at different times). The urethra consists of the following three regions:

    a. **Prostatic urethra.** The region surrounded by the prostate gland.

    b. **Membranous urethra.** The very short region within the urogenital diaphragm.

    c. **Spongy (penile) urethra.** The portion that courses through the penis and opens to the outside. The spongy urethra accounts for 75 percent of the urethral length (approximately 15 cm long), making it the longest portion.

Clinical Application

Vasectomy

The term "vas deferens" comes from the Latin word *vas*, meaning vessel, and *deferens*, meaning to carry away from. However, it is a misnomer, as the vas deferens is not a vessel. That is why the proper anatomical term is "ductus deferens." Despite this misnomer, the term has persisted with a form of male birth control called a vasectomy, meaning "to cut the vas." A vasectomy is a relatively straightforward procedure in which each ductus deferens is surgically severed and the cut ends are ligated or cauterized. A man who has had a vasectomy continues to produce sperm, but the procedure inhibits sperm from moving from the testis to the ejaculatory duct. As such, the man becomes sterile and the sperm are phagocytized.

Accessory Glands

The male reproductive tract consists of three exocrine glands: seminal vesicles, prostate gland, and bulbourethral glands. Together, these glands produce the bulk of semen.

1. **Seminal vesicles.** The seminal vesicles are located posteriorly on the urinary bladder and each one is approximately 5 to 7 cm in length. The excretory duct of each seminal vesicle opens into the corresponding ductus deferens as it enters the prostate gland to become the ejaculatory duct. Seminal fluid is yellowish in color, alkaline in composition (to help buffer the acidic environment of the vagina), and accounts for 60 percent of the volume of semen. In addition, the secretions are rich in fructose (to nourish spermatozoa), vesiculase (a coagulating enzyme), and prostaglandins (to oppose immune response against the semen). Sperm and seminal fluid mix together in the ejaculatory duct and enter the prostatic urethra during ejaculation.

2. **Prostate gland.** The prostate is a single, chestnut-sized gland that encircles the region of the urethra below the urinary bladder, called the prostatic urethra. The prostate consists of 20 to 30 compound tubuloalveolar glands embedded in a stroma of dense irregular collagenous connective tissue and smooth muscle. During ejaculation, the smooth muscle surrounding the prostate gland contracts and forces prostatic secretions into the urethra via several ducts. These secretions help liquefy the semen and account for approximately one-third of the seminal volume.

3. **Bulbourethral glands.** The bulbourethral glands are pea-sized glands located within the urogenital diaphragm. Prior to ejaculation the bulbourethral glands secrete a clear, viscous fluid containing abundant sugars that neutralize traces of acidic urine, facilitate the passage of sperm, and provide lubrication for sexual intercourse.

The Penis

The penis is an organ that delivers sperm into the female reproductive tract.

The external skin loosely covers the penis, which enables it to stretch during erection. Distally, the skin forms the prepuce, or foreskin, around the glans penis. Surgical removal of the foreskin is a procedure known as circumcision. Internally, the penis contains the spongy urethra and three parallel, cylindrical masses of erectile tissue that are enclosed in a capsule of dense irregular collagenous connective tissue. Erectile tissue resembles a sponge, consisting of connective tissue and smooth muscle riddled with vascular spaces. The penis contains the following two erectile tissue structures:

1. **Corpus spongiosum.** The midventral erectile body that surrounds the spongy urethra and expands distally to form the glans penis. The bulbospongiosus muscle covers the base of the corpus spongiosum.
2. **Corpora cavernosa.** The paired dorsal erectile bodies make up most of the penis. The ischiocavernosus muscle surrounds the base of the corpora cavernosa.

During sexual excitement, blood fills the vascular spaces within the erectile tissue (corpus spongiosum and corpora cavernosa), causing the penis to enlarge and become rigid and erect. An erection enables the penis to penetrate the vagina during intercourse.

Clinical Application

Prostate Health

The prostate gland has two main growth periods. The first occurs during adolescence and the second begins at the age of 30 and continues for the rest of life. Around the age of 50, hyperplasia of tissue in the prostate's periurethral zones produces nodules that compress the urethra, making it difficult to void the bladder. This is called benign prostatic hypertrophy (BPH), a non-cancerous increase in the size of the prostate gland.

Speed Racer
There are approximately 200 million sperm in one teaspoon of ejaculate, and semen exits the urethra between 30 and 40 miles per hour during ejaculation.

Clinical Application

Nocturnal Penile Tumescence

Nocturnal penile tumescence (NPT) describes how the penis spontaneously erects during sleep or when waking up (usually three to five times during REM sleep). One reason for NPT is that the kidney continues to filter blood and produce urine as we sleep. As such, the bladder fills and expands, stimulating the nerves in the penile region and causing the formation of an erection. Nocturnal penile tumescence is also used to determine if erectile dysfunction (ED) is caused by psychological or physiological origin. If a patient with ED achieves an erection during the night, it is presumed that their ED is due to psychosomatic problems and not a physiological problem.

1 Using your textbook or online resources, write a definition for each of the terms in Table 19.2.

TABLE **19.2** Anatomy of Male Reproductive Organs

Term	Definition
Epididymis	
Ductus deferens	
Ejaculatory duct	
Urethra	
Seminal vesicle	
Prostate gland	
Bulbourethral gland	
Penis	
Corpus spongiosum	
Corpora cavernosa	

2 Identify, label, and color the following structures on Figure 19.2.
- ☐ Peritoneal cavity
- ☐ Parietal peritoneum
- ☐ Visceral peritoneum of the bladder
- ☐ Urinary bladder
- ☐ Seminal vesicles
- ☐ Prostate gland
- ☐ Bulbourethral gland
- ☐ Ejaculatory duct
- ☐ Prostatic urethra
- ☐ Membranous urethra
- ☐ Spongy urethra
- ☐ Corpora cavernosa
- ☐ Corpus spongiosum
- ☐ Glans penis
- ☐ External urethral orifice
- ☐ Rectum

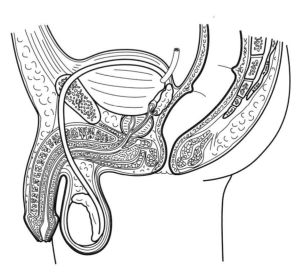

FIGURE **19.2** Organs of the male reproductive system, midsagittal section through the pelvis.

UNIT **5:** The Next Generation   CHAPTER **19:** Male Reproductive System ■ **363**

3 Label Figures 19.3 through 19.8 using the list of structures provided for each figure.

- ☐ Tunica albuginea
- ☐ Septae
- ☐ Rete testis
- ☐ Seminiferous tubules
- ☐ Lobules

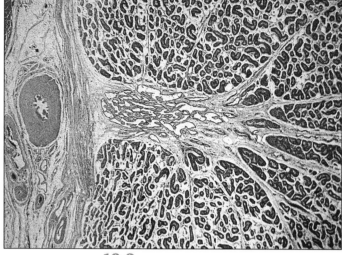

FIGURE **19.3** Testis (low magnification).

- ☐ Seminiferous tubule
- ☐ Lumen
- ☐ Interstitial (Leydig) cell

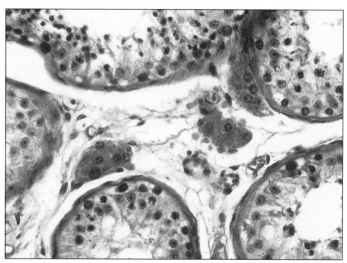

FIGURE **19.4** Testis (higher magnification).

- ☐ Sertoli cells
- ☐ Spermatogonia
- ☐ Spermatids
- ☐ Primary spermatocytes
- ☐ Boundary of seminiferous tubule

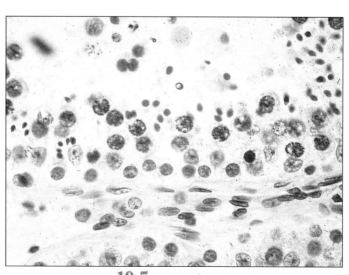

FIGURE **19.5** Seminiferous tubule.

364 ■ *Discovering Anatomy: A Guided Examination of the Cadaver*

- ☐ Lumen containing sperm
- ☐ Pseudostratified columnar epithelium
- ☐ Stereocilia

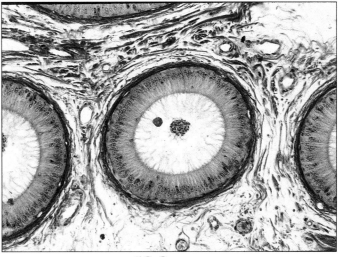

FIGURE **19.6** Epididymis.

- ☐ Pseudostratified columnar epithelium
- ☐ Lamina propria
- ☐ Muscularis externa

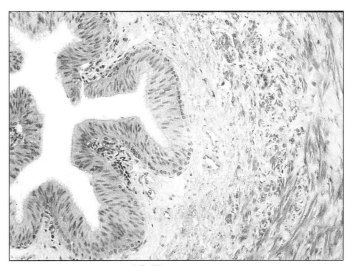

FIGURE **19.7** Ductus deferens.

- ☐ Prostate concretions
- ☐ Glandular acini
- ☐ Fibromuscular stroma

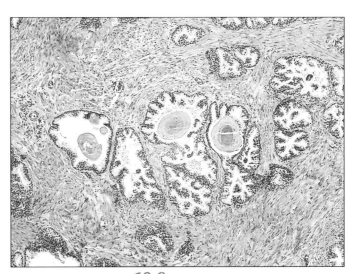

FIGURE **19.8** Prostate gland.

OBSERVING

Complete the following hands-on laboratory activities to apply your knowledge of the male reproductive system.

MATERIALS
Obtain the following items before beginning the laboratory activities:

- ❏ Colored pencils
- ❏ Light microscope
- ❏ Prepared microscope slides of the testis, ductus deferens, and prostate
- ❏ Cadaver
- ❏ Gloves
- ❏ Probe
- ❏ Protective gear (lab coat, scrubs, or apron)

ACTIVITY 1 Microscopic Structures of the Male Reproductive System

1 Obtain microscope slides of the testis, ductus deferens, and prostate from your instructor.

2 Place each slide under the microscope and start with the lowest power and advance to higher magnification to see more detail.

3 Once you have a view of the tissue use your colored pencils and draw what you see in the field of view. Color in the different cells or layers and label the terms indicated.

a. Testis (low power)

- ☐ Tunica albuginea
- ☐ Rete testis
- ☐ Septae
- ☐ Lobules
- ☐ Seminiferous tubules

b. Testis (high power)

- ☐ Seminiferous tubules
- ☐ Spermatozoa
- ☐ Interstitial cells

366 ■ *Discovering Anatomy: A Guided Examination of the Cadaver*

c. Ductus deferens

- ☐ Lumen
- ☐ Pseudostratified colunar epithelium
- ☐ Lamina propria
- ☐ Muscularis externa

d. Prostate gland

- ☐ Glandular acini
- ☐ Prostatic concretions
- ☐ Fibromuscular stroma
- ☐ Urethra (if present)

ACTIVITY 2 Gross Anatomy of the Male Reproductive System

1 Identify the following structures in Figure 19.9 on the cadaver.

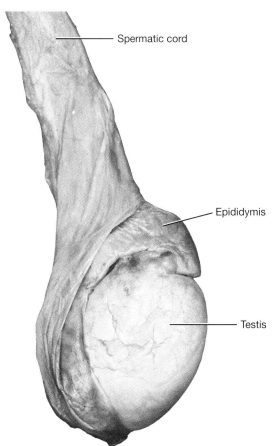

FIGURE **19.9** Lateral view of the spermatic cord and testis with scrotal sac removed.

2 Identify the following structures in Figure 19.10 on the cadaver.

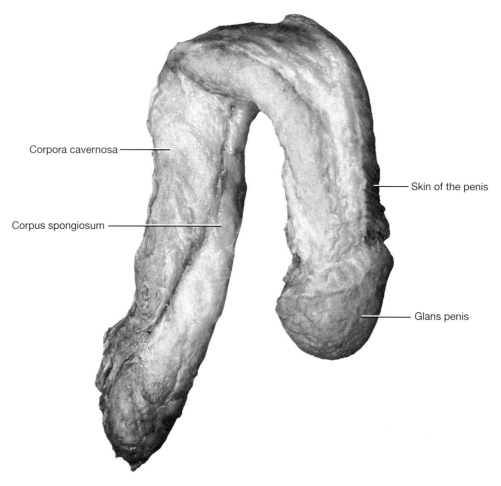

FIGURE **19.10** Structure of the penis, lateral view.

WRAPPING UP

Complete the following additional activities to help retain your knowledge of the male reproductive system.

Name _____

Date _____ Section _____

1 Match the male reproductive structure in the right-hand column with its correct description in the left-hand column.

Description	Male reproductive structure
____ Walnut-sized gland inferior to the bladder	a. Seminal vesicle
____ Sac of skin that houses the testicles	b. Bulbourethral gland
____ Contributes fructose to sperm for energy	c. Scrotal sac
____ Produces sperm	d. Ductus deferens
____ Male copulatory organ	e. Seminiferous tubules
____ Secretes mucus to line the lumen of the urethra	f. Penis
____ Erectile tissue that fills with blood resulting in an erection	g. Prostate gland
____ Tube that transports sperm from the epididymis to the ejaculatory duct	h. Corpora cavernosa

2 Trace the pathway for sperm, starting from its origin and ending outside of the body.

3 Provide an explanation for why the testes, rather than the penis, are considered the primary male sex organs.

4 A 16-year-old male jumps into a cold lake. He notices that his scrotum is shrunken and wrinkled. Provide an explanation for this observation.

5 A histological examination of 52 infertile men from a developing country reveals parasitic infection of the seminal vesicles by *Schistosoma haematobium*. This infection results in the inactivity of this gland. Identify the most likely component of semen that would be in low yield as a result of this infection and explain why.

6 In one sentence, explain what a vasectomy procedure is. Will a man who undergoes a vasectomy still produce ejaculate? Explain.

7 Upon palpation of a 71-year-old man's prostate gland, the examiner documents her findings as follows: "The prostate is smooth, without nodularity, size seems enlarged." She suspects that this patient has benign prostatic hypertrophy (enlargement of the prostate gland). Explain why asking if the patient has difficulty initiating the urination process and voiding his bladder would help confirm a diagnosis of benign prostatic hypertrophy.

8 Explain why the term "urogenital system" is more appropriate for males than it is for females.

Female Reproductive System

CHAPTER 20

At the completion of this laboratory session, you should be able to do the following:

1. List the major functions of the female reproductive system.
2. Describe the structure and function of the ovaries.
3. Outline the location, structure, and function of female accessory organs.
4. Describe the anatomy of female external genitalia.

The female reproductive system consists of the primary sex organs (ovaries) and secondary sex organs (oviducts, uterus, vagina, and external genitalia). All of these organs remain immature during the first decade or so of life. Within the following years, sexual development occurs and menses first appear; the first cycle of menstrual bleeding is referred to as menarche. The menstrual cycle has an average length of approximately 28 days. Menstrual cycles cease at about the fifth decade of life when women typically undergo menopause.

GETTING ACQUAINTED
Complete the following activities to become familiar with the female reproductive system.

MATERIALS
Obtain the following items before beginning the laboratory activities:
- ❏ Textbook or access to Internet resources
- ❏ Colored pencils

ACTIVITY 1 Female Reproductive Organs

The Ovaries
The ovaries (female gonads) are considered the primary sex organs of the female reproductive system because they produce eggs (oocytes, ova) and the female sex hormones estrogen and progesterone. Each ovary is approximately 4 × 3 × 2 cm in size. The paired ovaries are located on each side of the uterus in the pelvic cavity and are held in place by the broad ligament. The overall organization of the ovary is as follows:

1. **Tunica albuginea.** Each ovary is surrounded by a dense collagenous connective tissue capsule called the tunica albuginea. This tissue is covered with a simple cuboidal epithelium called germinal epithelium.
2. **Cortex and medulla.** The ovary consists of a cortex and a medulla. The cortex is the outer cellular region containing thousands of follicles embedded in connective tissue. The medulla is the central region consisting of loose connective tissue, vascular supply, and nerves.
3. **Follicle.** Each ovary contains follicles, which are the structural and functional unit of the ovary. Each follicle consists of a single oocyte surrounded by one or more layers of epithelial follicular (granulosa) cells, which produce sex hormones. During the reproductive years, follicles at various stages of maturation are present in the ovarian cortex, ranging from the immature primordial follicles to the mature vesicular (Graafian) follicle, from which an oocyte is released during ovulation.

Broad Ligament
The ovaries, uterine tubes, and uterus are covered by a layer of peritoneum on the anterior, superior, and posterior surfaces. Inferior to the uterine tube and lateral to the uterus, the peritoneal membrane is fused into a double layer called the broad ligament. The broad ligament is divided into the following three portions:

1. **Mesovarium.** The ovary is partially covered by a separate posterior fold of the broad ligament called the mesovarium.

371

2. **Mesosalpinx.** The mesosalpinx is located between the uterine tube in the free border of the broad ligament and the posteriorly projecting mesovarium.
3. **Mesometrium.** The mesometrium is the largest portion of the broad ligament. It is situated below the ovary and ovarian ligament. The mesometrium supports the uterus within the pelvic cavity and transports the uterine vessels.

The Female Duct System

The female reproductive system possesses the following accessory organs:

1. **Uterine tubes.** The uterine tubes, also known as the fallopian tubes and oviducts, consist of a pair of thin muscular tubes, each about 10 to 12 cm long, which stretch between the ovary and the uterus. The lumen of the uterine tube is very narrow (about as wide as a human hair) and its opening provides open communication between the uterine tube and peritoneal cavity. The uterine tubes join the superolateral portion of the uterus. Uterine tubes have the following three parts with specific functions:
   a. **Infundibulum.** Possesses fimbriae, which help direct the oocyte into the lumen of the uterine tube.
   b. **Ampulla.** Region of fertilization of egg and sperm.
   c. **Isthmus.** Area of zygote division or reabsorption of unfertilized secondary oocyte.

   Note: Unlike the male reproductive tract, the tubule system of the female tract is not continuous. Therefore, the oocyte is actually released into the peritoneal cavity, where the fingerlike extensions of the uterine tube, called fimbriae, must catch the oocyte and bring it into the uterine tube.

2. **Uterus.** The uterus is a pear-shaped organ that functions to support and nourish the growing fetus. The uterus is divided into the following regions:
   a. **Fundus.** The superior dome-shaped region.
   b. **Body.** The large central area where implantation occurs.
   c. **Cervix.** The inferior narrow portion that continues with the vagina. The external os of the cervix provides the communication between the uterus and vagina. The wall of the uterus has three layers:
      i. **Perimetrium.** Consists of a thin layer of mesothelium and a thin layer of loose connective tissue beneath it.
      ii. **Myometrium.** Consists of several layers of smooth muscle. The central layers contain large arcuate arteries, which give rise to both the straight and spiral (coiled) arteries of the endometrium.
      iii. **Endometrium.** Simple columnar epithelium that internally line the uterus. Some cells are ciliated cells and some are secretory in nature. The underlying lamina propria contains additional simple tubular mucus-secreting (endometrial) glands.

3. **Vagina.** The vagina is a muscular tube that extends from the external os of the cervix inferiorly to the vaginal orifice. The recesses between the cervix and vaginal wall are known as the fornices. The vagina is the receptacle for the penis during sexual intercourse and the passageway for menses during the menstrual cycle. It is also part of the birth canal, as it serves as the passageway for the fetus during childbirth. Flanking the vaginal orifice are the greater vestibular (Bartholin's) glands, which secrete mucus to lubricate the vaginal canal during sexual intercourse.

Clinical Application

Tubal Pregnancy

During ovulation, the egg leaves the ovary and typically enters the adjacent uterine tube. If intercourse has occurred, and the sperm have traveled into the correct uterine tube, and fertilization occurs, then a zygote is formed. If a typical pregnancy occurs, the zygote travels into the uterus where it implants in the endometrium. However, the zygote occasionally implants in tissues other than the endometrium. The most common location is in the ampulla of the uterine tube, producing a tubal (ectopic) pregnancy. Tubal pregnancy is a painful, potentially life-threatening condition because the tube can rupture, resulting in hemorrhage and shock. Chronic inflammation and scarring of the uterine tube, which can result from a sexually transmitted disease such as gonorrhea or chlamydia, increase the chance that a tubal pregnancy will occur.

Clinical Application

Uterine Size and Pregnancy

The uterus is a small organ and usually measures 7.5 cm long and 5.0 cm wide (about the size of a small pear). However, pregnancy changes all of that. By 20 weeks of pregnancy the fundus of the uterus expands to the level of the umbilicus. By 36 weeks the lateral edges of the fundus reach the inferior border of the costal angle. By 40 weeks (full term) the fundus is in contact with the diaphragm muscle.

Clinical Application

Endometriosis

Endometriosis is a condition in which endometrial tissue that normally lines the uterine lumen is growing outside the uterus in abnormal locations like the uterine tubes and peritoneal cavity. Approximately 10 percent of women of reproductive age are significantly affected by endometriosis, which can cause infertility, dysmenorrhea (menstrual cramps), and pelvic pain. Severe cases can result in hemorrhage and adhesions between the uterine tubes, ovaries, and other structures in the pelvis.

1 Using your textbook or online resources, write a definition for each of the terms listed in Table 20.1.

TABLE **20.1** Female Reproductive Organs

Term	Definition
Ovaries	
Follicle	
Broad ligament	
Uterine tube (oviduct)	
Uterus	
Cervix	
Perimetrium	
Myometrium	
Endometrium	
Vagina	

2 Identify, label, and color the following structures on Figure 20.1.

- ☐ Ovary
- ☐ Ovarian ligament
- ☐ Follicles
- ☐ Graaffian follicle
- ☐ Ovum (being ovulated)
- ☐ Uterine tube
- ☐ Uterus
- ☐ Mesosalpinx
- ☐ Mesovarium
- ☐ Mesometrium
- ☐ Endometrium
- ☐ Myometrium
- ☐ Perimetrium
- ☐ Cervix
- ☐ External os
- ☐ Vaginal fornices
- ☐ Vaginal rugae

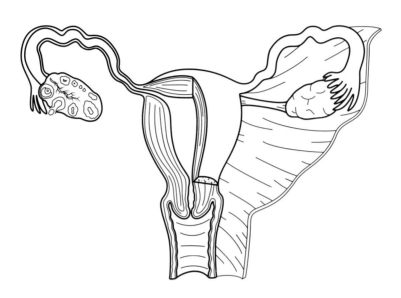

FIGURE **20.1** Posterior view of female reproductive system, coronal section.

ACTIVITY 2 External Female Genitalia

The female reproductive structures within the perineum are called the external genitalia (vulva). These structures include the following:

1. **Mons pubis.** The fatty, rounded area overlying the pubic symphysis is called the mons pubis. After puberty this area is covered with pubic hair.
2. **Labia majora.** The labia majora are two elongated, hair-covered fatty skin folds running posteriorly from the mons pubis.
3. **Labia minora.** The labia majora enclose two thin, hair-free skin folds called the labia minora. The labia minora enclose a recess called the vestibule, which contains the external openings of the urethra and the vagina. Flanking the vaginal orifice are the pea-sized greater vestibular glands, which secrete mucus into the vestibule and help to keep it moist and lubricated, facilitating intercourse.
4. **Clitoris.** The clitoris is at the anterior junction of the labia minora folds and is composed largely of erectile tissue. It is richly innervated with sensory nerve endings sensitive to touch (~8,000 nerve endings—more than any other tissue in the body). The clitoris becomes swollen with blood and erect during stimulation, contributing to a female's sexual arousal. Like the penis, the body of the clitoris has dorsal erectile columns (corpora cavernosa) attached proximally by crura, but it lacks a corpus spongiosum that conveys a urethra.

Note that in males the urethra carries both urine and semen and runs through the penis, but the female urinary and reproductive tracts are completely separate.

Weird and Wacky

Who Named That?

There are numerous peripheral nerves in the body. For example, the sciatic nerve provides innervation of the lower limb, the median nerve innervates the thenar muscles and skin over the lateral hand, and so forth. The perineum is no different. The pudendal nerve is the name of the nerve that innervates the perineum in men and women. It arises from the S2–S4 ventral rami and provides sensory innervation to the perineum (i.e., clitoris and labia minora) and motor innervation to perineal muscles. When early anatomists named this nerve, they said, "What should we call this ever so important nerve? How about the nerve that innervates tissue we should be ashamed of?" The discussion of reproductive parts, it seems, has always had some social discomfort associated with it. Perhaps this is a reason so many statues and paintings have fig-leafs covering the external genitalia. And so the pudendal nerve (Latin for "thing to be ashamed of") has persisted in anatomy books for centuries.

1 Using your textbook or online resources, write a definition for each of the terms listed in Table 20.2.

TABLE **20.2** Female Perineum

Term	Definition
Mons pubis	
Labia majora	
Labia minora	
Clitoris	
Urethral orifice	
Vaginal orifice	
Anus	
Perineum	

2 Identify, label, and color the following structures on Figure 20.2.

- ☐ Peritoneal cavity
- ☐ Parietal peritoneum
- ☐ Visceral peritoneum of the bladder
- ☐ Urinary bladder
- ☐ Ovary
- ☐ Uterine tube
- ☐ Uterus
- ☐ Cervix
- ☐ Vagina
- ☐ Rectum
- ☐ Anus
- ☐ Clitoris
- ☐ Labia minora
- ☐ Labia majora
- ☐ Urogenital diaphragm
- ☐ Pubic symphysis

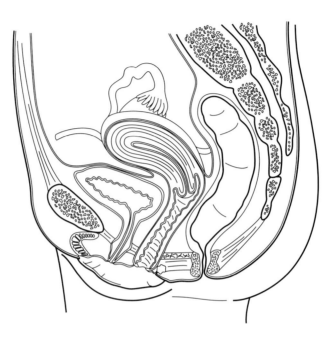

FIGURE **20.2** Organs of the female reproductive system, midsagittal section through the pelvis.

3 Label Figures 20.3 through 20.6 using the list of structures provided for each figure.

- ☐ Primordial follicles
- ☐ Primary follicles
- ☐ Atretic follicle
- ☐ Tunica albuginea

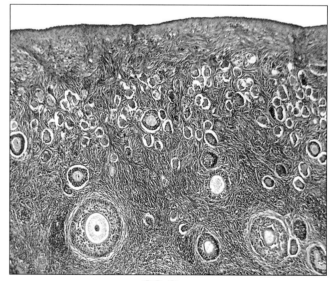

FIGURE **20.3** Ovary.

- Lumen
- Mucosal folds
- Muscularis externa
- Serosa

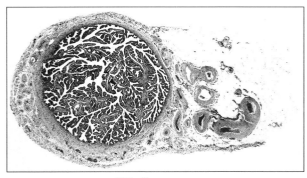

FIGURE **20.4** Uterine tube.

- Endometrial stroma
- Endometrium (lamina functionalis)
- Endometrium (lamina basalis)
- Endometrial uterine glands
- Myometrium

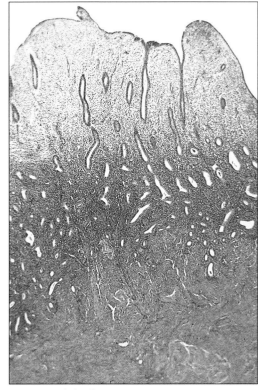

FIGURE **20.5** Uterus (proliferative phase).

- Lumen
- Nonkeratinized stratified squamous epithelium
- Lamina propria
- Capillary

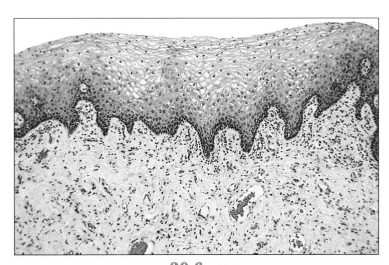

FIGURE **20.6** Vagina.

OBSERVING

Complete the following hands-on laboratory activities to apply your knowledge of the female reproductive system.

MATERIALS

Obtain the following items before beginning the laboratory activities:

- ❏ Colored pencils
- ❏ Light microscope
- ❏ Prepared microscope slides of the ovary, uterine tube, uterus, and vagina
- ❏ Cadaver
- ❏ Gloves
- ❏ Probe
- ❏ Protective gear (lab coat, scrubs, or apron)

ACTIVITY 1 Microscopic Structures of the Female Reproductive System

1 Obtain microscope slides of the ovary, uterine tube, uterus, and vagina from your instructor.

2 Place each slide under the microscope and start with the lowest power, advancing to higher magnification to see more detail.

3 Once you have a view of the tissue, use your colored pencils to draw what you see in the field of view. Color in the different cells or layers and label the terms indicated.

a. Ovary

- ☐ Tunica albuginea
- ☐ Primordial follicle
- ☐ Primary follicle
- ☐ Secondary follicle
- ☐ Graafian follicle

b. Uterine tube

- ☐ Mucosa
- ☐ Submucosa
- ☐ Serosa

c. Uterus

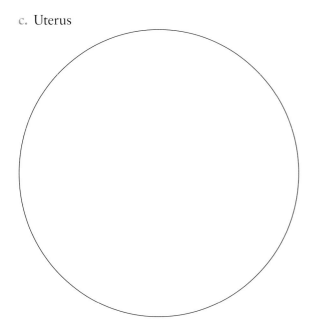

d. Vagina

☐ Endometrium
☐ Myometrium
☐ Endometrial glands

☐ Lumen
☐ Nonkeratinized stratified squamous epithelium
☐ Lamina propria

ACTIVITY 2 Gross Anatomy of the Female Reproductive System

1 Identify the following structures in Figure 20.7 on the cadaver.

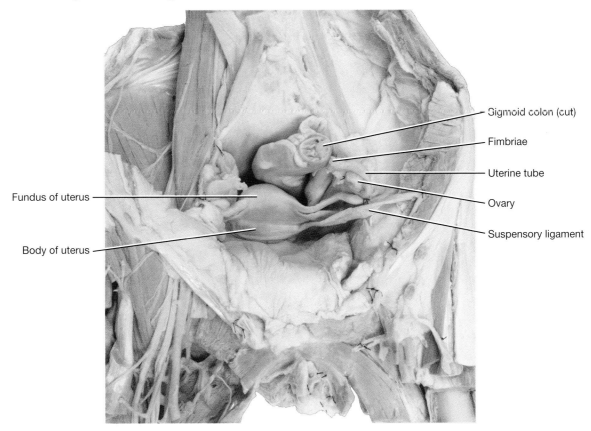

FIGURE **20.7** Female reproductive system.

UNIT 5: The Next Generation   CHAPTER 20: Female Reproductive System ■ **379**

2 Identify the following structures in Figure 20.8 on the cadaver.

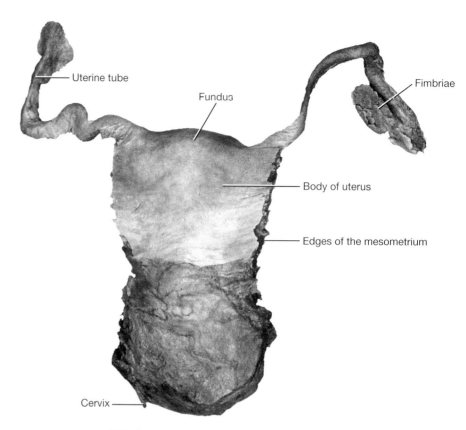

FIGURE **20.8** Uterus removed from pelvic cavity.

3 Identify the following structures in Figure 20.9 on the cadaver.

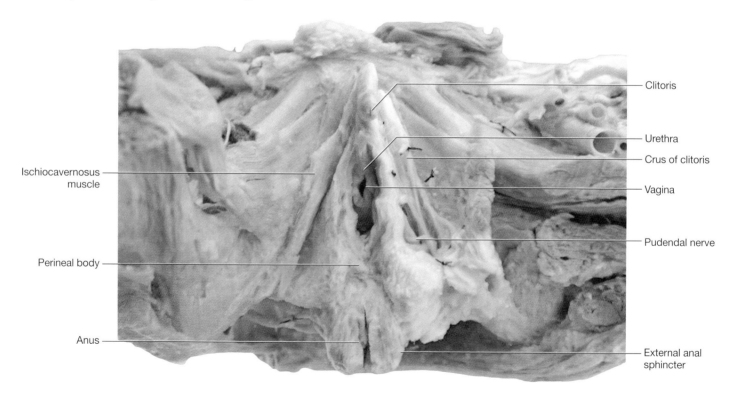

FIGURE **20.9** External genitalia of the female reproductive system (skin removed from perineum; right side shows muscles of the perineum and left side muscles are removed).

WRAPPING UP

Complete the following additional activities to help retain your knowledge of the female reproductive system.

Name _____

Date _____ Section _____

1 Match the female reproductive structure in the right-hand column with its correct description in the left-hand column.

Description

_____ Location of ovulation

_____ Birth canal

_____ Double-layered membrane that covers ovaries, fallopian tubes, and the uterus

_____ Erectile tissue body that flanks the vestibule

_____ Oviduct

_____ Erectile tissue equivalent of the glans penis in the male

_____ Perineal structure with pubic hair

_____ Lubricates the vaginal canal

Female reproductive structure

a. Clitoris
b. Vestibular gland
c. Labia majora
d. Uterine tube
e. Ovary
f. Vaginal canal
g. Labia minora
h. Broad ligament

2 Trace the pathway of a sperm from the seminiferous tubules of the male to the uterine tube in a female.

3 Occasionally the zygote implants at sites other than the uterus. The most common location for this to occur is the uterine tube. This condition is called a tubal (ectopic) pregnancy. Provide an explanation on why a tubal pregnancy can become a serious medical condition.

4 Endometriosis is a condition in which endometrial tissue flourishes outside the uterine cavity (most commonly in the peritoneal cavity). Provide an explanation for how endometrial tissue gets into the peritoneal cavity. Propose an explanation on why endometriosis may result in female infertility.

UNIT 5: The Next Generation

CHAPTER 20: Female Reproductive System ■ **381**

5. Fibroids (leiomyomas) are growths in the walls of the uterus and vary in size from as small as a seed to as large as a grapefruit. Fibroids are often referred to as tumors but they are not cancerous. They are smooth muscle growths. Explain from which layer of the uterus leiomyomas arise.

6. A 42-year-old woman presents with a painful, 3 cm paraovarian cyst located in her broad ligament. Ultrasonography reveals the cyst to be adjacent to the uterine tube, and laparoscopic surgery is scheduled. During surgery a small incision is made in the anterior abdominal wall, and the scope is inserted into her peritoneal cavity. What part of the broad ligament would most likely be incised to reach the tumor? Provide an explanation.

Index

Note: All *italicized* page numbers refer to figures and tables.

abdomen, 3, 8, 121, 266, 267, *337*
abdominal cavity, *252*, 294, 327, 332, *354*
abduction, 102, *103*, 114, 146, 161
abductor pollicis longus muscle, 151, 153, *156*
acetabulum, 86, 87, *92*, 161
Achilles tendon, 172, 174
acromioclavicular (AC) joint, 82, *110*, *111*
acromion, 81, 82, 83, *90*
adduction, 102, *103*, 114, 146, 161
adductor brevis muscle, 161, 162, *177*
adductor longus muscle, 161, 162, 165, *176*, *177*
adductor magnus muscle, 161, 162, 165, *176*, *177*
adipose tissue, 27, 28, 41, 50, 51, *257*, 340
adventitia, 297, 324, 348
alveolar sac, 300, 301, 305, 306
alveoli, 297, 298, 300, 301, 302, 303, *303*, 306
ampulla, 136, 316, 321, 322, 345, 372
anatomy, language of, 3, 4
anconeus muscle, 151, 153, *156*
ankle, 10, *97*, 101, *112*, 161, *170*, *173*
   muscles of, 169, *169*, 171, *172*, *179*, *180*
anterior inferior iliac spine, 86, 87, *93*
anterior superior iliac spine, 86, 87, *93*, *176*, *177*
anus, 317, 375, 376, *380*
aorta, 16, *16*, 43, 259, 260, 261, *294*, *307*, 327, *333*, 342
   abdominal, *252*, 266, 268, 272, 281, *354*
   ascending, 263, 277
   aortic arch, *253*, 263, 264, 265, 266, 268, 270, *275*, *276*, 281
appendicular skeleton, 59, 81, 90
appendix, 318, 319, *334*
arachnoid mater, 191, 192, 193, *197*
arm, 9, 12, *12*
arrector pili muscle, 50, 51, *55*
arteries, 22, *195*, *213*, *257*, 265, 267, 268, 278, 290, 293
   abdominal, 268, *272*
   anterior descending, 262, 263
   anterior tibial, 268, 273
   arcuate, 340, 346
   axillary, 266, 270
   brachial, 266, 270
   brachiocephalic, 266, 268, 270
   circumflex coronary, 264
   common carotid, *253*, 263, 268, *276*
   common iliac, 268, 272, 342
   coronary, *257*, 261, 262, 263, 264, 361

   dorsalis pedis, 266, 273
   external carotid, 268
   external iliac, 268, 272, 273
   femoral, 265, 266, 268, 273
   gonadal, 266, 268
   hepatic, 320, 321, 327, 329
   inferior mesenteric, 266, 268, 272
   intercostal, 123, *126*, *214*
   interlobular, 340, 346
   internal carotid, 266, 268
   internal iliac, 268, 272
   popliteal, 268, 273
   posterior descending, 264
   posterior tibial, 273
   pulmonary, 260, 261, 263, 264, 265, *276*, *308*
   radial, 266, 270
   renal, 265, 266, 268, 272, 340, 342, 343, *354*
   sinoatrial nodal, 261, 262, 263
   subclavian, 263, 266, 268, 270, *276*
   superior mesenteric, 266, 268, 272
   testicular, 360
   thoracodorsal, 139
   ulnar, 266, 270
   vertebral, 268
arterioles, 320
   glomerular, 265, 266, 267, 340, 343, 344, 345, 346, 347, 351
   pulmonary, 301, 302, 306
atlantoaxial joint, *111*
atlanto-occipital joint, *111*
atrioventricular (AV) bundle/node, 262, 263
atrium
   left, 259, 260, 264, 265, *276*
   right, 259, 260, 263, 264, 265, 267, *276*, 277
auditory canal, 228, 233, 237
autonomic nervous system (ANS), 203, 211, 212, *212*
axial skeleton, 59, 65–66, 70, 72
axillary, 8, 9, 12, *12*

back
   muscles of, 121, 122
   skin of, 208
ball-and-socket joint, 104, 108, 138, 161
basement membrane, 21, 302, 306
biceps brachii muscle, 143, 146, *155*, *156*, *215*
biceps femoris muscle, 166, 168, *175*
bladder, 323, 344, 352
   urinary, 43, 339, 341, 342, 348, 349, *349*, 352, 353, 363, 376
blind spot, 229, 236, *236*
blood, 26, 31, 33, *34*, 40, 265, 287

bones, 26, 30, 33, 108, 440
   identifying, 59
   micrograph of, *34*
   number of, 87
Bowman's capsule, 340, 343, 346, 347, 350, 352
Bowman's space, 343, 346, 347, 350, 352
Boyle's law, 304, 305
brachial plexus, 203, 207, 209, *209*, *215*, 223
brachialis muscle, 143, 146, *155*
brachiocephalic trunk, 263, 276
brachioradialis muscle, 151, 153, *155*, *156*, *157*, *158*
brain, *194*, 266
   CNS and, 187
   parts of, 187–88
   views of, *190*, *191*, *192*, *193*, *195*, *196*, *206*, *213*
brainstem, 187, 189, 212
breathing, mechanics of, 297, 304
bronchi, 43, 297, 298, 299, *307*, *308*
bronchiole, *300*, 302, *302*
   respiratory, 300, 301, 303, 306
   terminal, 298, 301, 303, *303*, 305, 306
buccal region, 7, 15, *15*
buccinator muscle, 119, 123, *125*
burns, 49

calcaneus, 90, *94*, 172, *179*, *180*
capillaries, 38, 266, 267, 340, 377
   glomerular, 341, 343, 350
   lymphatic, 285, 288
   peritubular, 341, 343, 345, 346, 351
   pulmonary, 265, 300, 301, *303*, 306
capitulum, 84, 85, 86, *91*
cardiac muscle, 35, 36, 41, 42, 139, 262
cardiovascular system, 3, 11, 245, *259*, 261, 263, *275*, 285
carina, 297, 299, *307*
carpal bones, 85, *110*
carpometacarpal joints, *110*
cartilage, 26, 40
   articular, 108
   costal, 71, 76, 121, *126*
   cricoid, *307*
   elastic, 28, 30, 33, 232
   hyaline, 30, 33, 302, 306
   micrograph of, *34*
   thyroid, 245, *253*, *307*
cartilaginous joints, 101, 104, *107*
cauda equina, 192, *194*, *197*
caudate lobe, *334*
caudate nucleus, *197*

cecum, 317, 318, 319, *334*
celiac trunk, 266, 268, 272
central canal, 31, *198*
central nervous system (CNS), 188, 194, *194*, 203, 204, 210
cerebellum, 187, 189, 190, 191, 193, *195*, *196*, *213*, *252*
cerebral aqueduct, 187, 189, 191, *196*
cerebrospinal fluid (CSF), 187, 191, 192
cerebrum, 187, 188, 192, 193, *252*
cervical plexus, 107, 207, 209
cervical region, 8, 14, 15, *15*, *253*
cervix, 372, 373, 374, 376, *380*
chemoreceptors, 227, 228, 230, 233
chordae tendinae, 259, 260, 277
choroid, 228, 230, *235*
chronic obstructive pulmonary disease (COPD), 298
ciliary body, 228, 230, *235*
clavicle, 81, 82, *125*, 135, *155*
clitoris, 374, 375, 376, *380*
club cells, 300, *300*
collagen fibers, 21, 26, 27, 40, 348
colon, 318
   sigmoid, 317, 319, *379*
   transverse, 319, 323, 327, *330*, *332*, *334*
compact bone, 30, *31*, *46*
conduction system, *257*, 262, *263*, 297–98
condyles, 59, 60, 63, 73, 88, 89, *93*
condyloid process, 61, 64, *72*
conjunctiva, 228, *235*
connective tissues (CT), 21, 26, 27, 28, *43*, 203, 207, *257*, 324, 340
   classification of, 28, *28*, *33*
   dense irregular collagenous, 28, 40, 50, 105, 106
   dense regular collagenous, 28, 41
   irregular collagenous, 362
   loose, 26, 28, 40, 50
   micrographs of, *29*
   types of, 30, 40
conus medullaris, 192, *197*
coracobrachialis muscle, 138, *155*, *155*
coracoid process, 82, 83, *90*
cornea, 228, 229, 230, *235*, 299
coronary circulation, 259, 261, 262, 263
coronoid process, 61, 64, *72*, 84, 85, *91*
corpus callosum, 188, 191, *196*, *197*, *252*
corpus cavernosa, 362, 363, *368*, 374
corpus spongiosum, 362, 363, *368*, 374

corpuscles
  Hassals, 286
  Meissner's, 48, 50, 51
  pacinian, 48, 50, 51, 52, 55
  tactile, 48, 50
  thymic, 286
corrugator supercilli muscle, 119, 120
cortex, 243, 291, 371
  adrenal, 245, 246, 248
  cerebral, 188
  lymph node, 292
  olfactory, 230
  renal, 340, 341, 343, 346, *350*, 352, *354*
cranial bones, 59, 50
cribriform foramina, 61, 64
cribriform plate, *73*, 228, 231
crista galli, 61, 64, *73*
crus, 10, *13*
cuboid, 90, *94*

deep, 4, *5*
deltoid muscle, 124, *127*, *128*, *129*, *130*, 138, 143, *154*, *155*, 155
deltoid tuberosity, 84, 85, *91*
depressor anguli oris muscle, 119, 123, *125*
depressor labii inferioris muscle, 119
dermis, 47, 48, 50, 51, 52, 53, 54, 55
  tattoos and, 49
  terms for, *50*
detrusor muscle, 349, 352, 353
diaphragm, 16, *16*, 252, 266, *275*, *293*, *294*, 299, 304, 305, *307*, 316, *330*, *333*, 342
  views of, *309*
diastole, 257, 259, 261
diencephalon, 187, 188, 189
digestion, 315, 316, 320, 321, 328
digestive system, 3, 11, 285, *319*, *320*, *325*
digits, *12*, *13*, 152
distal convoluted tubule (DCT), 344, 345, 346, 347
distal phalanx, 86, 90, *92*, *94*
dorsal ganglion, *214*
dorsal interosseus muscle, 1st, *157*
dorsal root, *198*, 207, 208, *214*
dorsiflexion, 102, *103*, 114, 169
ducts
  alveolar, 300, 301, 303, 306
  bile, 320, 321, 322, 327, 329
  collecting, 344, 345, 346, 351, 353
  common hepatic, 322
  cystic, 322
  ejaculatory, 359, 361, 363
  lymphatic, 285, 288, 289
  nasolacrimal, 62
  pancreatic, 250, 321, 322
  seromucous, 302, 306
  sweat, 51
  thoracic, 285, 286, 288, 289
ductus deferens, 359, 360, 361, 363, *365*, *366*, *367*
duodenum, *294*, 316, 318, 319, 320, 321, 322, *322*, 323, *333*, 327
dura mater, 191, 192, 193, 214, *214*
dust cells, 300, 301, *303*, 306

ear, 18, 220, 227, 228
  coronal section of, *233*
  external, 232, 233, *236*
  internal, 232, 233, *237*
  middle, 232, 233, *237*
elastic fibers, 40, 301, 324
elbow, 96, 101, *110*, *111*, 135
  movements of, *144*, *145*, *148*
  muscles of, 143, *144*, *154*, *155*
endocrine system, 3, 11, 243, *248*, 250, *252*, 255
  terms for, 244, *246–47*
endometrium, *372*, 373, 374, 377, 379
epicondyle, 59, 84, 88, *91*, 151, *158*
epidermis, 48, 49, 50, 51, 52, 55
  appendages of, 47, 53, 54
  terms for, *50*
epididymis, 359, 360, 361, 363, *365*, *367*
epiglottis, 234, 307, 316
epiploic appendages, 317, *334*
epithelium, 21–22, *23*, *24*, *25*, 38, 43, 324
  germinal, 371
  pseudostratified columnar, 23, *24*, *365*, *367*
  respiratory, 302, 306
  simple columnar, 23, *24*, *39*, 292
  simple cuboidal, 23, *24*, *39*, 343, 371
  simple squamous, 22, *23*, *24*, *39*
  stratified squamous, 23, *24*, *39*, 50, 324, 326, 328, 377, 379
  transitional, 23, 348, 351, 352, 353
erector spinae muscles, 121, 122, 124, *130*
esophagus, 316, 318, 319, 324, *326*, 328, *332*
estrogen, 245, 247, 371
ethmoid bone, 64, 65, *73*, 229, 371
extension, 102, *103*, 114, 143, 146, 147, 161, 166
extensor carpi radialis brevis muscle, 151, 153, *156*, *157*
extensor carpi radialis longus muscle, 151, 153, *156*, *157*
extensor carpi ulnaris muscle, 151, 153, *156*, *157*
extensor digiti minimi muscle, 151, 153, *156*, *157*
extensor digitorum communis muscle, 151, 153
extensor digitorum longus muscle, 169, 171
extensor digitorum muscle, *156*, *157*
extensor digitorum tendons, *156*, *157*
extensor hallucis longus muscle, 169, 171
extensor indicis muscle, 151, 153, *156*, *157*
extensor pollicis brevis muscle, 151, 153, *156*, *157*
extensor pollicis longus muscle, 151, 152, 153, *156*, *157*
extensor retinaculum, *157*

external genitalia, female, 371, 374, 375
external oblique muscle, 121, 122, 124, *126*, *127*, *128*, *129*, *130*
extracellular matrix (ECM), 21, 26, 27, 30, 40
extracorporeal shock wave lithotripsy (ESWL), 344
eye, 18, 228, 230, 235
  anatomy of, 227, 229
eyelids, 229, *235*

face, anterolateral view of, *15*
facial bones, 59, 61–62
facial muscles, *125*, 221
falx cerebri, 191, *197*
femeropatellar joint, *112*
femur, 81, 88, 89, *89*, 93, 161, 166
fibrocartilage, 30, 33, 40
fibula, 81, 88, 89, *89*, 94, 106, *106*, 175
fibularis, 169, 171, *179*
filum terminale, 192, *197*
fimbriae, *379*, 380
fingers, 135, *149*
flexion, 102, *103*, 114, 138, 143, 146, 147, 161, 166, 169
flexor carpi radialis muscle, 146, 150, *158*
flexor carpi ulnaris muscle, 146, 150, *158*
flexor digitorum brevis muscle, *180*
flexor digitorum longus muscle, 171, 172, 174, *179*, 180
flexor digitorum profundus muscle, 146, 150, *158*
flexor digitorum superficialis muscle, 146, 150, *158*
flexor hallucis longus muscle, 171, 172, 174, *180*
flexor pollicis longus muscle, 146, 150, *158*
flexor tendons, 85, *179*
follicles, 249, 250, 371, 373
  atretic, 376
  Graafian, *372*, 374, 378
  hair, 50, 51, 52, 54, 55
  lymphoid, 317
  primary, 376, 378
  secondary, 378
  thyroid, 250
  vesicular, 372
foot, 89, 101, 161
  views of, *13*, *90*, *94*, *111*, *179*, *180*
foramen, 59
  incisive, *73*
  infraorbital, 62, *72*
  interventricular, 187, 189
  intervertebral, 65, 208
  jugular, 60, 63, *73*
  lacrimal, 229
  mental, 64, *72*
  obturator, 86, 87, *92*, 93
  stylomastoid, 60, 63
  supraorbital, 60, 62, *72*
  transverse, 66, 69, *75*, *76*
  vertebral, 65, 69, *75*, *76*

foramen lacerum, 61, *73*
foramen magnum, 60, 63, 64, *73*, 74, *196*
foramen ovale, 60, 64, *73*
foramen rotundum, 61, 64, *73*
foramen spinosum, 61, 64, *73*
forearm, 9, *12*, 146, 147, 151
  views of, *12*, *147*, *150*, *151*, *153*, *156*, *157*, *158*, *159*, *160*
fossa, 59
  antecubital, 9, *12*
  coronoid, 84, *91*
  cranial, *73*, 74
  iliac, 86
  infraspinous, 83, *90*
  mandibular, 63, *73*
  olecranon, 84, *91*
  subscapular, 82, 83, *90*
  triangular, *236*
frontal bone, 50, 62, 64, 65, 72, *73*, 74, 105
frontal lobe, 188, 190, 191, *195*, *196*, *197*, 229
frontal region, 7, *15*, 15
frontalis muscle, 119, 120, 122, 123, *125*
fundus, 332, 372, *380*
furculae, 82, *83*

gallbladder, 17, *17*, 319, 320, 321, 322, *322*, 330, *332*, 333
gastrocnemius muscle, 171, 172, 174, *175*, *179*
gastrointestinal (GI) tract, 267, 297, 315–17, 322, 324, *324*, 335
glands, 22, 249, 252, 286, *294*, 333, *354*, 361
  adrenal, 243, 244, 246, 250, 251, 327, 342
  Bartholin's, 372
  bulbourethral, 359, 361, 362, 363
  endocrine, 243, 246, 321
  endometrial, 377, 379
  esophageal, 326, 328
  exocrine, 243, 321, 361
  mammary, 245
  parathyroid, 243, 244, 247, 248, 255
  parotid, 319, 320
  pituitary, 188, 190, 191, 244, 246, 248, *248*, 250, 251, *252*, 253, 255
  salivary, 22, 320, 321
  sebaceous, 47, 50, 51, 52, 54, 55
  seromucous, 302, 306
  sublingual, 319, 320
  submandibular, 319, 320
  sweat, 47, 50, 51, 52, 53, 54, 55
  thyroid, 243, 244, 247, 248, *249*, 250, *253*, 307
  vestibular, 372, 374
glans penis, 362, 363, *368*
glenohumeral joint, 81, 96, *110*, *111*, 114, 135, 161
  movement of, *140*, *141*, *142*
  muscles of, 139, *139*, 143, *154*, 155
glenoid cavity, 82, 83, *90*, 161

glomerulus, 340, 341, 343, 344, 345, 346, 347, 350, *350*, 352
gluteal region, 14, *14*, 163, *165*
gluteus maximus muscle, 161, 162, 163, 165, *165*, *175*, 176, *176*, 217
gluteus medius muscle, 161, 162, 165, *175*, 176, *176*, 217
gluteus minimus muscle, 161, 162, 165, *176*, 217
goblet cells, 302, 306, 316
goiter, 245, *245*
gracilis muscle, 161, 162, 165, *176*
greater curvature, *330*, *332*
greater sciatic notch, 86, 87, 92, 93
greater trochanter, 88, 89, 93, *175*, *176*
gyri, 188, 191, *195*

hair, 18, 47, 55
hamate, 86, 92
hamstrings, 166, 167, 210
hand, 12, 85, 101, 135
    views of, *12*, *84*, *86*, *92*, *110*, *149*
haustra, 318, 319, *334*
Haversian canal, 31
head, 84, 85, 88, 267, 327
    muscles of, 119, 121, 122, *122*, 123, *123*, 125, *131*
    views of, *7*, *18*, *91*, *96*, *197*, *199*, *200*, *201*, *231*, *252*
hearing loss, 232, 237
heart, 16, *16*, 22, 257, 258, 259, 261, *275*, *293*
    views of, *263*,*264*, *276*, *277*
heart attacks, 261, 317
Henle's ampula, 345
Henle's fissures, 345
hinge joint, 104, 108
hip, 101, *112*, 114
    movement of, *163*, *164*
    muscles of, 161, 162, *175*, 176, *177*, *182*
histology, 21, 38, 47
hormones, 243, 265
    adrenocorticotrophic, 244, 246
    antidiuretic, 244, 246, 344
    follicle-stimulating, 244, 246
    growth, 244, 246
    luteinizing, 244, 246
    parathyroid, 247
    pituitary, 244
    sex, 245, 371
    thyroid-stimulating, 244, 245, 246
humerus, 78, 84, 85, *91*, 143
hypodermis, 47, 48, 49, 50, *50*, 51, 52, 55
hypothalamus, 188, 191, *196*, 243, 244, 248

ileum, *291*, *292*, 316, 318, 319, 327, *333*, *334*
iliac crest, 86, 87, 92, 93
iliacus muscle, 161, *216*
iliopsoas muscle, 161, 162, 165, *176*, *177*
iliotibial band, 165, *175*, *176*, *177*
ilium, 86, 87, 92, 93, *176*

immune system, 285, 286, 316
inferior gemellus muscle, *176*
inferior vena cava, *252*, 259, 260, 263, 264, 265, 267, 269, 272, *294*, *327*, *333*, *334*, 342, *354*
infraspinatus muscle, *129*, *130*, 138, 140, 143, *154*
inguinal region, 8, *12*, *13*, 14
integumentary system, 3, 11, *50*, 52
intercarpal joints, *110*
intercostal muscles, 16, *16*, 120, 122, 123, *126*, *128*, 208, 304
intermediate cuneiform, 90, *94*
internal oblique muscle, 121, 122, 124, *128*
interphalangeal joints, *110*, *111*, *112*
interstitial cells, 359, 360, 364, 366
intertarsal joints, *111*, *112*
intertubercular groove, 84, *91*
interventricular septum, 277, *277*
intervertebral discs, 30, 66, 101, 104, 107, *110*, 208
intestines, 17, 316, 317, 323, 328, *330*, *331*, *332*, *334*
iris, 228, 229, 230, *235*
ischial spine, 86, 87, 92, 93
ischial tuberosity, 86, 87, 92, 93, *175*, *176*, 217
ischiocavernosus muscle, *380*
ischium, 86, 87, 92, 93
islets of Langerhans, 245, 247, 248, 250, 251

jejunum, 316, 318, 319, 326, 327, 329, *333*
joints, 102, 110
    classification of, 101, *104*
    movements of, *133*
juxtaglomerular apparatus, 344, *347*

keratinocytes, 47, 50, 55
kidney, *252*, 267, 327, *333*, 339, 340, 342, *354*
    anatomy of, *333*, *339*, *343*, *354*
knee, 101, *112*, 161
    movement of, *167*, *168*
    muscles of, 166, *166*, 176, *177*
    posterior surface of, *13*
    synovial joint in, *108*

labia majora/minora, 374, 375, 376
lacrimal bone, 62, 64, 72
lacrimal caruncle, *235*
lacrimal papilla, *235*
lamina, 65, 69, *75*, *76*, 377
lamina propria, 17, 324, 326, 329, 365, 367, 372, 377
larynx, 298, 299, *307*, 316
lateral canthus, *235*
lateral cuneiform, 90, *94*
lateral malleolus, 88, 89, *94*, *179*, *180*
lateral rotation, 102, *103*
lateral supracondylar ridges, *91*
latissimus dorsi muscle, 121, 122, 124, *127*, *128*, 129, *130*, 138, *138*, 143, *155*, 215

leg, 10, 13, 169
    muscles of, *169*, 171, *171*, *174*
    views of, *13*, *223*
lens, 228, 230, *235*
lesser curvature, *330*, *332*
lesser omentum, 323, *332*
lesser sciatic notch, 86, 87, 93
lesser trochanter, 88, 89, 93
levator labii superioris muscle, 120, 123, *125*
levator scapulae muscle, 121, 122, 129, 130, 135, 136, 138
Leydig cells, 359, 360, 364
ligaments, 263
    anterior cruciate, 30
    broad, 371–72, 373
    falciform, 322, *330*, *332*, *333*, *334*
    Henle's, 345
    inguinal, *127*, *128*, 176, *177*
    ovarian, 374
    patellar, 210, 221
    periodontal, *113*
    sacrotuberous, 217
    suspensory, 230, 379
linea aspera, 88, *93*
liver, 17, *17*, 21, *252*, *293*, *294*, 307, 319, 320, 322, *322*, 323, 328, 329, *332*, *354*
    views of, *330*, *331*, *332*, *333*, *334*
lobules, *236*, 359, *327*, *329*, 364, 366
long bone, anatomy of, *32*
longitudinal cerebral fissure, 188, *195*, *213*
loop of Henle, 340, 341, 344, 345, 346, *351*, 353
lower limb, *13*, 78, 88, 266, 267
    joints of, *112*
    muscles of, *165*, *168*, *175*, *181*
    terms for, *10*
    veins of, *274*
lumbar plexus, 203, 207, 209, 216
lumen, 17, 317, 325, 347, 353, 364, 365, 367, 372, 377, 379
lunate, 86, 92
lungs, 16, 22, 123, *293*, 298, 300, *307*
    views of, *16*, *294*, *307*, *308*, *313*
lymph nodes, 136, 285, 286, 288, *288*, 289, *291*, *292*, 294
lymphatic nodule, 290, 291, 292, 293
lymphatic system, 3, 11, 292, 293, 285
lymphatics, 285, 286, 324, 360

macula densa cells, 344, 347, 350
macular degeneration, 229
major calyces, 341, 343, *354*
mammillary bodies, *195*, *213*
mandible, 61, 62, 63, 64, 65, 72, *74*, *113*, 114
manubrium, 71, *76*, *126*
masseter muscle, 120, 123, *125*
mastoid process, 60, 63, 64, 72, 73, 237
maxilla, 62, 63, 64, 65, 72, 73, *74*, *113*, 229

meatus, 59
    acoustic, 60, 64, *72*, *74*
    external auditory, *236*
medial malleolus, 88, 89, *94*, *179*, 180
medial rotation, 102, *103*
medial supracondylar ridges, *91*
medicine, language of, 3
medium cuneiform, *94*
medius muscle, posterior view of, *165*
medulla, 190, 191, 243, 291, 292, 371
    adrenal, 245, 246, 248, 249, 251
    renal, 340, 341, 343, 346, *351*, 352, 353
medulla oblongata, 189, 192, *195*, *196*, *213*, 252
melanocytes, 47, 48, *50*, 55
meninges, 187, 191, 192, *192*
mental region 7, 15, *15*
mesentery, 322, 323, 328, *331*
mesometrium, 372, 374, *380*
metacarpals, 78, 85, 86, *92*, *110*
metatarsals, 81, 89, 90, *94*, *111*
metatarsophalangeal joints, *111*, *112*
microvilli, 22, 316, 317, 325
midbrain, 189, 191, *196*, 252
middle phalanx, 86, 90, *92*, *94*
minor calyces, 341, 343, *354*
mouth, 18, 315
mucosa, 291, 292, 297, 317, 324, 326, 329, 377, 378
    muscularis, 17, 324
    nasal, 348
    oral, 290, 293
    stomach, 316
mucosa-associated lymphatic tissue (MALT), 285, 286, 287, 288
muscle tissue, 35, *35*, 36, 43
    micrographs of, *36*
    types of, 21, 41
muscles, 3, 11, *123*, 135, *294*
    movement of, *132*, *133*
muscularis externa, 17, 291, 292, 297, 324, 325, 326, 328, 329, 348, 351, 353, 365, 367, 377
myocardium, 139, 257, 258, 259, 261, 263, 277
myometrium, 372, 373, 374, 377, 379

nails, 47
nasal bone, 62, 64, 65, 72
nasal conchae, 61, 62, 231
nasal region, 7, 15, *15*
nasalis muscle, 123, *125*
nasopharynx, 232, 287, 299
navicular, 90, *94*
neck, 3, 8, 88, 93, 119, *196*, 220, 327
    dissection of, *253*
    muscles of, 119, 120, *123*, *125*, *131*
nephron, 339, 340, 341, 344, 345, 346
    structures of, 343, *345*

nerves, *197, 232, 324, 360*
   abducens, 205, 206, *213*
   axillary, 207, 209, *215*
   cardiac splanchnic, 263
   chorda tympani, 228
   common peroneal, 207
   cranial, *195*, 203, 205, 206, *213*, 224, 230, 233, 237, 316
   dorsal scapular, *215*
   facial, 205, 206, *213*, 233
   femoral, 210, *216*
   fibular, 207
   glossopharyngeal, *195*, 205, 206, *213*, 233
   gluteal, 217, *217*
   greater splanchnic, *214*
   hypoglossal, 206, *213*
   iliohypogastric, *216*
   iliolingual, *216*
   inferior rectal, *217*
   intercostal, 123, *126*, 203, 207, 209, *214*
   long thoracic, *215*
   median, 207, 209, *215*
   musculocutaneous, 207, 209, 215
   obturator, *216*
   oculomotor, *195*, 205, 206, *213*
   olfactory, 205, 206, 228, 230, 231
   optic, 205, 206, *213*, 227, 229, 230
   pectoral, 215, *215*
   peripheral, 203, 218–21, 224
   phrenic, 305, 309
   posterior cutaneous, *21*
   pudendal, 217, *380*
   radial, 207, 209, *215*
   sciatic, *175, 176, 179, 194*, 208, *217, 375*
   spinal, 193, *194*, 197, 198, 203, 206, 207, 208, *209*, 214, *214*, 224, *225*
   subcostal, *216*
   suprascapular, *215*
   thoracodorsal, 139, *215*
   tibial, 207
   trigeminal, *195*, 205, 206, *213*
   trochlear, 205, 206, *213*
   ulnar, *84*, 207, 209, *215*
   vagus, *195*, 205, 206, *213*, 316
   vestibulocochlear, 205, 206, *213*, 232, 233
nervous system, 3, 11, 243
nervous tissues, 21, 37, *37*, 41, 42, 204, *204*
neurons, 37, 231
   sensory, 204, 210, 227, 230, 233, 317, 318
nose, 18, 227, 228

oblique fissure, *308*
obturator internus muscle, 176
occipital bone, 63, 64, 65, 72, 73, 74, 105
occipital lobe, 188, 190, 191, *195, 196, 197,* 227
occipital region, 7, *14, 15, 15*
occipitalis muscle, 119, 123
olecranon process, 84, *91, 158*

olfactory bulb, *195, 213,* 231
olfactory tract, *195, 213,* 231
optic canal, 60, 62, 64, 73
optic chiasma, *195*
optic disk, 228, 230, *235*
oral cavity, 299, 315, 319
oral region, 7, *15, 15*
orbicularis oculi muscle, 120, 122, 123, *125*
orbicularis oris muscle, 120, 122, 123, *125*
orbital region, 7, *15*
organs, 3, 11, 16, 19, 21, 43
   accessory, 359, 371
   central nervous system, 187
   classifications of, 315
   digestive, 215, 315, *318*, 320, 321, 322, 327
   functions of, *11*
   intraperitoneal, 327, 328
   lymphatic, 285, 286, 288, *288*, 289
   lymphoid, *289*
   peritoneal, 315
   retroperitoneal, 327, 328
   sensory, 227
   sex, 359, 371
   tissues in, *43*
   urinary, 342, *342*
os coxa, 81, 87, *92, 93,* 107
osteology, terms of, *31, 59*
otontoid process, 76
ovaries, 243, 245, 248, 371, 373, 374, *376,* 376, 378, 379
oviducts, 371, 372, 373
ovulation, 244, 371, 372, 374

palatine bones, 62, 63, *73,* 74
palmar aponeurosis muscle, 150
palmar arches, 270, 271
palmaris longus muscle, 146, 147, 150, *158*
palpebra, *235*
pancreas, *17,* 17, 22, 243, 245, 247, 248, 250, *250,* 251, 252, 294, 319, 321, 322, *322,* 323, 327, *333*
papillae
   circumvallate, 228, 233, 234, 316
   dermal, 50, *55*
   filiform, 316
   foliate, 316
   fungiform, 234, 316
   tongue, 233
papillary layer, 48, 52, 53, 54, 55
papillary muscles, 257, 260, 277
parasympathetic nervous system (PNS), 203, 211
pericardial cavity, 257, 322, 327
pericardial sac, 16, 258, *275,* 307
parietal bone, 60, 62, 63, 64, 65, 72, 73, 74, 105
parietal layer, 257, 304, 343
parietal lobe, 188, 190, 191, *195, 196,* 197
parietal peritoneum, 322, 323, 327, 328, 363, 376
parietal pleura, 299, 304, 305

patella, 13, *17,* 81, 88, 166, 168, *175, 176, 177, 179,* 210
patellar surface, 89, *93*
pectineus muscle, 162, 165, *176, 177*
pectoral region, 12, *12*
pectoralis major muscle, 120, 122, 124, *127, 128, 136, 138, 143, 155,* 215
pectoralis minor muscle, 120, 122, *128,* 135, 136, 215
pedicles, 65, 69, *75*
pelvis, 3, 8, *95,* 266, 267, 341, 343, 344
penis, 359, 362, 363
   views of, *368*
pericardium, 257, 258, 275
perichondrium, 30, 33, 302, 306
perimetrium, 372, 373, 374
perineal body, *380*
perineum, 3, 8, 266, 267, *375,* 375
periosteum, 30, 31, 192
peripheral nervous system (PNS), 203, 204, 227
peristalsis, 316, 318, 324
peritoneal cavity, 322, 323, *323,* 327, 328, 363, 376
peritoneal layers, 322, *323*
peritoneal membranes, *323,* 339
peritoneum, 322, *323,* 324, 327, 328, 349
peroneus brevis muscle, 169, 171, *179,* 180
peroneus longus muscle, 169, 171, *179,* 180
peroneus tertius muscle, 169, 171
Peyer's patches, 92, 291, *291,* 316
phalanges, 9, 10, 81, 85, 89, *110, 111*
pharynx, 316, 318, 319
pia mater, 191, 192, 193
piriformis muscle, *175, 176,* 217
plane joint, 104, 108
plantar aponeurosis muscle, *179*
plantaris muscle, 171, 172, 174, *179*
plasma, 27, 33, 288, 234, 339, 343
pleural space, 299, 304, 305, 322, 327
plicae circulares, 325, 326, 329, *333, 333*
pneumocyte, 300, 301, *303,* 306
pons, 189, 190, 191, *195, 196, 213,* 252
popliteal region, 10, 13, *13*
popliteus muscle, 171, 172, 174, *180*
position, anatomical terms of, 3, *4, 5,* 12
posterior chamber, 229, 230
posterior compartment, 172, 174
posterior inferior iliac spine, 87, *92*
posterior mediastinum, anterior view of, *214*
posterior superior iliac spine, 86, 87, 92, 93, 163
pouter, 82
pregnancy, 208, 372, 373
pronation, 102, *103,* 114
pronator quadratus muscle, 146, 150, *158*

pronator teres muscle, 146, 150, *158*
prostate, 359, 361, 362, 363, *365,* 365, 366, 367
proximal convoluted tubule (PCT), 343, 344, 345, 346, 347
proximal phalanx, 86, 90, *92, 94*
psoas major muscle, 161, *354*
pterion, *113*
pterygoid processes, 60, *74*
pubic, 8, *12*
pubic bone, 107, 121
pubic symphysis, 30, 87, *376*
pubis, 87, *92, 93,* 121
pulmonary circulation, 265, *265,* 266
pulmonary trunk, 263, 265, *276*
pupil, 228, 230, *235*
purkinje fibers, 262, 263
pylorus, *333*

quadratus femoris muscle, 166, *176*
quadratus lumborum muscle, *216*
quadriceps muscle, 210

radial notch, 85, *91*
radial tuberosity, 85, *91*
radiological imaging, 3, *18*
radioulnar joint, *110*
radius, 78, 85, 106, 143
   views of, *85, 91, 106*
ramus, 203
   communicating, *214*
   dorsal, 207, 208, *214*
   ischial, 86, 87, *92*
   mandibular, 64, *72, 74*
   pubic, 87, *92, 93*
   spinal, 207
   ventral, 207, 208, *214, 215, 216, 375*
rectum, 317, 318, 319, 323, 327, 363, 376
rectus abdominis muscle, 121, 122, 124, *128*
rectus femoris muscle, 166, 168, *176, 177*
rectus sheath, *127, 128*
reflex arc, *210*
renal capsule, 340, 343, *354*
renal column, 341, 343, *354*
renal papilla, 343, *354*
renal pyramid, 340, 343, *354*
reproductive system
   female, 362, 371, *373, 374, 375, 376, 378, 379, 380*
   male, 359, 361, *363, 366, 367*
respiratory distress syndrome (RDS), 300
respiratory system, 3, 11, 297, 300
   anatomy of, *298, 299, 301, 305,* 307
rete testis, 359, 360, 364, 366
reticular layer, 48, 50, 52, 53, 54, 55
retina, 228, 229, 230, *235*
rhomboid major muscle, 121, 122, 124, 129, *130,* 135, 136, 138
rhomboid minor muscle, 121, 122, 124, *130,* 135, 136, 138
rib cage, 70, *71,* 121, *307*
   views of, *76, 107, 110, 111*

ribs, 16, *70, 74,* 123, *126*
    costal, *16*
    false/true, 70, 71, 76
    floating, 70, 71, 76, 342
rotator cuff muscles, 138, 140, 143
rugae, *332,* 349, 374

sacral, 8, 14, 65
sacral plexus, 203, 207, 209, *217*
sacral vertebrae, *14,* 67, 68, 74
sacroiliac joint, 66, *110,* 112
sacrum, 66, 87, 317
sartorius muscle, 166, 168, *176, 177*
scaphoid, 86, *92*
scapula, 81, 82, 83, 114, 135, 136
    views of, *83, 90, 154*
scapular, 8, *9, 14,* 130
scapulothoracic joint, 135, *136,* 137, *137, 138*
sclera, 228, 229, 230, *235*
scrotum, 359, 360, *360*
sections, anatomical planes of, 3, 6, 18
sella turcica, 60, 64, 65, *73, 74,* 244, 252
semimembranosus muscle, 166, 168, *175*
seminal vesicles, 359, 361, 363
seminiferous tubules, 359, 360, 364, *364, 366*
semitendinosus muscle, 166, 168, *175*
serosa, 17, 324, 326, 329, 377, 378
serous membranes, 257, 304, *305*
serratus anterior muscle, 120, 122, 123, 124, *127, 128, 139, 130,* 135, 136, 138, *215*
shoulder, 82, 101, 220
sinoatrial (SA) node, 262, 263
sinuses, 61, 62
    coronary, 259, 262, 264
    dural venous, 191, 192, 267
    frontal, 50, 65, *74*
    sphenoid, 60, 65, *252*
skeletal muscle, 35, 36, 41, *42,* 119
skeleton, 3, 11, 59, 101, *106*
skin, 18, 43, 192, 208
    anatomy of, *51,* 53–54, *55*
    histology of, 52
    thick/thin, 47, 53, 54
skull, 18, 60, 192
    bony landmarks of, *72, 74*
    disarticulated, *77*
    joints of, *105, 113*
    orbit of, *229*
    sagittal section of, *65, 74*
    step dissection of, *192*
    views of, *27, 62, 63, 64, 72, 73, 78, 105*
smooth muscle, 35, 36, 41, *42,* 301, 302, 306, 352
soleus muscle, 171, 172, 174, *179, 180*
somatic nervous system, 203, 204
special senses, 234–35, *236*
    terms for, *227, 228*
sperm, 359, 361, 362, 365, 366, 372
sphenoid bone, 60, 62, 63, 64, 65, *72, 73, 229, 252*

sphincter
    cardiac, 316
    external anal, 317, 318, *380*
    external urethral, 341, 348, 349
    gastroesophageal, 316, 318
    internal anal, 317
    internal urethral, 341, 348, 349
    pyloric, 318
spinal cord, 190, 191, 192, 193, *195, 196, 197,* 207, 208, *213,* 214, *214*
    axial section through, *198*
    cervical, *194*
    CNS and, 187
    coronal sector through, *193*
    lumbar, *194*
    schematic of, *225*
    step dissection of, *193*
spine, 59, 82, 83, 90
spinous process, 65, 69, 75, 76
spleen, *252,* 267, 285, 287, 288, 289, *290, 290,* 292, 293, *294, 333*
splenius capitis muscle, 122, 124, *129,* 130
sternal body, 71, 76, *126*
sternal region, 8, *12, 12*
sternoclavicular joint, *10*
sternocleidomastoid muscle, 120, 122, 123, *125, 127, 128, 129,* 130
sternocostal joint, *110*
sternum, 12, 70, *127, 128,* 135
stomach, 22, 43, *252, 294, 307,* 316, 318, 319, *326,* 328, 329, *331, 333*
    body of, *330, 332*
stratum basale, *48, 50, 51, 53, 54, 55*
stratum corneum, 48, 50, 51, 53, *54, 55*
stratum granulosum, 48, 50, 53, 54
stratum lucidum, 48, 50, 53, 54
stratum spinosum, 48, 50, 53, 54
styloid process, 60, 63, 64, *73, 74,* 84, 85, *91*
submucosa, 291, 292, 297, 317, *324,* 325, 326, 328, 329, 348, 351, 352, 353, 378
subscapularis muscle, 138, 140, *155, 215*
sulcus, 188, 191
    central, 190, *195, 196*
    coronary, 261
    lateral, 190, *195*
    posterior median, *198*
superficial, 4, *5*
superior orbital fissure, 60, 62, 72
superior vena cava, 259, 260, 263, 264, 265, 269, 271, 275, 276, 307
supination, 102, *103,* 114
supinator muscle, 151, 153, *156*
supraciliary arches, 60
supraspinatus muscle, *130,* 138, 140, 143, *154*
suture, 59, 63, 101, 104
    coronal, 50, 61, 64, 65, *72,* 105, *113*
    lambdoid, 60, 61, 63, 64, 65, 72, 74, 105, *111, 113*

sagittal, 60, 61, 63, *74,* 105, *111*
squamous, 60, 61, 63, 64, 65, 72, 113
sympathetic nervous system (SNS), 203, 211
symphysis joint, 101, 104, *107*
syndesmosis joint, 101, 104, 106
synovial joints, 101, 104, 114, 119, 143, 161, 166, 169
    motions of, 102, *109*
    types of, *108*
systole, 257, 259, 261

taeniae coli, 318, 319, *334*
talus, 90, *94,* 169
tarsal bones, 81, 89, *111,* 169
tarsometatarsal phalangeal joints, *111*
tarsus, 10, 13, *13*
taste buds, 228, 233, 234, 318
teeth, 315, 318, 319
temporal bones, 60, 62, 63, 64, 65, *72, 73, 74,* 105, 233
temporal lobe, 188, 190, *195*
temporalis muscle, 120, 123, *125*
temporomandibular joint (TMJ), 61, *113,* 120
tensor fasciae latae muscle, 161, 162, 165, *176, 177*
teres major muscle, *129, 130,* 138, 143, *155*
teres minor muscle, *130,* 138, 140, 143, *154*
testes, 243, 245, 248, 255, 360, 361, *364, 366*
    structure/function of, 359
    views of, *360, 367*
thalamus, 188, 191, *196, 197*
thigh, 10, 13, *13,* 162, 267
    muscles of, 166, 182
    views of, *181, 208*
thoracic muscles, anterolateral view of, *126*
thoracic region, 14, *14,* 304
thoracolumbar fascia, *129, 130*
thorax, 3, 8, 119, 120, *196,* 286, 304
    cross section of, *208*
throat, 220
thumb, 114, 135
    movement of, *149, 152*
thymus, 285, 286, 288, *293*
tibia, 81, 106, 166, 169, *179*
    condyles of, medial/lateral, 88, 89
    views of, *89, 94, 106*
tibial tuberosity, 88, 89, *94*
tibialis anterior muscle, 169, 171, *179*
tibialis posterior muscle, 171, 172, 174, *179, 180*
tibiofibular joint, *112*
toes, 172
    movement of, *174*
tongue, 219, *234, 238,* 316, 318, 319
    parts of, *227, 233, 238*
tonsils, 285, 288, *289,* 292, 293
    adenoid, 287, 289
    lingual, 287
    palatine, 234, 287, *287,* 289, 290, *290*
    pharyngeal, 287, 289

tonsillectomy, 287, *287*
trabeculae, 31, 257
trachea, *253,* 297, 298, 299, 305, 306
    anterior view of, *307*
transverse abdominis muscle, 121, 122, 124, *128,* 345
transverse process, 65, 69, *75, 75,* 76
trapezium, 86, *92*
trapezius muscle, 120, 121, 122, 123, 124, *125, 129, 130,* 135, 136, 138, *138,* 143
trapezoid, 86, *92*
triceps brachii muscle, 143, 146, *154, 155*
triquetrum, 86, *92*
trochlea, 84, 85, *91*
trochlear notch, 84, 85, *91*
trunk, *8, 12, 12,* 209, *215,* 309
trunk muscles, 119, 121, 122, 123, 125
    views of, *122, 124, 127, 128, 131, 159*
tubercle, 59
    adductor, 88
    greater/lesser, 84, 85, *91*
    infraglenoid, 82, *90*
    pubic, 121
    supraglenoid, 82, *90*
tunica albuginea, 359, 360, 364, 366, 371, 376, 378
tympanic membrane, 228, 232, 233

ulna, 78, 84, 106, 143
    views of, *85, 91, 106*
umbilical cord, 293
umbilical region, 8, *12, 12*
umbilicus, 124, *127, 128,* 317
upper limb, 78, 84, 266, 267
    joints of, *110, 111*
    muscles of, 146, 154
    regional terms for, 9
upper skeleton, 67, *74*
ureters, *327, 339,* 341, 342, *343,* 349, *351,* 352, 353, *354*
    female, *349*
urethra, 339, 341, 342, 348, 349, 359, 367, 374, 375, *380*
    ejaculation and, 361
    female, *349*
    membranous, 361, 363
    penile, 361
    prostatic, 361, 363
    spongy, 361, 362, 363
urinary system, 3, 11, 339, 352
    anatomy of, *341*
    pyelogram of, *355*
uterine tubes, 371, 372, 373, 374, 376, *377, 378, 379, 380*
uterus, 22, 43, 371, 372, 374, 376, 378, *379, 380*
    expansion of, 373
    fundus of, *379*
    proliferative phase, *377*

vagina, 362, 371, 372, 373, 374, 375, 376, *377,* 378, *380*

valves, 258, *260*
  aortic, 259, *260*, 277
  atrioventricular, 257, 259, 261, 277
  bicuspid, 259, 260, 264, 277
  heart, 257, 259
  mitral, 259, 264
  pulmonary, 259, 260, 264
  semilunar, 259, 261, 264
  tricuspid, 259, 260, 264, 277
vasa recta, 341, 345, 346, 351, 353
vastus intermedius muscle, 166, *177*
vastus lateralis muscle, 166, 168, *177*
vastus medialis muscle, 166, 168, *176, 177*
veins, 257, 265, 266, 269, 278, *294*, 320, 327, 329, 340
  abdominal, 272
  anterior tibial, 274
  arcuate, 346
  axillary, 271
  basilic, 267, 271
  brachial, 271
  brachiocephalic, 271, *275*
  cardiac, 257, 261, 263, 264, 267
  cephalic, 267, 271
  common iliac, 267, 269, 272, 342
  external iliac, 267, 269, 272, 274
  femoral, 267, 269, 274
  gastric, 272
  gluteal, 163
  gonadal, 269, *333*, 354
  great saphenous, 267, 269, 274
  hepatic, 267
  hepatic portal, 269, 272, 321, *334*
  inferior mesenteric, 269, 272
  intercostal, 123, *126*, 214
  interlobular, 346
  internal iliac, 269, 272
  jugular, *253*, 269
  median cubital, 267, 271
  mesenteric, 267, 269, 272
  pampiniform plexus of, 360
  popliteal, 267, 269, 274
  portal, 320, 321, 327, 329
  posterior tibial, 274
  pulmonary, 259, 260, 263, 264, 265, *276, 308*
  radial, 271
  renal, 269, 272, 341, 342, 343, *354*
  splenic, 267, 269, 272
  subclavian, 271
  superficial, 267
  suprarenal, 272
  testicular, 360
  thoracodorsal, 139
  ulnar, 271
  upper limb, *271*

ventral horn, 192, *198*, 210
ventral root, 207, 208, *214*
ventricle, 257, 261
  fourth, 187, 189, 191, *196*
  lateral, 187, 189, *197*
  left, 43, 259, 260, 263, 264, 265, *276, 277*
  right, 259, 260, 263, 264, 265, *276, 277*
  third, 187, 189, 191
ventricular system, 187, *189*
venules, 266, 267, 340
vertebrae, 193
  atlas, 66, *69*, 76
  axis, 66, *69*, 76
  cervical, *14*, 66, 67, 68, 69, *69, 74, 75*
  coccygeal, 65, 67, 68, *74*
  fused, 66
  lumbar, *14*, 66, 67, 68, 69, *69, 74, 75, 197*
  thoracic, *14*, 66, 67, 68, 69, *69, 74, 75 197*
vertebral body, 65, 69, *75*
vertebral column, 65–66, 68, 192, *196, 197*, 214
vertebrocostal joint, *111*

vessels, 257, 258, *260*
  blood, 51, 55, 266, 324, 325
  coronary, 261
  lymphatic, 285, 286, 288, *294*, 301
  pulmonary, *307*
villi, *325*, 326, 329
viruses, 47, 286, 287
visceral peritoneum, 320, 322, 323, 327, 328, 363, 376
visceral pleura, 299, 301, 304, 305
vomer, 62, 63, 65, *72, 73*

wrist, 9, 12, *12*, 85, 101, 135, 146, 151
  movement of, *148, 149, 152, 153*

xiphoid process, 71, *76*, 121

zona fasciculata, 245, 246, 249, 251
zona glomerulosa, 245, 246, 249, 251
zona reticularis, 245, 246, 249, 251
zygomatic bone, 62, 63, 64, *72, 73*, 119, *229*
zygomaticus major muscle, 119, 120, 123, *125*